Slanted Truths

Lynn Margulis Dorion Sagan

Slanted Truths

Essays on Gaia, Symbiosis, and Evolution

Foreword by Philip and Phylis Morrison

C

COPERNICUS
An Imprint of Springer-Verlag

© 1997 Springer-Verlag New York, Inc.
All rights reserved. No part of this publication may be reproduced, stored in a retrieval system, or transmitted, in any form or by any means, electronic, mechanical, photocopying, recording, or otherwise, without the prior written permission of the publisher.

Published in the United States by Copernicus, an imprint of Springer-Verlag New York, Inc.

Copernicus
Springer-Verlag New York, Inc.
175 Fifth Avenue
New York, NY 10010
USA

Library of Congress Cataloging-in-Publication Data
Margulis, Lynn, 1938–
Slanted truths : essays on Gaia, symbiosis, and evolution / Lynn Margulis, Dorion Sagan.
p. cm.
Includes index.
ISBN 0-387-94927-5 (hardcover : alk. paper)
1. Biology. 2. Gaia hypothesis. 3. Symbiosis. 4. Evolution (Biology) I. Sagan, Dorion, 1959– . II. Title.
QH311.M37 1997
570—dc21 97-2160

Manufactured in the United States of America.
Printed on acid-free paper.

9 8 7 6 5 4 3 2 1

ISBN 0-387-94927-5 SPIN 10557766

QH
311
.M37
1997

111797-2376B2

To the memory
and legacy
of Carl Sagan

1934–1996

Foreword

This enticing collection is as devoted to the profound power of figures of speech as any oration of old Hellas. Metaphor reigns as we encounter the identification of our Earth as a single, integrative organism, a tale told in image and passion. The discoveries and conjectures upon which this grand view rest are here as well. I have enjoyed these two dozen pieces hugely.

No family tree of animal life but must somewhere disclose a cousinly infolding. We are compact of life past, and the looped handing down is more complex than the Mendel–Morgan dance of chromosomes and genes. That dance is essential, certainly, but it is the vital editing of an epic and many-rooted work, a book more like the Bible than like one great artist's *Remembrance of Things Past*. We follow a few old shelves of bound DNA, not just a single book.

Our major biochemical package for oxidative metabolism was described in a small DNA manual, somehow engulfed to become an organelle within a lucky ancient ancestral anaerobe, and passed ever since from mother to offspring outside of the chromosome shuffle. The sperm do carry half the compact genetic message of the human DNA, but they are too small to transfer this equally essential symbiotic one that comes down from the mothers, within roomy egg after roomy egg.

Thirty years ago a bold young biologist, building a new molecular biology out of many early hints, entered by experience and insight into the poorly known and bizarre variety of microorganisms and their habitats. The visible tiny packaged cell components, she saw, must once actually have been free-living bacteria. Such vital organelles have long been enfolded into the larger nucleated cells of all animals and plants. She is our senior author, Lynn Margulis. Most of her work was shared with her longtime partner, her oldest son, the artist-philosopher Dorion Sagan.

It is implausible that one-by-one assembly of hundreds of interrelated mutations made the leaves green or all animal cells oxygen users. That was rather the state of the bacterial arts very long ago, and the coded recipes worked out early on were stored in the bacteria that held the methods. They are still there to be read, if abridged. Once the larger cells incorporated such adept bacteria, mutuality of interest maintained the new physical relationship. Every individual plant or animal—you, too, reader—is a genuine mix of cells whose lineage is far more diverse than the implied parents of any centaur or chimera. Now we know that all the animals—mites to whales—like all the plants—duckweed to sequoia—are symbiotic chimeras, with an architecture more like that of the lichens than like the offspring of one single family tree.

Over the eons the host cells managed to bring the maintenance of their symbionts under control, so the union lost various walls and divisions that once isolated the partners. Unrelated residual DNA instructions, much abridged by now, still accompany the organelles, far from the nuclei of the enclosing larger cells. Our biology is thus a mine of synecdoche: we name them plants and animals only by naming the whole for its part.

The greenness of our countryside was a contribution of the cyanobacteria to an antique little-oxygenated world. The fuel-making seat of oxygenic energy harvest came as well from bacteria already able to use the waste product of photosynthesis, oxygen, at once powerful, toxic, and novel. Some large free cells move now only because they are manifestly rowed by symbiotic oarsmen, the once-free spirochetes. These essays speculate that the internal motility of cells with nuclei comes from the remains of spirochetes of long ago, lately perhaps disclosed by some little pools of strange DNA. Additional multiple legacies may have entered into the early stages of life's passage to oxygenation—by no means yet ubiquitous. Mudbanks, reef edges, bogs, and swamps still show colors that signal the vigor of anaerobic microbial life. These discoveries will continue.

The atmosphere and the waters unite all surface life. Life is visibly a geological force, from the locust flight that carries thousands of tons of carbonaceous compounds and water across the desert, to the slow accumulation and burial of carbon and sulfur within the limestone and gypsum flats of the world, mediated by the long, slow work of marine plankton. It is hard to doubt that as a principle. Its quantitative role is less surely known. Once accepted, it is tempting to try another step.

That step was made first by James E. Lovelock, a scientist of great imaginative strength, out of concern with the place of organics in the vast flow of the geochemical cycles. Because life is a force at such scales, can it inhibit as well as add? The answer is surely yes. Then a nonlinear system is pretty likely to develop feedback. An ecology that feeds back from life to rock and air, and from rock and air to life again, is no very big leap. That notion, developed beyond our present knowledge, may imply a life-generated stability able to maintain life itself. A strong greenhouse atmosphere might have kept the Earth from freezing when the old Sun was cooler, during the first tenth or two of terrestrial history. Was that a coincidence, or a built-in response stumbled upon by the living world soon enough to help survival by favoring some special properties of its molecules?

Here we are at full metaphor: regard the Earth and all its life as a new stabilizing entity, Gaia, named by Lovelock after an ancient goddess. Part III of this collection is centered on this issue, and both authors say an enthusiastic *yes* to it at many levels. A physicist is a little slower to accept this response. The Sun's output and the Earth's portion of it are mainly questions beyond life's reach. We are possibly survivors by good fortune, like many products of natural selection. (Consult the dinosaurs at the K-T boundary for the other side of the story.)

We should not multiply even metaphors without necessity. At the same time, we cannot ignore them but need to tease out the detail. If anything is clear these days, it is the erasure of old disciplinary boundaries, the merger of opposing points of view, the increased number of examples of what once seemed unique. These arguments are here, delightfully written, full of luminosity, and always honored by a firm stance on the facts.

We return to part I of the pieces compiled here. One of the writers of this foreword is a little troubled by reading his own name and experiences in the first essay, written some time ago. This is no mere coincidence; the places and topics are mine, engraved by experience in mind and heart. Lynn's narrative is cogent and not to be dismissed. At the same time, it is not simply a record but rather a serious docudrama woven of feelings and friendships. Its insights are those of the poet, perhaps as true as all else we can learn about old human dilemmas. Let me respond to the old piece by announcing the current optimism of an elderly physicist and his wife, long time friends of the authors. The tide of nuclear war and its possible winter have conspicuously receded. It is open to us all to prevent any revival of that threat by continued attention, deep concern, and unity of purpose.

This book of science and philosophy by a distinguished scientist at the bench and with the pen, by a gifted mother and a gifted son, is an erasure of a dark pattern of prejudice against women in science, one hard now even to credit. Yet sixty years ago that was the rule in many serious intellectual centers in the United States. The times are a-changing, not evenly, not fast, not always wisely, but they change. These pages witness such progress, by inclusion, in equity, through aesthetics, and in simple humanity. Here is good reason to hope.

Philip and Phylis Morrison
Cambridge, Massachusetts
January 1997

Acknowledgments

This work is grounded in the deciduous forests, university campuses, and cities of the northeastern United States, mainly Boston, Woods Hole, and Amherst, Massachusetts. In those places and others, we have conversed with hundreds of thoughtful scientists, scholars, naturalists, philosophers, social critics, editors, and business people. We are indebted to them all. Those whose work and ideas particularly influenced ours include the late E.S. Barghoorn, the late Alan McHenry, the late G.E. Hutchinson, the late Lewis Thomas, the members of the Commonwealth Book Fund Committee, David Bermudes, Daniel Botkin, Ricardo Guerrero, John Hall, H.D. Holland, Wolfgang Krumbein, James Lovelock, the late Heinz Lowenstam, David Luck, Mark McMenamin, Harold Morowitz, Kenneth Nealson, Eric Schneider, J.C.G. Walker, and Peter Westbroek.

We thank Connie Barlow, Andrew Blais, Gerard Blanc, Lois Brynes, Peter Bunyard, Michael Chapman, Carmen Chica, Al Coons, Kathryn Delisle, Michael and Sona Dolan, Deborah Fort, Teddy (Edward) Goldsmith, Steve Goodwin, Aaron Hazelton, Jeremy Jorgensen, Robin Kolnicki,

Christie Lyons, Kelly McKinney, Jennifer Margulis, Zachary Margulis, Claude Monty, Kenneth Nealson, Peter and Ann Nevraumont, Lorraine Olendzenski, Mercé Piqueras, Brian Rosborough, Jeremy Sagan, Landi Stone, William Irwin Thompson, Jorge Wagensburg, Peter Westbroek, Tom Wakeford, and especially Donna Reppard for aid with the innumerable tasks required to complete these essays.

We are grateful to Bruce Wilcox of the University of Massachusetts Press and William Frucht and Lesley Poliner of Copernicus Books for helping us shape an amorphous mass of words into a coherent book. Lewis Lapham, editor-in-chief of *Harper's Magazine,* contributed immensely when he encouraged us to publish the first selection here, "Sunday with J. Robert Oppenheimer," in a book of essays.

The scientific work in the laboratory of Prof. Margulis has been supported since 1972 by NASA. Essential financial support for students and scientific colleagues from the Richard Lounsbery Foundation, New York; the Boston University Graduate School; and the University of Massachusetts Amherst College of Natural Science and Mathematics is gratefully acknowledged.

Our writing partnership began in the spring of 1981, when we were visited in Lynn's Boston apartment by a colorful character in a pimp hat and three-piece suit, the literary agent John Brockman. As the host of an intellectual salon called the Reality Club, John sought out the most cutting-edge ideas in science, and he had come to persuade Lynn to popularize her work on symbiosis. She adamantly refused; she was strictly an academic and did not even read newspapers or watch television, let alone write popular books. "Go ask my ex-husband," she said. "That's his specialty."

At this point Dorion, then a college senior, launched into a smug, if verbose monologue, claiming among other things that extraterrestrial intelligence (Carl Sagan's field) was merely a replacement for religion in a secular age.

"That's good," said the agent when Dorion had wound down. "Why don't you write a book? You can write about growing up with your father."

"OK," Dorion replied, "but I'll have to make it all up because my parents divorced when I was three."

"Fine," said John, "as long as it's nonfiction."

Thus began our writing partnership as well as our association with John Brockman. Our debt to John, our beloved, sassy, intrepid agent, who stimulated so many of these essays, and to his marvelous wife, Katinka Matson, is immeasurable.

"Sunday with J. Robert Oppenheimer" was brought to a publishable stage through the skillful writing and organizational help of Jennifer Margulis. Philip and Phylis Morrison, Chick (Ernest) Callenbach, T.N. Margulis, Zachary Margulis, Dorion Sagan, Lois Brynes, William Lawren, Kameshwar Wali, Dennis Wepman, Freeman Dyson, and Michael Chapman also helped complete this four-decade-old work.

"The Red Shoe Dilemma" originally appeared in *A Hand Up: Women Mentoring Women in Science,* edited by Deborah C. Fort (The Association for Women in Science, 1993), and was written with the help of Deborah C. Fort, Dorion Sagan, Landi Stone, and the Richard Lounsbery Foundation.

"Marriage of Convenience" originally appeared in *The Sciences,* September/October 1990.

"Swimming Against the Current" originally appeared in *The Sciences,* January/February 1997.

"The Uncut Self" originally appeared in Volume 129 of the *Boston Studies in the Philosophy of Science,* Alfred I. Tauber's Symposium "Organism and the Origins of Self," Kluwer Academic Publishers, Boston.

"Power to the Protoctists" originally appeared in *Earthwatch,* September/October 1992, and was prepared with the help of Landi Stone and Nini Bloch.

"From Kefir to Death" originally appeared in *How Things Are: A Science Tool-Kit for the Mind,* edited by John Brockman and Katinka Matson (William Morrow and Company, New York, 1995).

"Kingdom Animalia: The Zoological Malaise from a Microbiological Perspective" originally appeared in *American Zoologist,* volume 30 (1990). The article was prepared with the help of Rene Fester, Tom Lang, and Gregory Hinkle, and employed ideas from Gail Fleischaker, Francisco Varela, and Humberto Maturana.

"Speculation on Speculation" originally appeared in *Speculations: The Reality Club,* edited by John Brockman (Prentice Hall Press, Englewood Cliffs, NJ, 1988).

"The Atmosphere as the Circulatory System of the Biosphere—The Gaia Hypothesis" was finally accepted by Stewart Brand for the *CoEvolution Quarterly,* Summer, 1975.

"Gaia and Philosophy" originally appeared in *On Nature,* edited by Leroy Rouner (University of Notre Dame Press, 1984).

"The Global Sulfur Cycle and *Emiliania*" originally appeared as "Sulfur: Toward a Global Metabolism" in *The Science Teacher,* January 1986.

"Descartes, Dualism, and Beyond" is based on Chapters 2 and 9 of *What Is Life?* by Lynn Margulis and Dorion Sagan (1996). *Quark* (a magazine published in Barcelona) published a Spanish version on pages 13–21 in their April–June 1996 issue dedicated to René Descartes, entitled "Descartes: El Sueño y la Razon" (Descartes: Dream and Reason). Translated by biology professor Mireia Artís, who works at the AMU of Mexico at Iztapalapa, it was extended and modified by Prof. Ricardo Guerrero at the Department of Microbiology of the University of Barcelona, Barcelona, Spain.

"What Narcissus Saw" originally appeared in *Speculations: The Reality Club,* edited by John Brockman (Prentice Hall Press, Englewood Cliffs, NJ, 1988).

"A Good Four-Letter Word" originally appeared as "Gaia: A Good Four-Letter Word" in *The Gaia Magazine,* Issue 3, 1991.

"The Biota and Gaia: 150 Years of Support for the Environmental Sciences" originally appeared in *Scientists on Gaia* edited by S.H. Schneider and P.J. Boston (MIT Press, Cambridge, MA, 1991).

"Gaia and the Colonization of Mars" originally appeared in *GSA Today,* November 1993. Dorion Sagan co-authored a first draft; E. Moores and David Snoeyenbos provided encouragement, editorial assistance, and useful discussion; Donna Reppard and Landi Stone helped with manuscript preparation.

"Futures" was adapted from material appearing in *Microcosmos: Four Billion Years of Microbial Evolution* (see p. 328), by Lynn Margulis and Dorion Sagan (Summit Books, 1986), and in "Gaia and the Evolution of Machines," by Dorion Sagan and Lynn Margulis (*Whole Earth Review,* volume 55, 1987).

"A Pox Called Man" originally appeared in *Science for the Earth,* edited by T. Wakeford and M. Walter (John Wiley and Sons, New York, 1995).

"Big Trouble in Biology: Physiological autopoiesis versus mechanistic neo-Darwinism" originally appeared in *Doing Science: The Reality Club,* edited by John Brockman (Prentice Hall Press, Englewood Cliffs, NJ, 1990).

"The Riddle of Sex" originally appeared in *The Science Teacher,* volume 52, March 1985.

"Words as Battle Cries—Symbiogenesis and the New Field of Endocytobiology" originally appeared in *Bioscience,* October 1990.

CONTENTS

Introduction

This collection of essays, a chorus of two voices with occasional guest artists, spans the forty-year period from 1956 until the end of 1996. While this book is a work of science in that all but the memoirs, the first two essays, are based on scientific investigation, we have taken the liberty of speaking out on issues widely considered taboo in scientific circles. Our motives are to show how scientific ideas should influence the world outside of science and to help lead the way toward a new, perceptive, critical, and responsible outlook for research science in the next century.

As both of us have talked over the years with scientists and other thoughtful people, we have come to see twentieth-century science and technology in the light of religion. By this we mean neither the secular Judaism and agnosticism of our youth nor today's commercial Christianity. Nor do we refer to the chanting Buddhism (Zen or SG International) that some of our formerly Christian friends have recently embraced with such fervor. We certainly are not speaking of the scary anti-intellectual and sexist beliefs of Creationist or Islamic fundamentalists nor the simplistic animism-paganism of nature worshippers. We do not refer to any kind of dogma; rather, we mean approaches to the questions that are traditionally asked by the great faiths. We mean religion in its original sense of *re-ligio:* re-ligate or bind together again. How did the world begin? What is the world? Who are we and where are we going? What are the relations between us? What is the purpose of life and death?

Although these questions are posed by religion, to answer them one must turn to science. But science is pragmatic: research investigations are limited to the physically verifiable. Scientists tend to return the large, unverifiable questions to philosophers and theologians. Isolated in their tiny "disciplines," professional research scientists must pay extraordinary attention to detail, a stance that not only makes science inscrutable to most readers but nearly precludes concern with the "big questions."

On rare occasions when science addresses a big question directly, for example, in documenting the origin and history of life, it provides far more insight than does revelation, meditation, or any other way of knowing. Thus if the nineteenth-century conflict between the Book of Genesis and the account of Earth history pioneered by James Hutton, Charles Lyell, and Charles Darwin has been definitively resolved, it is because on this issue traditional religion was vanquished. This, to us, is not the loss for religion that many assume. Rather, dogmatism was diminished while the urge to seek satisfying descriptions, provisional but cosmopolitan truths by verifiable means, was gloriously vindicated.

Science is an intrinsically flexible way of knowing requiring the activity of many talents. No single individual can accomplish a scientific feat, whether of insight, invention, or global discovery. Science is interactive and international, and in spite of delays caused by greed, private interests, intellectual laziness, or political barriers (to name a few obstacles), the results of science cannot be privatized: ultimately they belong to the world. The flexibility of science is a corollary of its worldwide dissemination and its pragmatism. Though errors may for a time be accepted as truth, eventually they are revealed and purged from the world's accumulated scientific literature.

Although dismayed by the closed-mindedness of traditional religions and their insistence on a special relationship with truth, we note that science too has its dogmas and dogmatic adherents. When religious zeal is injected into science, and the role of human sensibilities in the pursuit of knowledge denigrated, the consequences are equally dismaying. We consider the claim by neo-Darwinists that parental investment into male children can be precisely calculated, for instance, or that the principle of group selection has been proved wrong, as absurd as claims of a flat Earth or an immaculate conception. The persistent denial of the existence of emotions in nonhuman mammals is another absurdity. We suggest that nonhuman mammalian sexuality directly relates to human sexuality; complex behavior and development patterns, and even our moral concerns, are potentially subject to meaningful scientific inquiry. Research into these sensitive issues cannot yield understanding and insight if not undertaken by thoughtful investigators. No scientist should be deceived by a pretense of objectivity, and no information can be garnered on delicate issues by single researchers acting alone. All science is a highly social, self-correcting, interactive enterprise.

Our essays are as scientific as we can make them in that they are based on responsible work by dedicated, if occasionally gullible and imperfect, investigators. Taken together, the "big question" they address is the extent of connectedness of the elements of the living world. The familiar abstraction of the unity of life and its environment on the Earth's surface has implications humanity has barely begun to understand. We can summarize our view in the three key words of our subtitle: Gaia, symbiosis, and evolution.

Gaia is Earth's physiology: the sum of the energy- and material-exchanging activities of the living network at our planet's surface. The concept of Gaia, formulated by James E. Lovelock, is arguably the major spinoff of the international space program. Gaia theory, put simply, views Earth's biosphere (the place where life exists) as a single, self-regulating entity: the Earth is alive. Gaia theory embodies the concept of "life as a geological force" and furthermore posits that the conditions essential for life to continue, such as an average temperature that permits liquid water and the chemical composition of the soil and atmosphere, are maintained by living matter itself, using energy from the Sun. Nothing mystical is meant here; we suggest no conscious, benevolent goddess or god. When we speak of mammalian physiology and the mechanisms by which body temperature or calcium, chloride, and potassium ion concentrations in the blood are kept within narrow bounds, these are scientific, not philosophical or theological, matters. Similarly, when we speak of geophysiology, of the mechanisms that for several billion years have kept Earth's atmospheric composition so far out of equilibrium yet stable, we speak about flows of energy and matter that can be scientifically investigated.

Symbiosis refers to the physical connection between organisms of different species. These partnerships are often very strange. Members of species that are only very distantly related may be intimately connected through their roots, through holes in their exoskeletons or skin, by blood ties, and in many other ways. Strictly speaking, to be symbionts individual members of at least two species must touch each other most of the time. If we relax this criterion slightly and allow contact at one remove, we immediately see that all living things on Earth are in physical contact through its water, atmosphere, and soils, and that they all dwell in a coating on the surface of a limited planet. Gaia theory provides the framework through which to investigate the details of how all organisms exchange gases and other chemical compounds with the water, atmosphere, and soils of Earth. As Gregory Hinkle has put it, "Gaia is simply symbiosis as seen from space."

A modern discovery of biology is that some symbioses are contingent: the partners, like guests, come and go as conditions dictate. Other symbioses have become permanent partnerships as once-flexible associations have stabilized. As the former guests stayed and became incorporated, these codependencies led to new composites. Like the chimeras of Greek mythology, the cells of all animals and plants are such composites, formed from permanently fused bacteria. Because all life is directly or indirectly connected with all other life, these symbioses—loose and tight, permanent and temporary—form the components of Gaia. Gaia is indeed symbiosis as seen from space.

Symbiosis challenges our views of two fundamental realities: individuality and death. Bacteria, which are true individuals, ordinarily do not die, although of course they can be killed. Given food, water, and energy, they simply continue to grow by cell division. The first time "programmed" death appears as a predictable feature of the life history is in organisms which evolved as symbiotic bacterial communities: eukaryotes. A budding (growing) yeast cell studied under an electron microscope, for instance, reveals a scar in the place where the new bud appeared. Further scarring forms at different sites on the cell surface as more and more buds appear and break off to form new cells. When the original parent cell has generated some thirty buds, it stops reproducing and dies. Unlike bacteria, then, which are potentially immortal, the yeast cell does not reproduce forever. Animals and plants that must reproduce sexually have carried cell death even further. Two kinds of cells can be distinguished: those of the body, which die out, and germline cells, which retain their ability to reproduce new organisms by cell division. What we call the individuality of these organisms is actually a complex symbiosis of many formerly free-living component organisms that entails constant adjustment and integration. Programmed death permits the incessant reiteration this mode of life requires. One of life's many inventions, reproduction through sexual fusion of mates and the continual imperfect reiteration of forms, is a key elements of our subtitle's third term, evolution.

Evolution is history: it is simply defined as change through time. Stars and galaxies, solar systems and planets, life-forms and societies all evolve. The existence of evolution is undeniable, and how it has occurred is worthy of careful study. The evolution of most concern to us in this book is environmental evolution: how the surface of the planet has changed in response to life, and how life has evolved in response to the evolution of Earth. The biosphere is very old. The phenomenon of the living Earth dates back nearly four billion years, almost to the very beginning of the planet's existence. The continuity and unity of Gaia is clearly evident in the genetic systems of its component organisms: molecular biology shows convincingly that all life on present-day Earth shares a common ancestry. Thus there is an intimate link between evolution and Gaia: evolution connects all life on Earth through time, Gaia connects all life through three-dimensional space.

We welcome discussion, disagreement, new information, and criticism. Perhaps some readers will vehemently take issue with our claims. Even they, we hope, will welcome our extensive bibliography. For ease of reference, we have compiled a single alphabetized reference list at the end of the book.

Lynn Margulis and Dorion Sagan
Amherst and Northampton, Massachusetts
May 1997

Tell all the truth but tell it slant—
Success in Circuit lies
Too bright for our infirm Delight
The Truth's superb surprise
As Lightning to the Children eased by explanation kind
The Truth must dazzle gradually
Or every man be blind

Emily Dickinson
1830–1886
Amherst, MA

PART I

MEMOIRS

1

SUNDAY WITH J. ROBERT OPPENHEIMER

Turning and turning in the widening gyre
The falcon cannot hear the falconer;
Things fall apart; the centre cannot hold;
Mere anarchy is loosed upon the world,
The blood-dimmed tide is loosed, and everywhere
The ceremony of innocence is drowned;
The best lack all conviction, while the worst
Are full of passionate intensity.

. . .

"The Second Coming"
William Butler Yeats

September, 1986

As the men talked quietly, I imagined Kaori, my college roommate's aunt. The family chronicle was recounted by her uncle—the deaths of his wife and daughter. Aunt Kaori, he said, had groped for her baby who had slipped away from her, but could not move because only stumps and cascades of dark blood remained where her legs had been. Her slippery child, still trying to scream as charred black peeled off her cheek, died. Kaori managed to touch its thigh. Like Kaori and her daughter, one hundred thousand others perished instantly after that first still moment. And like my roomate's surviving uncle, many more were never again entirely alive.

We were seated comfortably in Cambridge—surrounded by large format atlases printed in Italy, origami miniatures (cranes, butterflies, squares) strewn on the floor, and the clutter of a too-ample late supper—talking about Hiroshima, about that day forty-one years ago. Despite my vivid images of Aunt Kaori, I was listening intently and more quietly than

I usually do. David Hawkins, my friend, and Phil Morrison, our host, were reliving it again, Morrison grimacing. Hawkins, philosopher and educator, Los Alamos veteran, was the author of the official history of the making of the bomb. His self-accusations of pedantry and churlishness in his own writing—a solicited report so detailed he claimed it was unintelligible even to him—did not take the catch out of his voice as he spoke.

I had heard of Professor Morrison long before meeting him. Thirty years earlier my then boyfriend, Carl Sagan, an aspiring scientist, had told me more than once: "Morrison armed Fat Man at Tinian." Tinian was the island in the South Pacific from which the atom bomb to Japan was launched, Fat Man the nickname of the bomb: Morrison, Carl claimed, had helped load the bomb onto the plane.

As if mocking how finite the post-Apollo, plate-tectonic Earth had become, a three-dimensional globe—on a turntable pedestal, lighted from within—cast shadows over us. I fixed upon the mid-Atlantic ridge, visible on the glowing planet.

Phil and David were remembering arguments with the great leader whose passion it had been to build the bomb: J. Robert Oppenheimer. The issue was whether or not to deploy Fat Man—whether or not, after the bomb had been successfully developed and tested, to drop it.

Phil looked through me. "Oppenheimer wanted to drop it; he felt it *must* be dropped. It was his Fat Man, his son, his invention. But not his alone—it was not his possession—rather his shared glory. It was living proof that all the esoterica, all the equations, all the arcane formulae and the ineffable mystery was something. Something tangible. Something of economic and political value."

He paused, neatly cracking open an almond shell with a table knife. "All the effort, the great quantities of money, were worth it. Our activities were not just the poetry of meaningless mathematics. For nearly three years Oppenheimer labored ceaselessly; with him, for him, we worked as hard. It was our labor of love. After all, we were stopping carnage in Europe, the destruction of civilization, the Hitler madness. We had the means to make the difference. Of course, Oppie argued we cannot just *have* the bomb. We must *drop* the bomb."

Phil ate the almond.

"I was such an idiot. I admired Robert Oppenheimer. He was of course my senior and my superior. It would be presumptuous to say I loved him or even feared him—Oppenheimer filled me with angst. Whatever the case, I listened to him. He had many arguments," Phil ticked them off. "We had the most dangerous weapon in the history of mankind. This A-bomb was qualitatively different; it wasn't just more TNT. The A-bomb could generate apocalypse. It absolutely had to be contained by common consent. But how could the public restrain a monster they could not even imagine? We were responsible for public display of the A-bomb. We must unequivocally demonstrate the enormity of potential devastation." He paused.

"We had to end all war. We could end all war. Therefore, the bomb had to be dropped in an unequivocal manner—there must be thousands of eyewitnesses to its destructiveness."

We were silent as the shadows continued to spin.

"I agreed with Oppenheimer," Phil was nodding, slowly. "I wanted to drop the bomb to reveal indiscriminate horror; I was sure my fellow man would end war. All sane people, including even the most dedicated military men, would see that war where one bomb could obliterate New York or Paris was unthinkable. At first we had paper calculations proving that a single detonation might even burn the air itself, the potential conflagration of the entire atmosphere. OK, they proved incorrect . . . the probability of burning air was vanishingly low. Still I was nagged for years by a persistent nightmare of the final whole-Earth bonfire.

"Who was I? A young physics student. Very close to the action and passionately interested but, as usual, overassessing my own importance. Inspired by my own sense of moral imperative I spoke out with the assurance of the young. We met nearly every night then—any scientist or technician could come and, of course, I always went feeling swept by the big fact that we were making history. I always spoke up, even then. I insisted on a public A-bomb explosion. I knew Oppenheimer was right, we must demonstrate the power to as many military people from as many places as possible. This explosion—the very fact of it—would reveal how ludicrous was the continuation of weapons development. But I suggested a well-publicized display of the bomb in the Pacific Ocean. President Truman would invite not only our allies but Japan and Germany too. I found out, only later of course, that I was far from alone in advocating a grand public demonstration.

"Oppie vehemently opposed demonstration. He was emphatic: 'For us to stop all war, we have to stop this war now.' We must launch a surprise attack that counted, that would demolish a significant Japanese military target. We must demand the immediate surrender of Japan and Germany. Oppie had said spookily, I remember it so clearly: 'We must *use* Fat Man. We must bomb Berlin and Tokyo simultaneously.' "

Phil held the pedestal, spinning the lighted globe hard, throwing shadows. He cracked another almond.

"Why do you say you were such an idiot?" David asked.

Phil looked, distracted, past David, past me, past the globe. "After weeks of pondering, arguing, and debating we were all just about agreed: General Leslie Groves, the other physicists, everyone, even the wives—but mostly Oppie. We listened to Oppie that night after supper, in the big hall. The consensus was," he began to hold up his fingers:

"One, we must drop the bomb to show the world its power—to end war.

"Two, we'd drop it on Japan first, on as strictly a military target—like Hiroshima or Nagasaki—as we could find. If Japan didn't surrender after Fat Man was deployed, then we would have a contingency plan to hit Germany immediately afterward with the second bomb.

"Three, we must drop the A-bomb as soon after testing as possible.

"There were only a few remaining questions. How much prior notice should the world be given that the bomb attack was real and inevitable? Who should be notified? The entire literate world? How much notice? An ultimatum should be issued first. At the very least, there must be time for evacuation.

"And I was an idiot. In the beginning I argued, I wanted to see as much publicity beforehand as possible. In the end, I was for no notification of anyone. I was an idiot."

"Why?" David asked.

"I let one argument sway me," Phil was nodding again, "an argument to which I never should have listened. Can you guess what it was that convinced me to let us go ahead and drop the bomb without telling anyone?"

David answered, "Probably no one listened to you because Oppenheimer—in his typically ambivalent way—wanted to feel the full power of the moment: the bomb released at a target and on a schedule to which only he was privy. You were too young anyway, who would listen to you?"

Phil shook his head, "Yes, I was young but that wasn't it. I don't think Oppenheimer was such a power-seeker. That wasn't it. It was the crew. It was all about the American crew—I knew the pilot: Tibbets."

"What do you mean the crew? Who is Tibbets?" David asked softly. I held my breath.

"They pointed out—Oppie, Groves, and others—that if we gave any notice at all we would be rightly accused of murdering the crew. If warned, the enemy defense would get ready; the plane delivering the bomb would surely be shot down. They'd be dead, sacrificed. The crew would become a rallying cry, scapegoats of physicists' whims. There were, in fact, several potential pilots. All had the requisite technical training. Two or three had been at Los Alamos for years with us; I knew them all by sight.

"But Tibbets I knew best. One of his sons was already a superb pianist with an ambition to compose. I thought about the inevitable death of the father. I ended up agreeing to a secret bombing. I thought I was saving this pilot's life. One life against one hundred thousand."

Pain showed through Phil's eyes. David looked away.

"But if you had another chance," David asked, "what would you have done—if you knew what you know now? You were only twenty-nine then, after all. What would you have decided?"

"I don't know." There were shadows in Phil's face. "I do not know. The public still does not understand. No one fathoms the seriousness of these weapons. Sometimes I still think we need more demonstrations—not in New York City or Washington, D.C., but say in Fairfax County, Virginia or Bethesda, Maryland. Or here on the Cambridge Common, Westchester County, or Grosse Point, Michigan." He paused. "I'm not seriously advocating bombing residential areas.

"In fact, I guess if I had it all to do over I'd end up listening to Oppie again. He conceived the project, he brought us together to work like demons, he inspired us with the physics, he exuded ambience—it was his sweet victory. At that moment, I trusted his wisdom. It wasn't just loyalty to my leader. It was that JRO was sage. I still believe he believed his goal was to stop Hitler in particular and war in general. Yes," Phil's voice was sadly certain, "Yes, I probably would just support Oppenheimer again."

We dispersed shortly after, serious conversation reverting to the superficial as we descended the stairs—boom in health food sales of *Spirulina*, bacteria sold under the safer, plant-sounding name of "blue-green algae."

As we entered the sudden evening chill, I remembered, first with clarity—Aunt Kaori's charred daughter and then, joltingly, my version of J. Robert Oppenheimer. Oppie, the man who had changed the world in ways he never knew and could never have predicted. Until this conversation at the Morrisons', I had forgotten what I knew, thought I knew, of Professor Oppenheimer. A long winding of years, indeed, separated 1986 from 1955. I, too, had admired Oppenheimer.

The hand that signed the paper felled a city;
Five sovereign fingers taxed the breath,
Doubled the globe of dead and halved a country;
These five kings did a king to death.

"The Hand that Signed the Paper"
DYLAN THOMAS, 1936

March, 1955

I was literally sophomoric and had just turned sixteen. The winter quarter at the University of Chicago had just ended; schoolwork finished was a great relief. Overassessing my talents, unconscious of my self-centeredness, I was convinced that great insights were revealed in my latest brilliant paper: "Not 'Whether or Not?' but 'How?'—J.R. Oppenheimer and the Decision to Drop the Atomic Bomb." Of all the pages of analyzed quotations, Oppie's 1954 statement kept playing in my mind.

> It is my judgment in these things that when you see something that is technically feasible you go ahead and do it and argue about what it is only after you have had your technical success. That is the way it was with the atomic bomb. I do not think anybody opposed making it, there were some debates about what to do with it after it was made.

My boyfriend, one of many, coaxed my company for his spring-break trip East. The Princeton campus—a place I had always wanted to visit—was, he urged, only 900 miles from Chicago. Looked at another way, it was only half an hour drive to Rahway, New Jersey, Carl's home town. Reluctant to drive thirty hours alone, Carl was equally ambivalent about taking me the final thirty minutes to Rahway. He wanted to avoid the inevitable flurry were he to arrive home with a young woman.

Carl, less sophomoric than I, enjoyed an exaggerated sense of his own importance. He had only superficially understood Copernicus' lessons: Galileo's idea had been taken in logically but not emotionally. Despite centuries of thought succeeding Ptolemy, my aspiring astronomer functioned as though the Earth (Rahway in particular) were the dead center of the universe. His mother, incessantly orbiting around her son, would ask unanswerable questions:

"Where did you sleep on the road?"
"What are your intentions toward your young lady friend?"
"Are you going to interrupt your studies?"
"What kind of family does she come from that they permit her to travel such a long distance with a young man—unaccompanied?"
"Do they have money?"
"What does her father do?"

Resolved. I was invited, but not quite for the distance. Nevertheless, I assented, out of Princeton curiosity and Carl lust. Leaving on a mid-March Friday night, we shared the burden of the drive: Indiana, Ohio, and, especially trying, Pennsylvania's interminable new turnpike. Finally, Princeton, New Jersey; with a peck on the lips, Carl dropped me at the Nassau Tavern and accelerated, alone and unembarrassed, homeward.

To pay for the Nassau Tavern, I carefully counted out money from meager after-school earnings. Since September I had set up pins in a bowling alley for the women's Physical Education classes. Counting out the bills, I could hear smooth balls echoing down shiny alleys and shocked pins bouncing. Sheeeeeeeeeen, cluey. The tavern was not pompous. As Aunt Kaori's husband would say, it was *shibui*—studied, quiet elegance; *shibui* in a familiar Yankee American way.

It was dusk when I settled into the spare, comfortable, maple-furnished room. I telephoned home. As expected, my father answered—he hated Saturday night crowds so had made going out only on Fridays part of his religion. I tried my plan on him: "I'm in Princeton. I'm thinking of visiting J. Robert Oppenheimer."

"Do it!" Pa responded straightaway. "Now is the time. You may never have another chance. Tell him you admire him." Pa was excited. "Tell him—now that he's been put down, now the publicity's faded, tell him lots of us respect him. Do it now," he shouted with characteristic enthusiasm. "He's Jewish too, you know. At least," the slow sigh, "he used to be. That's what making waves does to *Yekehs*." (My father, not without a tinge of envy, always called German Jews, whom he supposed to be his cultural—but not genetic—superiors, *Yekehs*). "Go on—it's now or never. Go see him Lynnie. Now. You'll never have a chance again."

The media had tired of Oppenheimer. Maybe because the public had villainized him, I thrilled at every mention of him. His history, the irony of his life's work, the extraordinary blue eyes all fascinated me.

Did he have trouble sleeping at night? Could he have sensed at all, worrying through theoretical equations, that their application would lead, finally, not only to his own fall but to the destruction of thousands of people? Was it as clichéd as climbing a mountain because it's there? Could it possibly be human nature to do what the tools enabled, the "technologically sweet," no matter what the consequences? Was Dr. Oppenheimer, as he grappled in Los Alamos, thinking of the stinking meat smell of gassed Jewish flesh, was he trying to stop the Nazis? Or was he just glorying in the chance to push his intelligence to its limits? What did he believe; rather, what did he believe he believed?

Oppenheimer's story obsessed me. Could I ever lose myself like he did? I loved science, could I ever hunt and then be haunted by its applications, forgetting about the knowledge in and for itself?

By March of 1955, Oppenheimer's fame had metamorphosed in a way I could imagine terrifying to such a public person. He did not matter any more. No one cared. The FBI had accumulated voluminous files for many years, then, abruptly, "case closed." He might as well have been dead. The man revered as an American hero, then beaten by scandals, had retreated, withdrawn from the public eye.

The weather when I awoke was delicious—cold, bright. I still remember the distinct brown and yellow sunbeams, the sweet smells of eastern spring as I gazed across Palmer Square from my second floor window. Just as the man selling newspapers blew into his cupped hands, I saw the headline printed in red: "We Meet the Oppenheimers."

Columnist and owner of the *Post,* Dorothy Schiff's "Dear Readers" column took the front page, an attempt to wring some news from a case closed since the summer of 1954. Schiff was "surprised to hear Mrs. Oppenheimer," the German-born "femme fatale" who had led a dramatic life, "speak American so well." "She sounded," Schiff crooned, ". . . like a Park Avenue socialite; her face shows the ravages of strong emotion rather than time." Katharine Oppenheimer reminded Schiff of a "flaming youth" of the 1920s; a heroine in an F. Scott Fitzgerald novel "capable of romantic and reckless action." And Oppie, "With his wide mouth and liquid blue eyes looked like a man who had been crying, but . . .

> I think he is too aloof, too disdainful, too philosophical to indulge in self-pity. Something about his ears and the way he moved reminded me of a fawn. Perhaps he is more like a chameleon. . . . That afternoon he was not the frustrated fa-

> natic I had seen on the Murrow program, the abstruse poet I had heard on a Columbia University broadcast, nor the trapped scientist-turned-politician who emerged from the testimony. . . . Someone at the luncheon had asked Oppenheimer if he thought the H-bomb would ever be used. For the first time he laughed, but not merrily. 'Only a sphinx could answer that,' he said, adding something about the bomb's 'limited use,' which, to my dismay, he seemed to advocate.

I grew angrier as I read. Schiff's article reflected both the attitude of the public and their flagging interest in the Oppenheimer case. It personified for me the society from which he had withdrawn: prying eyes, vicarious thrills, superficial judgments, facile criticisms of what was not understood, patriotism and parochialism, fear of imagination and culture, and abysmal, abysmal, ignorance of science.

What could Dorothy Schiff possibly know of science, of love, of the too human failure of judgment? At sixteen, with my term paper conflicts and subtleties completed, I clearly understood. He, personally, had had the power to do something about stopping war and stopping Nazis. Wasn't to play technologically, to explore and deploy as intrinsic to our nature as making love?

I put the *Post* away, ate breakfast, and then, simply, looked up the address of Dr. Oppenheimer in the Princeton telephone directory—Olden Lane. A short walk, five-minute hitchhike, and a little local aid brought me to the Olden Lane street sign.

I scrutinized every house on the Lane. The first faced another street, the second one too. I assumed they did not have Olden Lane addresses. The third house—charming, old, and weatherbeaten—directly across the wide lawn of the Institute, seemed likely. No name at the door. In front on a fence a sign: "Drive Carefully Children Playing." The fourth had a name, not Oppenheimer. The garish, tiny, fifth house looked improbable. The sixth had a name plaque. At the seventh I was approached by a menacing dog. The eighth looked too square and modern. I returned to the third house.

As I walked toward the door I saw Dr. Oppenheimer, his wife, and two children coming toward me. I mumbled that I was "here" in Princeton. Studying science at college. Glad to see you, excuse me, just a minute—well—from Chicago, yes, yes. Glad to meet you. Going downtown—I, with you? Certainly, thank you.

Joining them in the white Cadillac convertible, I tried to sound modest as I boasted that I was the last of the liberal arts students to be graduating The College under the Robert M. Hutchins A.B. plan. "You must be smart," said the son, Peter, born in August '41 and therefore just fourteen.

"Peter," Mrs. O. spoke slowly, too distinctly, "has just published his first article. It appeared in a column called 'Small Talk' in the television section of the university newspaper, ***The Princeton,*** with his byline." Silence.

"Oh," was the best I could do, not sure what to say. I wondered if Dr. O. felt as oppressed with his wife as I did.

We arrived at Palmer Square; I tagged along self-consciously. Mrs. O. looked in a shop window at Wedgewood china. "That," I offered "could not be Wedgewood; it has no wedges." She snapped a correction: "No, dear, I'm afraid you're wrong. I have bought Wedgewood sets several times without wedges."

Dr. O. went for tobacco and the newspaper, their ostensible reason for driving to town together every Sunday morning. There on the front page was Dorothy Schiff's "Dear Readers" with the dreadful banner: "We Meet the Oppenheimers—see page M7." I was embarrassed, for myself and for them, already feeling awkward in anticipation of the discussion it might provoke. I bought a carton of milk and walked across the street to drink it as Dr. O. and Peter were coming back. They waved. I offered pudgy Peter some milk, which he declined, saying that he only drank skim. My milk carton and I must look silly, I thought. When Dr. O. claimed he didn't mind I realized I had said it aloud. "Harvard," he told me, "is a finer school than the University of Chicago."

"Robert, we need your help," Mrs. O. had opened the delicatessen door and was calling to us.

"Mine too?" I asked, not realizing the help was for choice of tobacco brand. Peevish. "You needn't come in unless you like this store particularly."

I waited outside until Dr. O. joined me again in the street.

"About a month ago there was a screening of the longer version of the Murrow TV film about your work," I was enthusiastic, "Mandel Hall, the largest University of Chicago auditorium, filled up twice!"

His voice was contemplative, "Murrow's hour and a half film interview gives an audience a better idea of the discussion and even though the conversation, which lasted two hours, was cut, the film was much better than the forty-five minute TV show. But in both"—here he was emphatic—"the physics expositions that had had some continuity were lost on the audience. The cuts were unjust."

Then, he smiled. Those eyes looking directly, asked warmly if I would like to come back to their home again. Mrs. Oppenheimer, coming out of the deli, overheard the invitation; she grimaced.

"Would I be disturbing you?"

She replied, a purposefully unmistakable tension in her voice, "Quite frankly, you'd be welcome only if you'd stay *just a little.* This is the only chance we have to be with the children, isn't it, Robert dear?"

"About twenty minutes," he added encouragingly and quietly so his remark went unnoticed by her. "Come, do come home with us, I have something to show you." I was delighted.

On the way back in that luxurious car we talked more about Murrow. The first time he had seen the finished TV film it had been in lovely surroundings. "The Murrows invited me up to their home; the poet and Abraham Lincoln biographer, Carl Sandburg, was there, also a guest. I had never met Sandburg before. He was very impressive. Sandburg is all that we generally think he is and more. . . . He also is a real man."

"Murrow?"

"No, Sandburg. Murrow is too, but Murrow is just like his television appearance; it reflects him accurately." Dr. O. went on, clearly talking directly to me.

Peter was sheepishly buried in the Sunday papers, hunkering down in the back seat with sullen Toni, the ten-year-old daughter. Although he's not more than two years younger than I am, he's a child and I'm an adult, odd. Peter, half under his breath, apologized for his dedication to the funny papers spread around him. "I'm not an intellectual type," he nodded as I scanned the "Peter Rabbit" he was reading.

Dr. O. handed me Peter's "Coming Attractions" column. "Peter is to be a regular contributor with a byline to the TV section." Mrs. O. interceded, audibly mumbling "unfair." She requested, rather fiercely, that I give her the paper.

Climbing out of the car, Dr. O. took my frowsy jacket and hurried to show me the thing he had promised. It was a Van Gogh original—a striking, colorful farm at sunset; the stars amazing and amazed.

"This was part of my father's collection in San Francisco." He smiled wistfully and showed me another framed work, a pastel by a Frenchwoman, Belle Greene—was that the name of the person he mentioned? "I met the painter, a French contemporary of mine, not very well known.

Greene died recently." Then, after he had examined with approval my large acetate scarf, tied and adjusted as a book sack—Harold Blum's thermodynamics book, *Time's Arrow and Evolution,* and a frayed, secondhand Kafka *Metamorphosis*—he showed me into the study.

"I apologize," he said, "for the sad state of my polylingual bookcase." There were academic German books, mostly mathematics, some French novels, and even a few Russian ones on the science shelves, a two-volume Chinese art set from Taiwan, English and American poetry. He took up his pipe and opened a beer, obviously not his first of the day. He urged upon me a cigarette and whiskey or beer, which I refused. He pointed out books to me: a beautiful handwritten, illustrated Blake. Another, Eric Bentley's *In Search of Theater.* "Bentley is a professor at Bennington and Rutgers. He works occasionally at the Institute;" he explained, "we were classmates." His talk wandered dreamily.

I felt privileged.

"I've been at the Institute for seven years—I was in New Mexico during three of the war years. The name of this house is really Olden Manor. I just this week received a short-story collection—autographed," he added with a quick, proud smile, "from Carson McCullers."

"Is it *The Ballad of the Sad Cafe* by any chance?," I asked. "Yes, I think so." I told him I had read it and found it perfect, mythic. He asked me if I planned to write.

"How did you guess?" The question was silly, but it was already out of my mouth.

"Well—don't go to school for too long," he suggested, "and certainly don't go to study writing. You should go to Harvard." Then, as an aside, dryly, softly, and without much interest, he said to Mrs. Oppenheimer, "Kitty dear, you are not very entertaining."

She answered bluntly, "Well?"

"I imagine many people annoy you," I said, reserved.

"They do, they certainly do, don't they, darling?"

"You'd rather entertain your children, I'm sure," I said.

"We will, after you leave."

But then, Dr. O said, "Oh, no. Do stay and eat with us." He invited me with real feeling in his voice. He seemed sincerely to want me to stay.

"No, thank you very much"—it was hard, I was flattered—"but I'm not hungry now." Mrs. O. flashed him a triumphant smile.

Then, amid the tension that we all seemed to be feeling, we discussed the Schiff article. Dr. O. said that Mrs. Murray Kempton, whom they hardly knew, will more feel the repercussion of that trumped-up "luncheon meeting" that she will more bear the consequences than he. "The conspirators usually suffer more than I when they are responsible for my bad press." He paused. "They usually are."

He and Mrs. Oppenheimer had accepted the invitation and agreed to the social lunch without any notion, of course, that a working reporter would be there. "Schiff maneuvered an interview by using her friends, Lloyd Garrison" (I had read that he was Dr. O's chief counsel) "and the *Post's* editor, Murray Kempton. She obviously savored every minute of that contrived lunch visit . . ." He hadn't felt betrayed, really, only impatient as if he were outside of it, a spectator, perhaps, at a silly performance.

"I thought the article was stupid," I said.

"Yes," he went on. "I imagine it would seem confused and almost incoherent to the reader, although in fact it was just poor manners, crude."

Mrs. O. sat conspicuously reading the paper. With a sort of syncopated rhythm she kept glancing up at me. She eyed her watch, then glared forward into space. She looked harsh and determined. It was precisely 25 minutes since we had come in from the car; there was time for nothing more. Dr. O seemed to be dreaming, letting his thoughts unravel. Then, suddenly, noticing his wife's rigid posture, her eyes staring with intense, stifled impatience, he returned to alertness, not without revealing to me a twinge of regret.

"Well, goodbye," I said, politely shaking her hand. I turned to Dr. O. "It's been wonderful to meet you." He invited me, please, to come again soon—he must have guessed how I thrilled at his sincere invitation.

Furtively glancing at his tight-lipped wife (who was now, with some relief, leaning over the Sunday newspapers in concentration) he checked that she wasn't looking and smiled, warmly, deeply. He meant it; I tingled with pleasure. He shook my hand, gazed out affectionately at me, then blankly at the street, with melancholy, gleaming blue eyes.

I walked back to the Nassau Tavern in a dream, wondering about my foray into the Oppenheimer family. I was certain that although Mrs. O. still resided with him, she had long since deserted him. She obviously did not share his interests, needs, nor aesthetic discernment; she clearly wanted no part of his battles with himself.

He had wanted me to stay. I had been permitted to catch a glimpse of, to feel the "intellectual sex appeal" that had attracted physicists to Los Alamos in 1943. He seemed like an aging stallion staggering from a broken spirit.

Did he ever think of Jean Tatlock? Had he ever, with his infamous antifascist lover, enjoyed a relationship of a higher quality, a special plane of intimacy? Had he simply mismeasured Kitty's character, capabilities, during the exciting days of November 1940? Was her bitterness, the tension, the clash of wills, none of which they had made any effort to hide, all that was left between him and—as Dorothy Schiff had written—his "petite, chic, witty, tense and vivacious" wife?

Had he felt a sense of failure or frustration or pity after all those years of pacing the corridors of power when, in the end, his security clearance, a symbol of his former influence, was denied? Was he left to feel it alone?

As I turned down Olden Lane drunk with Dr. O. and scents of spring, grinning with self-satisfaction and vanity, scarf with books tied over my shoulder, I easily convinced myself that I had known him better in that precious hour than Kitty had in all her years with him.

The adolescent euphoria receded; I had second thoughts. By the time I reached the hotel it began to come through to me. How brash to think that in one hour of invasion I could know anything at all. I thought of Schiff and her shabby article. I began to feel a bit ashamed. "You never know," a fellow student had said to me once as we were leaving class together, "you never know about couples unless you sleep under their bed." Still it had seemed to me then, and even does now that Kitty and Oppie had lived days and years together in the same house but in separate worlds.

At the hotel I found a telephone message at the desk from Carl: "Please be packed and ready to leave by 3." I thanked the clerk and slowly climbed to my room, pondering my future. Should I consider marrying this cocksure scientist-to-be? Did I want a public life? Should I become a scientist's wife? Should I endure? Could I enjoy his certainty of future fame?

Of course, I'd ride back with him to Chicago, but I knew, just then, I'd leave my New Jersey hero to his own imposing future. Dr. O. suggested I go to Harvard. Was he right? I contemplated Oppie's kind advice, "Don't go to school for too long, and certainly don't go to study writing."

Whatever his words, I learned far more from his example—the foils of his wife's eyes, the trapped desolation of his own. Yes, I would reassure Carl that he could enjoy the ease as he bloomed into a world-class scientist: he would never have to present me formally in Rahway. He would be eternally excused from ever introducing me to his mother.

As we sped west, a horizon of complexity lay before me.

Were those who failed to ask whether or not the bomb should be dropped to be blamed for the deaths of Hiroshima children?

Everyone had conceded that the building and detonation of the bomb was "technically sweet." Could J.R.O. have ceded his pleasure in the sweetness of how? Was it in the domain of the possible to demand, instead, that the bomb never be dropped? Could the "whether or not" question be asked at all? Did Oppie and his friends have only a single option?

All bomb systems were go. These weapons were unlike any in the history of warfare. No civilian, no sage elder, no mother, no child was exempt. Raging fires burn without distinguishing the armed soldier, the uniformed sailor, the dauntless marine.

. . .
The darkness drops again; but now I know
That twenty centuries of stony sleep
Were vexed to nightmare by a rocking cradle,
And what rough beast, its hour come round at last,
Slouches toward Bethlehem to be born?

"The Second Coming"
WILLIAM BUTLER YEATS

Feeling the undetectable spin of our planetary home, this globe, the Earth, I sky-gazed before mounting our own front steps. Incredible autumn stars. Bold, like his personal Van Gogh, the study for "Starry Night," Oppie so proudly had shown me, I had so proudly seen.

How ironic that on August 7, 1945, a hundred thousand Hiroshimi died at the hands of these superscientists whose only wish was to stop war forever, "It was his Fat Man, his son, his invention," Phil had said.

Is it irony that today over fifty thousand nuclear fusion weapons are tucked away in caves, in submarines, in ship's holds, in airplane payload bays? Or, had the desire to end all wars gone awash in the capability of doing something "technologically sweet," carrying on without asking questions?

Was Oppenheimer correct or corrupting as he convinced a young Philip Morrison that a "mere demonstration" would not be enough? Settling that, that any warning to the Japanese would bring certain death to an American pilot, one death too many? It must be dropped over a populated area, it must be dropped with no warning. Sorting dark ironies is harder for me now than it was at age sixteen, Kaori and of all those other human beings nameless, to us, now dead.

Orchestrating the project, Oppie became famous for his "tendency to play ball" with the military, with the politicians, especially with the savvy and ambitious General Leslie Groves. Oppenheimer was mesmerized with the question "How?" never asking, "Whether or not?" With pain, he later broadcast that he and his physicist colleagues had known sin. He stuttered. He halted. He raised issues of conscience. He debated Edward Teller, father of the next generation of bombs, mastermind of the H-bomb. "Wasn't the A-bomb enough?" Oppie questioned publicly. The turn of the screw, accusations at the 50s hearings—"Is he, with his sudden severe attack of conscience, standing in the way of progress?"

I am still the person I was at sixteen, perhaps more cynical, certainly more circumspect. Could I ever lose myself as Oppie did? My love for science—but not for scientists—has only deepened, as has my disdain for its applications. I see the touted use of any scientific insight for improvement of human health or military prowess as profoundly distracting from the task of generating knowledge itself. So-called applied science—from molecular medicine miracles to hydrogen bomb detonations—is just more of the same talky-talky to thinly disguise greed: academic greed, scientist greed, corporate greed, governmental greed. Science is rationalized, confusion abounds, description obfuscated, havoc wreaked. Intrinsically hypocritical, unrecognized anthropocentrism breeds with itself, reducing exploration to snide self-justification. And then there are the victims, always the victims, the helpless-hopeless—the propaganda science writers, the memorizing students, the part-time technicians, the lab rats and guinea pigs, all the marginalized others.

Yes, my enthusiasm for knowledge is unabated, while my confidence in my fellow humans, especially fellow intellectuals, has eroded. Oppenheimer the man seems like the rest of them—far too accommodating, too malleable, too insubstantial, too indecisive to warrant serious reflection at this late date. Only, in some strange way, perhaps he, or at least the situation, *was* my mentor: the complex ambiguity of a persistent and vicious war, the Nazis and the disintegration of the great German culture, the racist comic caricatures of a tiny, slant-eyed half-people.

All were crystallized through Oppie's eyes and Phil's dilemma—the balance of one U.S. pilot against half a million Japanese citizens.

Should Phil have agreed to *not* warn targeted cities and their military installations? Most of the dissenting clergy, Communists, Jews, political activists, and intellectuals had departed or, in the final solution, had been permanently silenced. Should not Hitler's master race—so white, so Christian, so Saxon, so German—have been bombed? At least the allied victors would have been spared the sight of Auschwitz, Dachau, and Bergen-Belsen, which still offends so many sensibilities.

Nazi-haters certainly, but residually respectful of Teutonic efficiency, should not the Oppie and Groves' neat plan for destroying two cities of black-eyed Asians have been overruled? Should they have announced a political demonstration, dropping the A-bomb in full official view off some South Pacific rim city? And, after Hiroshima, whatever could have been Oppie's, Groves', anyone's rationalization for the second bomb on Nagasaki?

The evolutionary backdrop and the ecological setting, rather than the flashy on-stage protagonists of the drama, are more of my concern. Yet the answers are never simple, in part because the questions cannot be clearly phrased. The shadowed globe continues to spin. We are embedded in our history, in all our personal histories, and in theirs.

Term Paper Notes

Not "Whether or Not?" but "How?": J. R. Oppenheimer and the Decision to Drop the Bomb

Lynn Alexander
Social Sciences 2
February 1955

1907 Julius Robert Oppenheimer was born in San Francisco, USA.

1918 For some years the family of J.R.O. summers in wooded hills near The Ranch School at Los Alamos, New Mexico. His father, who had in his teens come to this country from Germany, has by now made a fortune as a merchant.

1926–7 Göttingen: the old and real world. Oppie lives on Geismaran Landstrasse in a granite villa owned by a practicing, well-educated physician. He is enrolled in the faculty of natural science to study physics with Max Born. He is talkative, popular, fascinating to his peers. Extremely industrious.

Jungk, author of ***Brighter than a Thousand Suns,*** writes ". . . excessive garrulity and eloquence began to cause irritation and envy among a number of his companions. They submitted a written petition to one of the professors suggesting a check might be put on the 'wunderkind.' " Paul Dirac, quiet mathematical physicist who also lives at the physician's villa, learning that Oppie writes poetry as well as equations describing physical laws, comments [this is a paraphrase] "How can you even attempt both at once? In science we write hoping to be understood by all whereas in poetic writing one tries to be the opposite—incomprehensible." [Some might maintain exactly the contrary.]

1927 J.R.O.'s petition to take his doctoral exam denied on the basis of faulty records. Because Oppie had never filed a curriculum vitae, he had never formally matriculated and hence was not, in fact, a member of the academic community. Max Born interceded, claiming that Oppie's contribution to physics was outstanding and that because of economic circumstances Herr Oppenheimer would be unable to remain in Gottingen after the current summer term. [We might say now, Professor Born requested that Oppie's residence requirement be waived.] The truth of Born's statement is questionable—Jungk feels that Oppie was far more impatient than he was impoverished.

May 11, 1927 J.R.O. received his doctoral degree in physics from Göttingen "with distinction."

1936–9 J.R.O. develops and sustains an intimate relationship with Jean Tatlock. She is the daughter of a Professor of English Literature at the University of California, Berkeley. Several times they make and break engagements to be married. Tatlock, an active antifascist, collects money and clothing for the Spanish Civil War relief. She joins the Communist Party. She is an avid reader of socialist and communist literature. Their relationship lasts nearly a decade.

1937 J.R.O. teaches physics at the University of California at Berkeley and at California Institute of Technology, Pasadena, California. Throughout this period, according to Jungk, "The many scientific papers he published in the periodicals of various countries unquestionably contributed valuable sections to the growing edifice of modern physics but they laid no new foundations for it."

Late November 1940 J.R.O. struck by lightning: he falls in love with Katharina Harrison (née Puening). She is married to someone else, an English physician.

August 1941 Peter Oppenheimer born to Katharine Puening Oppenheimer and J.R.O.

Autumn 1941 Oppie attends a meeting under the auspices of the National Academy of Sciences on the military applications of atomic energy. The quantity of the fissionable isotope of uranium, U235 (among other items required for explosion), is discussed. Soon thereafter J.R.O. publishes a paper on an electromagnetic method of separating U235 (the potentially explosive form of uranium) from U238 (the more abundant but unfissionable isotope). Eventually this piece of work leads to a 50% to 70% reduction in the cost of obtaining uranium needed to achieve critical mass.

Summer 1942 The international scientific community becomes aware of the potential destructive power of fissionable uranium. Taken as fact, even though later shown to be not true, Nazi Germany is well along in its efforts to develop atomic bombs. By mid-1942, President Franklin D. Roosevelt and Prime Minister Winston Churchill agree to concentrate in Canada and the United States atomic research teams of the Western allies. In the U.S., control of atomic research is taken out of the hands of the scientists and transferred to the Military Policy Committee: General Styer, Admiral Purnell, General Groves, Dr. Vannevar Bush, and Dr. James Conant. This group becomes the nucleus of power for the future coordinated atomic research project.

August 13, 1942 A code name is adopted for a correspondence bursting with ideas and plans for action among scientists, military brass, and high-ranking officials: DSM (Development of Substitute Materials) is baptized. This mental concoction also becomes known as the Manhattan Project, so named after a New York post office box.

November 25, 1942 The fateful train ride: J.R.O. embarks on the luxurious Twentieth Century Limited with Colonel K.D. Nichols; first-nameless Marshall, a member of Groves's staff; and General Leslie R. Groves himself. These four hold a floating meeting en route from Chicago to California. Atomic energy must be exploited for military uses in order to thwart the Nazi war offensive. The need to unite the basic scientific research effort is well argued by J.R.O. A crucial decision is made with Oppie's encouragement. The goal is firm: it is to be attained. A commitment for good is made. The Manhattan Project, at first a wispy conception, is now to be moved into the realm of the physical. A research effort is to be mounted where memories of the Ranch School still hover, at Los Alamos, New Mexico.

Spring 1943 J.R.O. begins a quiet campaign in his inimitable way to convince nuclear physicists to join the war effort. He looks at many colleagues straight, beaming his fabulous blue eyes into theirs. He explains that the task at hand is to develop a nuclear weapon that employs fissionable uranium. The ultimate goal is to stop the carnage in his beloved Europe. "With great cunning and intellectual sex appeal," Oppie rounds up into action many American physicists, corralling them in his glorious Southwest.

Early June 1943 Although he has continued to be responsive to the needs of his true friend, Jean Tatlock, by now J.R.O. has deliberately severed relationships with all others of his former life who are known to him to be Communists or sympathizers. J.R.O. also informs the F.B.I. of his outer fringe activities and his former generosity toward leftist causes. He admits an affair with Jean Tatlock, which he claims is over.

Saturday night, June 12, 1943 At her request, Oppie meets Jean Tatlock in San Francisco at the Top of the Mark for drinks. He stays the night with her. Oppie informs Jean he is going away with Katharina and Peter on government business. He expects to be out of touch.

Sunday morning, June 13, 1943 Jean drives J.R.O. to the airport. The movements of the couple are followed by the F.B.I.

July 1943 General Leslie R. Groves refuses to accept an adamant suggestion that Dr. E.U. Condon, irreverent powerful scientist that he is, be director of the top-secret Manhattan Project. Groves opts for J.R. Oppenheimer whom, he claims, he can far better control. Groves writes the following:

> District Engineer
> Office of the Chief of Engineers
> Manhattan District
> Station F., New York City
>
> July 20, 1943
>
> In accordance with my verbal direction of July 15, 1943, it is desired that clearance be issued for the employment of Julius Robert Oppenheimer without delay irrespective of the information you have concerning Mr. Oppenheimer. He is absolutely essential to the [Manhattan] project.
>
> Leslie R. Groves,
> Brigadier General

Summer 1943 J.R.O. is appointed as director of the Manhattan Project. He orchestrates one of the most intense scientific and engineer-ing ventures in history. The atomic bomb is created by these immense efforts.

January 1944 Jean Tatlock kills herself.

1943–1953 J.R.O.'s secretary, as instructed, classifies, files and stores every newsclipping, article, caricature, and photo of Oppie. During the periods of greatest volume, this devoted recordkeeping requires at least a few hours a day. When the decade closes J.R.O. has won many awards, perhaps the most publicized is President H.S. Truman's Medal of Honor, yet he has published only five scientific articles in the journals of professional physics. All are minor contributions.

May 3—June 1, 1945 Secretary of war, Henry Stimson, sets up the "Interim Committee." Politicians join physicists to discuss bomb testing: Vannevar Bush, Karl T. Compton, J.B. Conant, J.R.O., Enrico Fermi, A.H. Compton, and E.O. Lawrence. These seven scientists, with the possible exception of Fermi, had the reputation of a "tendency to play ball with the politicians and the military." A.H. Compton later remarked that he had never recalled being asked whether the new bomb should be used, but only how it should be used.

July 16, 1945, 5:30 a.m. The awesome explosion "brighter than a thousand suns" is seen by a Los Alamos crowd at dawn, over the New Mexico desert at Alamogordo, New Mexico, 120 miles south of Albuquerque. The fruit of their labor.

August 7, 1945, dawn General Kawabe, Deputy Chief of Japan's General Staff in Tokyo, receives a telegram from the senior civil official from Chugoku district. It reads:

THE WHOLE CITY OF HIROSHIMA WAS DESTROYED
INSTANTLY BY A SINGLE BOMB

August 9, 1945, Nagasaki The newspapers name Oppie the "Father of the A-bomb."

1947 J. Edgar Hoover, Director of the F.B.I., famous for his anticommunism and his perseverance, attempts to retract J.R.O.'s security clearance. He lacks enough evidence. Undaunted, Hoover assiduously collects evidence against Oppie. By 1953 the pile of Oppenheimer papers will measure 4 feet 6 inches in height.

January 31, 1950 The Special Committee of the National Security Council recommends that President H.S. Truman employ a crash program to build the hydrogen bomb. Truman announces this oligarchical decision, saying: "I have directed the Atomic Energy Commission to continue its work on all forms of atomic weapons including the 'hydrogen' or 'superbomb'!"

1951 The calculations by many physicists show the feasibility of fusion reactions. Talk of the Superbomb is now in the air. A rumor widely spread and known even to the lay public is that Dr. Edward Teller is father of the hydrogen bomb. (Many physicists and philosophers, however, people on the inside, know that mathematician Stanislav Ulam slept with the mother nine months before.)

June 1951 Gordon Dean, Chairman of the Atomic Energy Commission, later recalls, "We had at that meeting in June 1951 every person, I think, that could conceivably have made a contribution [to the Superbomb]. . . . I remember that everyone around that table without exception, and this included Dr. Oppenheimer, was enthusiastic now that you had something foreseeable. The discussions were pretty well ended and we were able within a matter of just about one year to have that gadget ready."

1951–1953 Oppenheimer questions. He begins to raise issues of conscience. He debates Edward Teller. "Wasn't the A-bomb enough?" he asks.

1953 J.R.O. is appointed to be Reith Lecturer, British Broadcasting Company. He receives his sixth doctorate degree (*Honoris causa*, Oxford University) on the same trip to England. U.S. Senator Brian McMahon is Chairman of the Joint Congressional Committee on Atomic Energy. He, as many others, feels that the prestige of the technical skill of all of America is at stake: The United States must produce the Super. It is also a question of our defense against Communism. Mr. William Borden, formerly an assistant to the Senator, sends McMahon a letter in which he states that J. Robert Oppenheimer is "probably a Soviet agent in disguise."

December 3, 1953 President D.D. Eisenhower meets with two of his cabinet members: Atomic Energy Commission Chairman Lewis Strauss and Robert D. Cutler of the National Security Council. The President orders that access by J.R.O. to all government secrets be denied.

December 21, 1953 Constitution Avenue, home of the Atomic Energy Commission. At a long conference table are Lewis Strauss, A.E.C. Chairman, and his general manager, Colonel K.D. Nichols, the same man who rode with L.R. Groves and J.R.O. on the Twentieth Century Limited to California on November 25, 1942. Strauss and Nichols make small talk with Oppie, whose presence they have so urgently requested. Strauss hands him, stretching his arm the long distance across the table, a draft of a many-paged letter written by Nichols in which the depth of Oppie's disloyalty to his country is detailed. The first twenty-three paragraphs outline Oppie's associations with leftists and his support of their activities. The twenty-fourth paragraph gets to the heart of the matter: it accuses J.R.O. of strong opposition to the construction of the Superbomb—not only before but even after President Truman's decision to proceed with all deliberate speed on its construction. The latter closes by insisting that, given its parade of facts as cited, one must raise "questions as to your veracity, conduct and even your loyalty."

April 1954 The private proceedings of the Atomic Energy Commission against J.R.O. are about to begin. Is he loyal? Does he deserve to retain his clearance? Is he, with his sudden severe attack of conscience, standing in the way of progress? Is he impeding the construction of the Superbomb? The public must be made aware of what is about to happen. Oppie's attorney, Lloyd Garrison, hands to James Reston of the Washington office of the *New York Times* the Nichols letter along with J.R.O.'s answer: 44 pages in defense of himself. All major newspapers publish his statement—at least in part.

November 12, 1954 The public hearings begin over the question of J.R.O.'s right to retain his security clearance. Roger Robb of the Atomic Energy Commission presides over them and "prosecutes" Oppie as if he were charged with high treason.

> Robb: But you supported the dropping of the bomb on Japan, didn't you?
>
> J.R.O.: What do you mean, "supported"?
>
> Robb: You helped pick the target, didn't you?
>
> J.R.O.: I did my job, which was the job I was supposed to do. I was not in a policy-making position at Los Alamos. I would have done anything I was asked to do, including making the bombs a different shape, if I thought it was technically feasible.

Asked about his role in the creation of the A-bomb, Oppie stated:

> "However it is my judgment in these things that when you see something that is technically feasible you go ahead and

do it and argue about what it is only after you have had your technical success. That is he way it was with the atomic bomb. I do not think anybody opposed making it, there were some debates about what to do with it after it was made."

Mid- to late-November, 1954 "Testimony evidence" is presented by many atomic scientists and others before the Security Board of the Atomic Energy Commission.

End of November, 1954 The Personnel Committee of the Security Board of the Atomic Energy Commission votes four to one to deny J.R.O. the reinstatement of his security clearance. They submit to him a statement that includes the following:

> "We have, however, been unable to arrive at the conclusion that it would be clearly consistent with the security interests of the United States to reinstate Dr. Oppenheimer's clearance and therefore do not so recommend.
>
> 1. We find that Dr. Oppenheimer's continuing conduct and associations have reflected a serious disregard for the requirements of the security system.
>
> 2. We have found a susceptibility to influence which could have serious implications for the security interests of this country.
>
> 3. We find his conduct in the hydrogen bomb program sufficiently disturbing as to raise a doubt as to whether his future participation . . . would be clearly consistent with the best interests of the country.
>
> 4. We have regretfully concluded that Dr. Oppenheimer has been less than candid in several instances in his testimony before this Board."

Although I feel like a cheat tampering with my old Soc Sci 2 paper, I add the following just for closure:

1960s–present Peter Oppenheimer becomes more and more reclusive; refuses to answer correspondence and denies requests for all interviews.

1967 J.R. Oppenheimer dies in Princeton.

1977 Toni Oppenheimer commits suicide; attributed by a J.R.O. biographer to an unhappy love affair.

2

The Red Shoe Dilemma

Lynn Margulis

For as long as I can remember, when someone asked me what I wanted to be when I grew up, I always answered "an explorer and a writer." Explorer of what? As a child, I didn't know: undersea cities, African jungle pyramids, unmapped tropical islands, polar caves. "Whatever will need exploring," I said without hesitation. Today, nearly incessantly, I explore with passion the inner workings of living cells to reveal their evolutionary history. And as soon as I learn something new about bacteria or insect symbionts that helps explain the history of life on the Earth's surface, I write about it.

So you see, I am, after all these years, an explorer and a writer. Science for me is exploration, and no scientific work is complete if it has not been described and recorded in an article by the scientist herself (the "primary literature") or in a book or paper by someone else (the "secondary literature"). Much of my day is spent in description: generating literature that speaks to fellow scientists and graduate students, talking in classes or lecturing to amuse the curious, writing notes and observations, collecting references, and jotting down the insights of others. I have become a mother (four children), a wife (twice), and a grandmother (once, so far).

Because no one in my early life ever even explained the existence of science, I never realized until adulthood that I could participate in the great adventure of science as a profession. Unlike many friends, neither as an adolescent nor as a young adult did I wait for "my prince to come." Rather I expected some—any—opportunity to join serious expeditions. Then, as today, I read nearly everything in sight: bottle labels, train schedules, recipes, Spanish poetry, and novels. Decades ago, on the south side of Chicago, I used to ride the "IC" (Illinois Central Railroad) some forty minutes, both in the stifling heat of summer and the freezing cold of winter, at least once weekly to the downtown Loop for ballet. Ballet classes (demanding, exhausting, French, and irrelevant) were sufficiently escapist to be captivating before scientists or exploratory missions were available in my life.

Choices

One film moved all of us dancers of those days: We all idolized redheaded Moira Shearer prancing in *The Red Shoes*. Set near Nice on the Mediterranean, close to a place with a marine station (Villefranche-sur-Mer) that I would get to know many years later, this romantic movie mesmerized my dancing classmates. The talent of this beautiful ballerina in the prima donna role was exhilarating, as was her true love for her sexy, handsome beau. I remember anger at the melodrama of that movie, however. I thought the dichotomy of her life that led to her self-instigated fate utterly ridiculous.

Why did there have to be "necessity to choose" between devotion to a man or a career? What generated the psychic dissonance that drove her to destruction? Obviously there was no reciprocity: if the star had been male, he would not have been driven to choose. He simply would have taken a wife. Instead, under relentless pressure to be the perfect dancer whose shoes run away with her, the ballerina yields to the dance master's demands that she remain in the spotlight, stage center of his world. But, equally enamored of her man, she is driven by another exigency: her lover demands that she marry him and have a family.

Why hadn't she simply married her lover, borne her children, and continued dancing? Hollywood resolved her dilemma tragically, making the young heroine jump to her death from the summit of a sea wall. What infuriated me was the idea that the healthy, beautiful, and ambitious ballerina had to accept the "either-or" notion imposed upon her by the two men who ran her life. Should she simply have opted for *everything*, however, she would have deprived the film of its trumped-up fatal conflict. Wasn't a strong family life *and* a career possible for Moira Shearer's character? Isn't such a full life even easier today in the age of food storage by deep freeze, the private automobile, the dishwasher, and the laundry machine?

At age fifteen I was certain that the ballerina died because of a silly antiquated convention that insisted that it is impossible for any woman to maintain both family and career. I am equally sure now that the people of her generation who insisted on *either* marriage *or* career were correct, just as those of our generation who perpetuate the myth of the superwoman who simultaneously can do it all—husband, children, and professional career—are wrong.

Today many students, especially women, ask me for enlightenment, how to combine successfully career and family. When they learn I have four excellent, healthy, grown children and never abandoned science for even a single day in over 35 years, they request my secret. Touting me as an example of an American superwoman, they label me a "role model" (a term I despise). But there is no secret. Neither I nor anyone else can be superwoman.

Aspiring to the superwoman role leads to thwarted expectations, the helpless-hopeless syndrome, failed dreams, and frustrated ambitions. A lie about what one woman can accomplish leads to her, and her mate's, bitter disappointment and to lack of self-esteem. Such delusions and self-deceptions, blown up and hardened, have reached national proportions. Rampant misrepresentation of feasibility abounds as everyone falls short of the national myth peopled with a happy family, educated children, and professionally fulfilled parents. Something has to give: the quality of the professional life, of the marriage, of the child rearing—or perhaps all—must suffer.

The unreality of such expectations, coupled with the gross inadequacy of our educational system, such as it is, often leads to despair temporarily relieved by mind-numbing drugs—marijuana, whiskey, cocaine—or other escapes.

Each husband, wife, and child in this sea of false hope suffers the crushing pain of inadequacy. In the United States, we value the beauty and strength of youth, but as a culture we disdain love for children as "touchy-feely" and denigrate home-making as trivial and unworthy. We marginalize or expel the elderly and ridicule life on communes. By no means are the homeless on the street the only ones without homes. Unwilling to care for our greatest resource and those in direst need—our infants and children—we, speaking through money, debase their instructors, despising the seriousness needed to acquire a fine education. Our culture laughs at the inquisitive while lauding the merely acquisitive.

I have not in any way overcome these stresses or resolved these problems. I have just ignored them, as if they were laws that do not apply to me. Looking beyond such social heartaches, I chose intellectual exploration as my way of life and allied myself with nonhuman planetmates, with the scientific quest, rather than devoting myself to an arbitrary integrity of family and human community.

And, of course, I never jumped off the ballerina's cliff; the thought of abandoning life itself has always been unthinkable. Be warned, though, I do not offer a recipe for personal fulfillment—superwoman does not exist, even in principle.

Mine is the story of scientific enthusiasm and enlightenment coming to a foolish and energetic girl who turned down dates on Saturday night and who *never* watched television. The point is that I was willing to work. This is not a statement of advocacy, as no single answer or easy path suits every woman. Probably, I have contributed to science because I twice quit my job as a wife. I abandoned husbands but stayed with children. I've been poor, but I've never been sorry.

Children, husband, and excellence in original science are probably not simultaneously possible. Yet women who feel the urge must be encouraged to pursue scientific careers. Such women need our help. If life does not pose its problems as melodramatically as a Hollywood movie, neither does it resolve them so cleanly or definitively.

Yes, women can, of course, be superb scientists, but only at great sacrifice to their social life and its obligations. Most critically productive women and girls must be surrounded by supportive and loving men and boys. We all need a cultural infrastructure that respects the deep needs of our young children and older family members. Let us hope that the provision of such enablers as scholarship monies, family leave opportunities, enlightened health insurance programs, imaginative and indulgent day care for preschoolers, and afterschool play programs will increase the probability that talented and determined women will contribute much more to the scientific adventure in the future than they ever have been able to in the past.

PART II

Symbiosis and Individuality

3

Marriage of Convenience

Did the movement of chromosomes in mitosis, the lashing of sperm tails and the sensing of odors and sounds evolve from an ancient symbiosis of swimming bacteria?

Lynn Margulis
and Mark McMenamin

In the warm muddy shallows that lap mangrove thickets along the shores of the Philippines, Malaysia, and Indonesia live twenty-one species of small silvery fishes, members of a single family called the leiognathids. Some species have upturned mouths, for feeding on insects at the surface; others scour the bottom, aided by mouths that turn down. A third group feeds on smaller fishes in the middle depths, and these species are equipped with jutting teeth for snaring prey. No more than five inches long, the leiognathids offer little else to catch the eye.

Unless, that is, one sees them in the dark. Then they shine with a ghostly blue-green light. Sometimes it pierces the water in narrow search-light beams, lighting a fish's way. Sometimes the entire underbelly of a fish gives off a diffuse glow, randomly patterned with darker patches like a landscape shadowed by clouds. The focused beams probably aid the fishes in their search for prey; the abdominal glow may camouflage them from predators lurking below by matching the sun-dappled water surface. When a predator does approach, a leiognathid emits a flash of light that may confuse the attacker. And the light may also serve as a sexual marking; male fishes flaunt distinct patterns of bright areas.

Under its nondescript exterior a leiognathid is a bundle of optical devices for controlling its luminescence. Three muscular shutters—sheets of opaque tissue that the fish can extend or withdraw at will—control the emission of light from an organ in the throat. The fish opens all three shutters to flash or to produce its protective glow. Two of them, below the light organ, let light escape directly through the translucent muscles and scales of the fish's underside, where pigment-filled cells contract or dilate to create the changing patterns of shadow. The third allows light to play off the swim bladder, a gas-filled sac above the light organ, which controls the fish's buoyancy. The bladder, coated with a silvery organic compound, acts like the reflector of a lamp. When the fish projects its searchlight beams, it closes this shutter and opens the first two only partway

Most startling of all, given the sophisticated optical mechanisms the leiognathids have evolved, is that the light itself is borrowed. The eerie glow comes from a dense population of luminescent bacteria, feeding, growing, and excreting inside a pouch attached to the upper gut. The fish's circulation supplies the bacteria with oxygen and nutrients, but otherwise the host contributes nothing to the production of the light. Each generation of fishes seems to take on new bacterial colonists; to start a new colony the fry simply swallow the luminescent bacteria that live freely in seawater. The shutters, mirrors, translucent windows, and the behaviors that go with them all exist to take advantage of the symbiosis—an intimate joining of different types of organisms.

Dozens of other kinds of luminous fishes have been identified worldwide; some generate their own light, and others, like the leiognathids, rely on bacterial symbionts. But the leiognathids have developed by far the most impressive array of adaptations for controlling biological light. The establishment of the symbiosis must have been an evolutionary turning point for these fishes—perhaps even a kind of founding event. The evolutionary changes that distinguish the leiognathids from other fish lineages in both anatomy and behavior may well have begun when luminescent bacteria first began to feed and grow in a pocket of gut, a precursor of the light organ.

Symbiosis has shaped the features of many other organisms. The great evergreen forests that spread across the northern latitudes would wither and die without the threads of symbiotic fungi that extract nutrients from rocks and soil, and convey them to the tree roots. Termites would be no threat to houses, except that their guts contain myriad protists, "large" microbial creatures capable of digesting the cellulose in wood. The giant tube worms that live near hot springs on the ocean floor lack mouths; they take nourishment from symbiotic bacteria that live in their tissues, metabolizing energy-rich sulfide compounds carried out of the Earth's crust by the springs. In these cases the union of two or more kinds of organism has yielded what is in essence a new organism.

But symbiosis may have had a still more profound role in evolution. It may have been critical to the emergence not just of specific groups of organisms but of fungi, plants, animals, and protoctists, all life forms made up of eukaryotic cells. These cells that, unlike bacteria, contain a nucleus and specialized subunits, or organelles. In the mid-1960s one of us (Margulis) pursued an explanation for strange genetic data—data suggesting the eukaryotic cell itself originated in a series of ancient symbiotic unions. By now it is widely accepted that two kinds of organelle were once free-living bacteria that became established within the confines of very different bacteria.

New genetic findings are providing further support for the symbiotic theory of the cell. A third crucial element of the original theory, far more controversial than the first two, holds that the remarkable dynamism of all eukaryotic cells—their ability to shunt material around internally, change shape, or lash whiplike appendages—was also acquired from a bacterial symbiont. If the eukaryotic cell owes this ubiquitous feature to symbiosis, biologists must begin thinking of the cell as a complex community of microorganisms, not merely as a unit in larger structures. And the merging of bacterial symbionts with other organisms, having been held responsible for a fundamental branching of the tree of life, will be firmly established as a key principle of evolution.

The idea that cells owe some of their complexity to symbiotic microorganisms originated long before biologists had the tools needed to explore the notion. Eighty years ago the Russian biologist Konstantin Sergeivich Merezhkovsky began to suspect that the chloroplasts of plant cells—the green speckles that capture sunlight and produce sugars and oxygen through photosynthesis—are interlopers. Merezhkovsky realized that the organelles resemble blue-green photosynthetic bacteria, called cyanobacteria, more closely than they do any other structure within the cell. He also knew that chloroplasts reproduce on their own, independent of the cell's division cycle. They simply split, or fission, as ordinary bacteria do, but they do so in the confines of the cell. Merezhkovsky and his colleague Andrei Sergeivich Famintsyn, who tried to grow isolated chloroplasts, proposed that the organelles are actually cyanobacteria that took up residence in an early ancestor of plant cells and eventually lost their autonomy.

That scenario was ignored or flatly rejected. But decades later, in the 1960s, the electron microscope showed that chloroplasts contain an intricate stack of internal membranes, similar to the ones in cyanobacteria. Investigators examining chloroplasts also spotted ribosomes, the molecular factories with which proteins are assembled. Ribosomes are a hallmark of independent cells.

Workers also began noticing that certain mutations that turned the chloroplasts of an affected plant white or yellow were not inherited in the same way as genes in the nucleus, the main repository of genetic information in a cell. Only a female plant could pass on the abnormalities to its descendants; the male parent was irrelevant. In many plants, chloroplasts travel from generation to generation in the female part of the flower; perhaps the chloroplasts themselves carried the genetic information determining their color. Eventually DNA, the molecule of heredity, was found in chloroplasts. In its organization the DNA had much more in common with the DNA of certain cyanobacteria than with nuclear DNA. These findings were accepted as clear confirmation of the former independence of the chloroplast's ancestors.

Evidence for the bacterial origin of another set of organelles, the mitochondria, accumulated in much the same way. These rice-shaped subunits are the power stations of the eukaryotic cell, where molecules from food react with oxygen during aerobic respiration, yielding a substance called ATP. Like the electricity generated by a power plant, ATP made in the mitochondria is a convenient and portable energy source for use elsewhere. Mitochondria, like chloroplasts, resemble bacteria and reproduce on their own; their appearance and behavior led Ivan E. Wallin, an anatomist at the University of Colorado Medical School in Denver, to conclude in the 1920s that mitochondria too originated as bacterial intruders.

Wallin, who went on to propose a broader theory of the symbiotic origin of species, was spurned by biologists of the time. But in 1966 decisive evidence for the validity of his proposal about mitochondria came with the discovery that these organelles have their own DNA. Later, comparisons of mitochondrial DNA with DNA from various kinds of bacteria revealed parallels to certain purple nonsulfur bacteria—photosynthetic microorganisms that can also carry out aerobic respiration. Their ancestors are the most likely precursors of bacteria that became mitochondria.

An image of piecemeal evolution emerges from these findings. The eukaryotic cell did not emerge from a single precursor cell—a bacterium of some kind—that gradually evolved more sophisticated features. Rather, it arose from several organisms that interacted closely. Each precursor contributed an entire module of genes, which specified a distinctive set of biochemical abilities.

The kind of bacterium that accommodated the other symbionts in its interior may have been one similar to *Thermoplasma,* a tough microorganism living in acidic hot springs. Like all other bacteria, *Thermoplasma* has DNA that floats freely in the cytoplasm, the jelly-like substance of the cell. (Eukaryotic DNA, in contrast, is bundled into dark rod-shaped structures, or chromosomes, which in turn are enclosed in the nuclear membrane.) Yet *Thermoplasma* and its relatives differ from most other bacteria. Their DNA is coated with a protein similar to the histones that form the scaffolding of chromosomes in eukaryotes and they lack cell walls. Histones are conspicuously absent from other kinds of bacteria. Some hardy bacterium resembling *Thermoplasma* may have been the ancestor, which acquired additional metabolic abilities wholesale by taking in other bacteria.

These microbial interactions took place at a critical juncture in the history of life. Before two billion years ago there was little oxygen in the atmosphere, but as photosynthetic bacteria (including the cyanobacterial precursors of the chloroplasts) spread, the concentration of this gas rose. Oxygen, a poison to most of the microorganisms that represented the universe of life at the time, spurred the evolution of respiration. Tough, wall-less *Thermoplasma*-like ancestors, now motile, took up respiring bacteria through their membranes, probably after surviving invasion by these aggressors. Thus these swimming consortia gained a way of removing any oxygen that penetrated their membranes and, in the long term, a new way of deriving energy. Equipped with the precursors of mitochondria, the new symbiotic complexes spread into environments neither component organism could have colonized. See Chapter 4, page 47, for details.

Later the metabolic repertoire of some of these compound cells was enlarged still further. They fed on carbohydrate-rich photosynthetic bacteria, but eventually some of the microbial prey resisted being digested. Surviving within the cell, the photosynthetic bacteria turned into dependent guests and ultimately into chloroplasts. The evolution of green algae (the precursors of the green plants) had begun.

This account is speculative, but the phenomena it describes—predatory relations in which microorganisms are the aggressors or the prey, followed by survival, coexistence, and symbiosis—are seen in nature today. And one need not look hard to find organisms that recently acquired new metabolic abilities when foreign microorganisms became incorporated into their own cells. Witness the tube worms, clams, and mussels that form oases of life on the ocean floor, surviving on food synthesized from carbon dioxide in seawater by the sulfide-oxidizing bacteria that live in their tissues. Just as the bacterial precursors of mitochondria inadvertently protected their hosts from oxygen, the sulfide-oxidizing bacteria convert into a benign form the sulfide that would otherwise poison the animals.

What other feature of the eukaryotic cell might have evolved through symbiosis? Eukaryotic cells have a dynamic quality that bacteria cannot match. Some can change shape by extending tentacles or even broad skirts of membrane. Others use whiplike appendages to swim through a watery medium or to sweep material across their own surfaces. Eukaryotic cells draw in food from their environment, shunting it around in membranous sacs. They rearrange their organelles individually and their cytoplasm in bulk. And during mitosis—the process of cell division characteristic of eukaryotes—they engage in a dance of the chromosomes. Typically the chromosomes line up at the center of the dividing cell, and the halves of each chromosome are drawn apart as two new cells form. (Bacterial cells simply duplicate their loop of DNA and then growing new membranes and walls, pinch apart.)

Bacteria lack all these forms of movement, though many bacteria can swim. They do so by means of flagella—minute rigid appendages that rotate like propellers. Eukaryotic cells such as sperm cells, the hair cells that sweep mucus up the lining of the throat and the cells of many protoctists (a kingdom that includes microscopic protists, algae, slime molds, and all other eukaryotes that are not plant, fungi, or animal) wield much more complex appendages. Flexible along their length, they lash and undulate. Given the differences between bacterial and eukaryotic flagella, it is better to revive a fifty-year-old term from the German and Russian literature and to refer to the latter appendages as *undulipodia* (waving feet).

Undulipodia are just one expression of a motility system that extends throughout the eukaryotic cell, animating the cell interior as well as its exterior. We believe that this motility system, like chloroplasts and mitochondria, shows signs of having originated in formerly independent microorganisms. The establishment of a symbiosis between these motility precursors and a *Thermoplasma*-like bacterium may have been the first step on the road to the protoctists, and ultimately to fungi, plants, and animals.

One hint that many of the forms of motility in eukaryotic cells might have a common origin is a structure called the centriole, of animal cells, plant sperm, and many protoctists.* Under the microscope this organelle resembles a small bundle of sticks adrift in the cytoplasm; higher magnification reveals nine smaller bundles of three sticks each. The centriole is best known for its behavior during animal-cell mitosis. It reproduces just before the chromosomes become visible, and the progeny migrate to opposite ends of the cell. The new centrioles sit at the far ends of the spindle, the array of fine cables along which the chromosomes migrate toward each pole of the dividing cell.

The same structure plays a role in another kind of motility. The lashing undulipodia of eukaryotes all grow from organelles called kinetosomes (also known as basal bodies). Located at the base of an undulipodium, a kinetosome has the same nine-times-three architecture as the centriole; indeed, in many organisms, kinetosomes and centrioles are manifestly the same thing. When their cells complete mitosis, the centrioles can be seen to migrate to the cell surface, where they seed the growth of new undulipodia. The ninefold structure of any kinetosome is mirrored in the internal structure of the undulipodium that grows from it: nine pairs of fibers running the length of the shaft, arranged around a central pair.

Besides being linked in appearance and behavior, all these structures are made of the same stuff: a protein named *tubulin* because its molecules naturally assemble themselves into hollow fiber, or microtubules. Under the electron microscope the rodlike elements of centrioles and kinetosomes, the cables of the mitotic spindle, and the internal fibers of undulipodia all are revealed as minute tubules. Microtubules give cells still other forms of motility, for example, by acting as a cell's dynamic scaffolding. The rapid assembly and disassembly of microtubules reshapes some protist cells—species of plankton, for instance—just as a tent is remodeled when poles are added or removed. Microtubules also serve as guideways for traffic within the cell.

*This is the [9(3)+0] centriole-kinetosome, see Chapter 4, p. 47 and Glossary, p. 348.

Not only do the different motility structures all seem to be related in any given eukaryotic cell, but also identical structures occur in very different cells. Undulipodia have the same ninefold structure in sperm from ginkgo trees and from humans; centrioles look the same in all animal cells. At the same time, these structures are oddly aloof from other cellular activities. Centrioles and kinetosomes in animal cells sometimes reproduce on their own, like mitochondria and chloroplasts; and kinetosomes in plant sperm seem to appear out of nowhere. It is as if the mechanisms of motility in one cell are more closely related to counterparts in other organisms than they are to processes within the same cell. One way of interpreting the evidence is to propose that the motility system of eukaryotic cells has its own history, distinct from that of the rest of the cell: it originated in ancient microorganisms, whose components have since been dispersed and put to new uses.

Again there is a Russian precedent. In 1924 the biologist Boris Kozo-Polyansky proposed that undulipodia are relics of active, motile bacteria that once clung to an ancient protocell, acting as a kind of outboard motor. Kozo-Polyansky did not develop the broader implications of his proposal. We argue that once the motile bacteria became an integral part of the protocell, they influenced internal processes, causing the evolution of mitotic cell division. Although the shared ninefold structure of the kinetosome and the centriole was unknown, investigators even before electron microscopy had already discerned a close relation between these two organelles.

Kozo-Polyansky's symbiogenesis work is still unknown in the West. Meanwhile, Margulis independently proposed the idea in a more modern form. In this view, early eukaryotic cells—immobile or slow moving at best—gained the ability to move rapidly when they were joined by slender, motile confederates. The kinds of organisms most likely to have filled this role are spirochetes—common spiral-shaped bacteria, some kinds of which cause syphilis or Lyme disease. Present-day spirochetes not only are slender and fast moving but also tend to associate with other cells, often grazing on their surfaces or even boring into their interiors.

These predatory relations can give way to symbiosis. One of the most vivid examples—what may be a replay of an event early in the history of life—comes from the hindgut of a termite that lives only near Darwin, Australia. Among the menagerie of organisms that help the insect digest the cellulose it eats is the protist *Mixotricha.** A large single-celled organism that ingests crumbs of wood and releases compounds the termite can digest, *Mixotricha* has a cluster of four undulipodia at one end. They serve only as

*See Figure 9.4, page 120, in Chapter 9.

rudders; *Mixotricha* gets all its forward impetus from spirochetes—hundreds of thousands of them, clustered as densely as hairs over its entire surface.

The flagella of most motile bacteria extend outside the cell wall, but spirochetes carry their flagella internally. As the flagella rotate, the entire length of the spirochete flexes back and forth, just as eukaryotic undulipodia do. Thus the spirochetes cloaking *Mixotricha* (the name means mixed-up hairs) are easy to mistake for undulipodia. They move in synchrony, like the oars of a galleon, because they are packed so closely.

Striking as this contemporary illustration of symbiotic spirochetes is, the ancestral association of spirochetes and *Thermoplasma*-like bacteria would have gone much further. The earlier symbiotic spirochetes must have gradually lost their metabolic self-sufficiency, since they could rely on the larger bacterium for food and protection against heat and acidity. They must have also lost genetic autonomy as genes were transferred from spirochete DNA to the histone-coated *Thermoplasma* DNA—the precursor of the nucleus.

Still, the attached spirochetes (now undulipodia) kept their ability to move and also a kind of reproductive independence. As we mentioned, the growth of each undulipodium is seeded by a kinetosome, and in animal cells each kinetosome duplicates itself on its own timetable, independent of the cell's division cycle. Indeed, we take the kinetosomes (and their alter egos, the centrioles) to be a key remnant of the original symbionts. They may represent relics of the spirochetes' original anchor points.

The capacity of the kinetosome to reproduce itself suggested it was there, if anywhere, that one might find the genetic signature of an ancient symbiont—a fragment of DNA passed down from what was once an autonomous organism. The discovery of DNA associated with chloroplasts and mitochondria convinced most biologists that those organelles were once free-living microorganisms; perhaps similar evidence for the motility system waited in the kinetosomes. One hint that kinetosomes, or at least something outside the nucleus, contain genetic information came from studies of paramecia, the slipper-shaped protists familiar to beginning biology students. In some strains of paramecia, the organism's fringe of undulipodia is abnormal; instead of swimming in a straight line, it twists and gyrates. The abnormality is passed on from generation to generation as the protists divide, but the inheritance follows an unusual route.

Paramecia engage in a kind of sex in which two individuals dock and exchange their nuclei. If the genetic abnormality lay in the nucleus, genetic exchange between a mutant and a normal paramecium should transfer the mutation so that later, when the normal party to the exchange divided, some of its offspring would swim abnormally. But none of the offspring ever does; only descendants of the paramecium that was abnormal in the first place show the defect. No exchange of nuclear genes ever transfers the abnormality; it seems to be controlled by genes outside the nucleus.

The possibility that the genes responsible for such mutations sometimes reside in the kinetosomes gained support when some microscopists reported that colored stains specific for DNA pick out material in the kinetosomes of paramecia. But other investigators disagreed, and definitive evidence for kinetosomal DNA was slow in coming.

The evidence is now at hand, in work reported late last year by a group led by David J.L. Luck of Rockefeller University. Luck, John L. Hall, and Zenta A. Ramanis studied an organism called *Chlamydomonas*—a single-celled green alga equipped with one large chloroplast and two undulipodia. The organism is subject to many mutations that truncate or eliminate one of its undulipodia and cripple its swimming ability. Pairs of sexually active *Chlamydomonas* cells regularly fuse to produce cells with twice the usual amount of genetic material; these diploid cells then divide to form haploid cells, parceling out half their DNA to each offspring. Genes on different pieces of DNA often end up in different offspring.

When a *Chlamydomonas* cell suffering from several motility mutations fuses with a normal cell and the resultant diploid cell then divides, the mutations might be expected to be distributed between the offspring, following the usual pattern in genetics. But the result is quite different for some of the motility mutations. Those mutations tend to stay together from generation to generation; they belong to the same linkage group. Geneticists infer that the mutated genes are on the same piece of DNA.

The Rockefeller workers developed a molecular probe for the distinctive DNA. They linked the probe to a fluorescent tracer and applied it to *Chlamydomonas* cells. Two distinct spots of fluorescence appeared, indicating large quantities of packed DNA. They were found at the base of each undulipodium, in the kinetosome. Here, vivid in a photomicrograph, was the kinetosomal DNA predicted by the symbiotic theory—perhaps a genetic relic of a spirochete that last swam freely two billion years ago.

Luck's result by itself does not prove our case. No one else has yet been able to reproduce the finding. What is more, some workers have argued that such a large quantity of DNA should be easy to see in electron micrographs, which show no sign of it. But if the result is borne out, additional evidence should be easy to find. If all eukaryotic cells are descendants of an alliance with motile bacteria that ended up being incorporated into the host cells, a DNA signature of the symbionts should be widespread. The signature may not always be inside the kinetosome; in many organisms—perhaps most—the telltale DNA may reside in the nucleus. Yet probes developed by the Luck group for *Chlamydomonas* should eventually track it down, regardless of its location in the cell. We would also expect the probes to recognize DNA in certain existing spirochetes—relatives of the original symbionts. When a probe for kinetosomal DNA also specifically recognizes DNA in spirochetes that resemble undulipodia in other respects, such as diameter, the origin of microtubule-based motility in symbiotic spirochetes will in our view be firmly established.

In the meantime, other kinds of supportive evidence are accumulating. In the symbiotic scenario the tubulin protein—the basic constituent of the motility system—is a molecular relic of the original symbionts. One might therefore expect to find it in existing spirochetes, even though the presence of tubulin has not yet been confirmed in any bacteria. Several workers, including David G. Bermudes of the University of Wisconsin at Milwaukee, Stephen P. Fracek of the University of North Texas at Denton, Gregory Hinkle of the University of Massachusetts at Amherst, Robert Obar of the Worcester Foundation for Experimental Biology in Massachusetts, and George Tzertzinis of Harvard University, have been searching for tubulin in extracts made from spirochetes. They have already found proteins that by several criteria—chemical makeup, response to changes in temperature, and reaction to antibody probes—are similar to tubulin from brain cells. In addition, several kinds of spirochetes contain long, thin structures reminiscent of microtubules.

Still, eukaryotic cells and spirochetes have quite different means of propulsion. A modern spirochete attached to another organism sinuously lashes its entire body, like an undulipodium. But the basic molecular mechanism—the bacterial rotary motor, powered by a flow of hydrogen ions—has little in common with the ATP-driven process that seems to underlie the whipping of undulipodia. The symbiotic theory would get a considerable boost if biochemical similarities were found between the power sources of spirochetes and undulipodia.

If the ability of eukaryotic cells to move and rearrange their contents turns out to be the legacy of symbiotic microorganisms, biologists will have gained deeper insight into how early cells crossed the evolutionary gulf separating bacteria and eukaryotes. The acquisition of motility, after all, must have preceded the advent of chloroplasts or mitochondria. All animal cells lack chloroplasts, and many obscure protists (*Mixotricha*, for example) lack even mitochondria. But nearly every nucleated organism has a motility apparatus based on microtubules, which accounts both for standard features of eukaryotic cells, such as chromosome migration during mitosis, and for so much of their wonderful diversity of shape and movement.

If the eukaryotic cell is viewed as a community of microorganisms, much of cell biology will be cast in a new light. One example is differentiation, the cellular specialization that goes on in many-celled organisms. The differentiation process that yields a heart-muscle cell packed with mitochondria, a human sperm with its undulipodium, or a chloroplast-laden photosynthesizing cell in a blade of grass might be viewed (as it was seventy years ago by Wallin) as the disproportionate growth of one or another of the microbial components of the nucleated cell. And if symbiosis gave rise to something as elaborate and unlikely as the eukaryotic cell, how many other key evolutionary advances may have come about through past symbiotic alliances?

Ordinary evolutionary change, incremental in nature, is hard put to account for some of the sudden advances in the fossil record. It also has trouble explaining how complex new structures and fine-tuned metabolic abilities could have arisen. An incremental step toward a new capability might handicap an organism by impairing an existing one. Only the full-fledged development, at the far side of some evolutionary barrier, might be viable. As existing symbioses—luminous fishes, tube worms, termites, and the like—make clear, partnership with microorganisms provides a ready way of tunneling through such barriers. The larger organism gains all the necessary genes at a stroke, packaged in a tamed microbial invader or in undigested microbial food.

Over time the more familiar processes of evolution—mutation and selection—tend to eliminate many of the distinctions between symbiotic partners. Thus the branches on the tree of life do not always diverge. One branch can merge with another, and from these unions new limbs can grow, unlike anything seen before.

4

Swimming Against the Current

Harold Kirby's Calonymphids and Centriole-Kinetosome DNA

Lynn Margulis
and Michael F. Dolan

On a bright August afternoon on the southern coast of Nova Scotia, we were zipping along Highway 3, the Lighthouse Route—Dolan at the wheel of the van, Margulis navigating, half a dozen students in the back—headed to Halifax for the 1994 meeting of the International Society for Evolutionary Protistology. We were a few miles out of Yarmouth when a road sign flashed past. Lynn's eyes widened.

"There's Tusket!" she shouted. The students stared blankly. Microbiology trips are not heavy with roadside attractions. "Harold Kirby's hometown," she added, not very helpfully. By now, even the old-timers in Tusket might not have been able to identify Kirby. His colleagues and students from the zoology department at the University of California at Berkeley had long since retired. To the two of us, however, Kirby was a living presence. The preceding year, we had spent months tracing the contours of his well-tuned, methodical mind, editing one of his unfinished manuscripts, a monograph on an arcane class of microorganisms. Kirby became our intellectual hero and, in a way that he could not have predicted, our posthumous champion in a war of scientific ideas.

The war centers on the way cells with nuclei—the cells that make up plants, animals, fungi, and many lesser known organisms—evolved from the non-nucleated cells of bacteria. On one side are the so-called neodarwinists, who assert that new organisms and organs evolve primarily through the accumulation of random mutations in DNA. In contrast, we and our allies maintain that a more important source of Darwinian evolutionary novelty in living beings is symbiogenesis, evolutionary change through long-term physical contact between members of different species. Symbiogenesis has been known since the nineteenth century as the means by which certain quite abundant and diverse organisms have originated: the lichens. (A lichen is a partnership of two entirely different kinds of life: a fungus and a photosynthesizer—either a green alga or a cyanobacterium.)

But, we believe, symbiogenesis has been instrumental not only for the evolution of the lichens, but also for all plant and animal cells—including, of course, the cells of human beings. Random mutations in DNA, so accredited with power by the neodarwinists, lead to small, mostly harmful changes. Mergers of symbionts lead to large, functional evolutionary jumps: new organs or major new groups of organisms, such as lichens. To account for the evolution of one of the most revolutionary developments in all biological history, the nucleated cell, we posit serial endosymbiosis theory, or SET. SET holds that all cells with nuclei are composites formed from the mergers of as many as four different kinds of bacteria. One kind of bacterium evolved into mitochondria, pill-shaped bodies inside the cell that act as organic batteries, converting carbon compounds and oxygen into energy. Another kind became organelles called plastids, including chloroplasts, the green-pigmented "solar panels" of algal and plant cells. The third bacterium is the host cell that absorbed the others. We hypothesize that the fourth partner—if it existed—played a subtler role. The earliest to join the host cell, it merged completely and created the first non-bacterial cell, a nucleated cell with many new powers of movement.

It is hard to know what Harold Kirby would have made of serial endosymbiosis theory. Three decades ago, when symbiogenesis was first put forward to explain cell organelles, mainstream (that is to say, neodarwinian) biologists ignored or belittled the idea. That cells might evolve by merging sounded like something out of a second-rate science-fiction movie. But SET had explanatory power. Mitochondria and plastids *look* like bacteria, and the closer microbiologists examined them, the more resemblances they saw. Starting in the mid-1960s, new techniques for analyzing DNA and RNA showed over and over again that mitochondria and plastids have the same structural details as bacteria do, including their own genes. Skeptics conceded the point, and now three of the four bacterial mergers of SET have become mainstream science, accepted by most biologists.

Today only one battle remains. It is the most important battle of all, because the merger of the fourth bacterium, if it took place, lies at the very heart of what it means to be a nucleated cell. It explains how cells move and how they reproduce. But did it take place? The critical evidence has eluded cell biologists for more than thirty years. Now we think we have seen it. And Harold Kirby, who never heard of serial endosymbiosis theory, gave us the clue that made it possible.

Kirby himself was no revolutionary. Like the founders of Tusket, Tory loyalists who fled New York in the wake of American independence, he preferred to plow a familiar furrow. But he plowed it well. He left Nova Scotia almost eighty years ago with a scholarship to Emory University in Atlanta, Georgia. There he developed an interest in invertebrate animals, and he earned his doctorate in the zoology department of the University of California, Berkeley. In 1928, after two years of postdoctoral work at Yale University, he returned to Berkeley and spent the rest of his life there.

The zoology department at Berkeley was a hotbed of research on the microscopic swimmers then called protozoa—single-celled, nucleated creatures that were then considered primitive animals (they have since been reclassified as members of a distinctive group of nucleated organisms called protoctists). Kirby quickly found his niche. He specialized in the biology of the protoctists that inhabit the hindguts of wood-eating termites. His quarries were beautiful, amazingly active swimming cells that spend their lives immersed in dank, dark, low-oxygen gut fluid, coursing incessantly upstream to keep from being excreted into aerated soil, where they would die within minutes on contact with air. More than 450 species of termite protoctists are known; any given species of termite harbors as many as a dozen different kinds—thousands of individuals—along with a rich community of bacteria.

Kirby spent his career studying insect gut communities. An indefatigable naturalist, he collected termites from North, Central, and South

America; the Caribbean; the Pacific islands; South Africa; Australia; and elsewhere, brought them back to his laboratory, eviscerated them, and studied their contents with meticulous care. The fruits of his observations filled the equivalent of ten volumes of published work, including thirty monographs and a handbook of laboratory methods. When he died in 1952, of a heart attack he suffered while chaperoning a Boy Scout skiing trip, unfinished manuscripts lay on his desk.

American biologists of Kirby's generation, like most of those practicing today, engaged very little in evolutionary speculation about how symbiont acquisition could lead to the appearance of new traits. It smacked too much of the nineteenth century, a freewheeling era when cell theory was in its infancy and biological heresies were rife. Rigor was what they wanted: hard data born of patient, methodical, sharply focused observation. Renegades paid a price.

One maverick was Ivan E. Wallin, an anatomy professor at the University of Colorado Medical School in Denver. At the same time that Kirby was beginning his scientific career, Wallin published a series of papers in the *American Journal of Anatomy* describing experiments aimed at proving that mitochondria originated as bacteria. In 1927 he published his magnum opus, *Symbionticism and the Origin of Species,* in which he propounded the ideas that major components of cells had arisen through symbiosis and that new species evolve by acquiring bacterial symbionts. The reviews were so scathing that Wallin withdrew from scientific discourse. For the last forty-two years of his life, although he published on other topics, he never wrote another word about cell evolution.

To mainstream biologists, Wallin was little better than a crank—an upstart from the middle of nowhere, working outside his specialty and employing unusual (and, to be honest, sometimes sloppy) experimental methods. He also had no discernible knack for cultivating powerful scientific allies. His natural allies, in any event, would have been overseas. Around the turn of the century, two Russian botanists—Konstantin S. Merezhkovsky of Kazan University and Andrey S. Famintsyn at the Academy of Sciences in Saint Petersburg—had promoted the idea that evolutionary change is driven by symbiosis, a process Merezhkovsky called *symbiogenesis.* After their deaths (Famintsyn died in 1918, Merezhkovsky in 1921) another botanist, Boris M. Kozo-Polyansky of Voronezh State University, refined their ideas and kept them in circulation until the 1950s as a respectable, if minor, strain of biological thought. But few of their writings were translated, and the theories remained obscure in English-speaking countries. When ideas about symbiogenesis resurfaced in the West in the 1960s, their Russian precursors were still unknown.

One of the early shots in the war for SET came in 1967. In the paper "Origin of Mitosing Cells" in the *Journal of Theoretical Biology,* one of us (Margulis) set out the central tenets of the theory: that certain present-day cell components were once free-living bacteria, and that any live being larger than a bacterium is a superorganism whose cells evolved by symbiogenesis through bacterial corporeal mergers. Since then, thousands of workers have embellished the theory with data and ideas. Now it is possible to match each class of cell structure with the kind of free-living bacterium from which it evolved. The host cell itself is probably related to *Thermoplasma,* a heat- and acid-tolerant "archaebacterium" that lacks a cell wall. Mitochondria are related to the proteobacteria, very common oxygen-breathing walled bacteria that inhabit water of all kinds. Chloroplasts began as photosynthetic bacteria, glistening blue-green bodies that live in microbial mats, muds, pools, and rivers and at the surface of the ocean. In the course of becoming cell organelles, they lost their cell walls and much other equipment needed for independent life. The DNA governing those features—now so much extra baggage—they shed as well, or relinquished to the nucleus of the host.

The fourth former bacterium in the nucleated cells, if it exists, is the most stripped-down symbiont of all. We think that it descended from a spirochete, a member of an agile, serpentine family of bacteria that often merge symbiotically with other organisms. But now, seen through the light microscope, the former spirochete is a tiny remnant: a little dark-staining dot called a centriole-kinetosome (see Chapter 3).

A hundred years ago most biologists thought centrioles and kinetosomes were different structures with unrelated jobs. The dots biologists called centrioles, for instance, appeared in animal cells about to undergo division, or mitosis; they marked the ends of a spindle of fibers that guide chromosomes into position in the offspring cells. Kinetosomes, in contrast, turned up in all cells with moving hairs. They lay at the bases of sperm tails, as well as at the bases of cilia in the oviduct; they underlay the specialized appendages of cells lining the sense organs, such as those of olfactory or taste organs. Kinetosomes were found at the bases of the modified rod and cone cells of the retina. They formed rows beneath the bristly linings of throats, lungs, and noses. To emphasize that such lively hairlike appendages have a common origin, we prefer to call them all by the generic name *undulipodia.* Biologists began to suspect centrioles and kinetosomes are the same in the first decade of the twentieth century, when microscopists observing cell division saw centrioles sprout fibers to become kinetosomes. Decades later, electron micrographs showed that they have the same structure as well—barrel-shaped arrays of protein fibers, about a quarter of a micrometer across and three times that long, with a distinctive nine-sided symmetry.

From the point of view of microbial evolution, the importance of centriole-kinetosomes is that they drive a wedge between the non-nucleated cells of bacteria and the nucleated cells of everything else. Most nucleated cells have them; non-nucleated cells always lack them. Once acquired, centriole-kinetosomes became the touchstone of virtually all subsequent cell evolution. They grew tails that gave cells the ability to swim more efficiently and to sense other organisms—in short, to become all that nucleated cells have since become, while bacteria remain bacteria.

The search for the genetic material of centriole-kinetosomes, so-called c-kDNA, is one of the most exciting and frustrating areas of research in cell biology. From 1966 until 1972, one of us (Margulis) searched for it in *Stentor,* a huge funnel-shaped protoctist whose large mouth is made up of thousands of wiggling undulipodia. When dropped into a solution of one part sugar to ten parts water, *Stentor* sheds those appendages and then grows new ones, manufacturing as many as twenty thousand new centriole-kinetosomes in an hour and a half. Margulis hoped to show that, in the process of forming new centriole-kinetosomes, *Stentor* also synthesizes new DNA. By the early 1970s, however, she concluded that it does not. Other cell biologists undertook similar searches in various organisms, with inconclusive results.

Then, at long last, came news. In 1989 three cell biologists at Rockefeller University in New York—John L. Hall, David J. L. Luck, and Zenta A. Ramanis—reported they had discovered c-kDNA in a one-celled protoctist called *Chlamydomonas,* a green alga that spends most of its time swimming about with the aid of two undulipodia. Naturally enough, some of the DNA in *Chlamydomonas* is devoted to genes that influence how the organism swims. By studying mutations in swimming behavior, Hall, Luck, and their colleagues identified about twenty "swimming" genes and showed them to be genetically linked, or inherited together.

They tagged the "swimming" DNA with fluorescent tracers. At first they reported that the special DNA was concentrated in two glowing spots, one at the base of each undulipodium, apparently inside the centriole-kinetosomes. But by 1995 they had changed their minds: the genes turned out to be clustered near the centriole-kinetosomes but on a chromosome inside the nuclear membrane.

Critics of Hall and Luck took the null result as evidence that c-kDNA does not exist, but we see it differently. Consider what Hall and Luck discovered: a block of specialized DNA that governs movement, passes from generation to generation as a unit, and lies in the outermost fringe of the nucleus, near the centriole-kinetosomes. To a symbiogeneticist those data are clues to assimilated remnants of c-kDNA—not the grand prize, but the next best thing to c-kDNA itself.

Hall and Luck's observations spurred us on to continue the search. We were not interested in *Chlamydomonas,* whose many DNA-containing mitochondria and single DNA-containing chloroplast could all too easily distract and confuse us. No, we needed an organism with a good chance of yielding evidence for c-kDNA. But which?

That was where Harold Kirby came in.

For perhaps the last two decades of his life, Kirby had studied a family of protoctists called calonymphids. Like most of his subjects, calonymphids dwell only in the swollen, oxygen-poor hindguts of termites. They are a diverse group of large microorganisms, some of them as big as giant amoebae—more than 100 micrometers long. With a good magnifying glass and a little imagination, swimming calonymphids can be observed as little bouncing dots; with a decent light microscope, their internal structures are relatively easy to see.

And calonymphids contain much worth seeing. For one thing, they have multiple nuclei—as many as fifty of them in some species. For another, the front ends of their pear-shaped bodies bristle with row upon row of undulipodia, which emerge from the front of the organism in clusters of four. In some kinds of calonymphids, each cluster is rooted in a nucleus; in others, only some of them are; and in one genus, *Snyderella,* the nuclei and the sets of four undulipodia are separate.

Each of the undulipodia sprouting from the bodies of calonymphids is endowed with a centriole-kinetosome. A single calonymphid may have hundreds of them, arranged in tidy rows; Hall and Luck's *Chlamydomonas* and most animal cells have only two. A cluster of undulipodia is known as a mastigont (from *mastix*, Greek for whip). Intrigued by what he called the "mastigont multiplicity" of calonymphids, Kirby sought to explain, in evolutionary terms, how the profusion of front-end undulipodia had come about. An unfinished monograph on that topic was among the papers on his desk when he died.

Kirby's papers—an assortment of drawings and photographic negatives and a manuscript in Kirby's neat handwriting—came into our hands in 1992. They were rescued from the wastebasket by Rhoda Honigberg, the widow of Kirby's former Ph.D. student Bronislaw Honigberg, who had been a professor emeritus of zoology at our university and editor-in-chief of the *Journal of Protozoology*. In 1993, thanks to a research fellowship from the university, we began preparing Kirby's unfinished manuscript for publication (it appeared in the journal *Symbiosis* in 1994). While annotating, editing, and illustrating it, we began to understand cell evolution from Kirby's point of view.

Kirby hypothesized that calonymphids and similar creatures, known collectively as trichomonads, had evolved from small swimmers with one nucleus and four undulipodia apiece. Those so-called basal trichomonads developed into modern species over hundreds of millions of years by making remarkable changes in the steps by which the cells reproduce themselves or grow new structures.

In most other protoctists with which Kirby was familiar, evolutionary changes were fairly straightforward. A single-celled organism might easily divide in such a way that the offspring cells would stay together, thus forming a multicellular organism. Such mechanisms, however, left each resultant cell with a single nucleus; they seemed inadequate to explain the galaxy of multinucleated, multiundulipodiated creatures descended from the original trichomonads.

Instead, Kirby discovered, calonymphids and their kin had hit on a hitherto undocumented kind of cell evolution: The nuclei and undulipodia of a cell had reproduced independently of the cell itself, proliferating at times when the cell had no intention of dividing. Even today, the jumble of reproductive schedules makes calonymphid cell division quite a spectacle. In most plant and animal cells, mitosis proceeds with the solemn regularity of a stately dance; in calonymphids it looks more like an orgy. We began to wonder: with so many fecund centriole-kinetosomes around, surely some of their genetic material might be in the neighborhood, too.

Another oddity soon caught our attention: Kirby's calonymphids, in contrast to most plant and animal cells, have no mitochondria. That lack is not a handicap. In anoxic environments, mitochondria (which need oxygen to function) are so much dead weight, and several lineages of organisms probably have lost them over the generations. Most biologists have assumed that calonymphids, nestled in their food-rich rear-end niche, also lost their mitochondria fairly recently in evolutionary history.

We disagree, profoundly. In our opinion, calonymphids and related protoctists without mitochondria represent an extremely ancient lineage. Their cell division is so unlike standard mitosis that they must have evolved before either mitochondria or mitosis did. All calonymphids live in oxygen-free environments—exactly what we would expect from organisms whose remote ancestors lived in anoxic mud flats and ponds at a time when the entire world was extremely low in oxygen. Furthermore, studies of the genes for trichomonad ribosomal RNA place the trichomonad ancestors of calonymphids early on the family tree of nucleated cells. Termite-dwelling calonymphids evolved fairly late; they probably coevolved inside termites a mere hundred million years ago or so. But their ancestors, Kirby's basal trichomonads, probably evolved a billion years ago, before nucleated cells had acquired mitochondria. The reason calonymphids lack mitochondria today is that their ancestors never had them.

In short, Kirby's legacy to us was a family of microorganisms studded with centriole-kinetosomes and dating back to the earliest nucleated cells. What better place to look for centriole-kinetosome DNA outside the nucleus?

Our organism of choice was *Calonympha grassii,* a beautiful protoctist with between thirty and fifty nuclei. *C. grassii* lives inside the powderpost termite, *Cryptotermes brevis,* a voracious wood-eater widespread in tropical countries. We took specimens of the insects and, drop by drop, squeezed out the contents of their hindguts. One of us (Dolan) treated the fluid with a series of fluorescent stains for DNA and created gorgeous microscope preparations for study. Then, amid growing excitement, we flicked off the laboratory lights, switched on our microscopes, and, after

FIGURE 4.1. Harold Kirby's drawing of *Calonympha grassii,* c. 1950. The more numerous akaryomastigonts with four undulipodia each are anterior (above) the two dozen nuclei. Each nucleus also is attached to its four undulipodia.

waiting for our eyes to adjust to the darkness, scrutinized the slides for evidence of c-kDNA (Figure 4.1).

We found it! There, amid glowing archipelagoes of nuclear DNA, were rows of smaller dots, many of them random but a few of them visible as clusters in foursomes outside the nuclei—right where the DNA-stained centriole-kinetosomes should be. To check that we were really seeing DNA, Dolan made more slides, but before staining them he dipped them in DNase, an enzyme that breaks down DNA. The dots disappeared.

Since then we have seen the rows of stained DNA about a dozen times. The next step will be to verify that the dots really do represent c-kDNA and not, say, spurious DNA from the many symbiotic bacteria that cling to *Calonympha*'s surface. After that we plan to track down c-kDNA in other species of *Calonympha* and in the other branches of the calonymphid family. In preliminary studies of c-kDNA in *Coronympha,* a

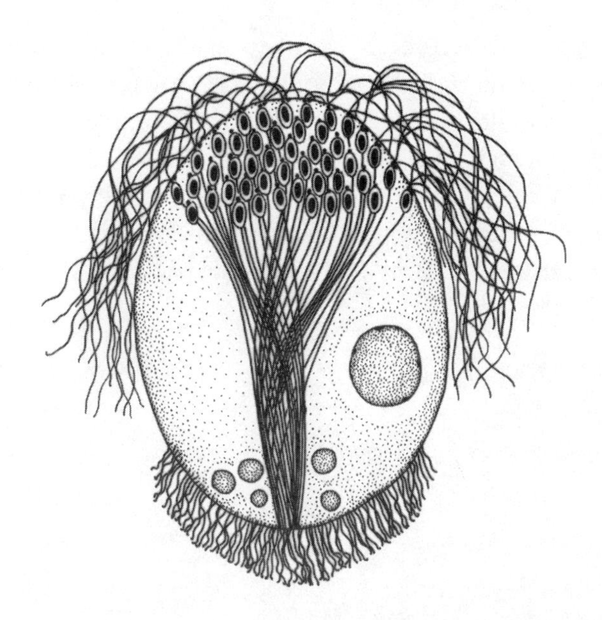

FIGURE 4.2. Harold Kirby's drawing of *Stephanonympha nelumbium*, c. 1950. Each nucleus is attached to its four kinetosomes of the undulipodia (above). Spirochetes fringe the posterior (below) region of this calonymphid.

smaller calonymphid with only eight nuclei, we have not detected centriole-kinetosome staining outside the nucleus. The remaining calonymphid genera—*Metacoronympha, Snyderella,* and *Stephanonympha*—lie ahead (Figure 4.2).

Our interpretation is that centriole-kinetosome DNA began as the DNA of once-independent swimming bacteria. The swimming bacteria became attached, but still swimming, symbionts. Early in the history of cells, the foreign bacterial DNA caused the nucleus to form. Even the DNA that codes for proteins involved in the movement of mitosis was incorporated into the newly formed nucleus. The merging of two kinds of DNA, from the host and from swimming bacteria, led to the evolution of nucleated cells, with their striking form of division by mitosis.

Nearly all subsequent nucleated organisms evolved from mitotic ancestors that had incorporated c-kDNA into the rest of the nuclear genes. In most organisms, as in the *Chlamydomonas* alga that Hall and Luck studied, centriole-kinetosome DNA is entirely integrated in the nucleus. Animal or plant cells might never reveal it. To uncover the treasure, cell biologists must descend into mud, and the hindguts of termites, and other dim redoubts where the old ways still prevail—where, inside forgotten amitochondriate relics such as Harold Kirby's beloved *Calonympha,* centriole-kinetosomes still reproduce as a unit tied to the nucleus and retain vestiges of their own DNA. Only from such legacies of ancient life can we hope to reconstruct the history of the ultimate bacterial symbiosis that led to our own mitotic cells.

5

The Uncut Self

Dorion Sagan
and Lynn Margulis

Speeches and books were assigned real authors, other than mythical or important religious figures, only when the author became subject to punishment and to the extent that his discourse was considered transgressive.

M. Foucault
Language, Counter-Memory and Practice

full circle, not based on the rectilinear frame of reference of a painting, mirror, house, or book, and with neither "inside" nor "outside" but according to the single surface of a Moebius strip. This is not the classical Cartesian model of self, with a vital ensouled *res cogitans* surrounded by that predictable world of Newtonian mechanisms of the *res extensa;* it is closer to Maturana and Varela's conception of autopoiesis, a completely self-making, self-referring, tautologically delimited entity at the various levels of cell, organism, and cognition (Maturana and Varela 1973). It would be premature to accuse us therefore of a debilitating biomysticism, of pandering to deconstructive fashion, or, indeed, of fomenting an academic "lunacy" or "criminality" that merits ostracism from scientific society, smoothly sealed by peer review and by the standards of what Fleck calls a "thought collective" (Fleck 1979). Nor would it be timely to label and dismiss us as antirational or solipsist.

All such locutions stem from the mundane reason, the ethnocentric conception of self that precisely comes under question here. "The philosophy of the subject," writes Jürgen Habermas, "is by no means an absolutely reifying power that imprisons all discursive thought and leaves open nothing but a flight into the immediacy of mystical ecstasy" (Habermas 1987). On the one hand we position ourselves beyond the sixteenth-century European Enlightenment, its faith in reason, the arrogance of its secular priests, and the later Darwinian smarm. In this sense we have a poststructuralist, postmodern, nonrepresentational view of self. On the other hand, we dialectically question this position, motionlessly turning it inside out, as it were, and paying heed to the successes of scientific positivism and biochemical reductionism—movements that philosophically, cannot (at least provisionally) be disentangled from the pervasive influence of IndoEuropean grammar, subject-verb-object structures, and the like. In this sense, our view of the organism is less ontological and more biological; the order of metaphysics and physics, the primacy of philosophy over biology, undergoes a reversal more in keeping with the academic notions of self, and the anthological effort to enclose in a coherent, comprehensive, rectilinear manner. Membrane-bounded indeed.

But the membrane is no concrete, literal, self-possessed wall; it is a self-maintained and constantly changing semipermeable barrier. The idea of the semipermeable membrane permits us to jump organizational levels, from intraorganismic cell, to cellular organism, to organismic ecosystem and biosphere. Whether we are discussing the disappearing membranes of endosymbiotic bacteria on their way to becoming organelles, or the breakdown within the global human socius of the Berlin Wall, we must revise this rectilinear notion of the self, of the bounded I. Alan Watts pejoratively referred to it as the "skin-encapsulated ego"; indeed, even though so deeply entrenched, this bounded sense of "self" seems to us to be thoroughly natural—it is neither an historical or cultural universal. For example, the Melanesians of New Caledonia, known in French as the Canaque, are unaware that the body is an element that they themselves possess; the Melanesians cannot see the body as "one of the elements of the individual" (Leenhardt 1979). So, too, the Homeric epics never make mention of a body—the flesh-enclosed entity we today take for granted as the definable material self—they speak only of what we would think of as the body's parts, for example, "fleet legs" and "sinewy arms" (Snell 1960). "The idea of the 'self in a case'. . . ," writes Norbert Elias, "is one of the recurrent *leitmotifs* of a modern philosophy, from the thinking subject of Descartes, Leibniz' windowless monads and the Kantian subject of knowledge (who from his aprioristic shell can never quite break through to the 'thing in itself') to the more recent extension of the same basic idea of the entirely self-sufficient individual" (Elias 1978).

Psychoanalytically, the sense of self on the level of personhood has been construed to be a convenient fiction, an effect of infantile representation that is jubilant but essentially ersatz. (Etymologically, the word *person* means "to sound through"; coming from the Greek *persona,* it refers to a dramatic mask with a speaking hole.) According to Lacan (1977), the jubilation that creates the essentially false and paranoid ego in the infant occurs when its gaze confronts the image of a fully contoured and coordinated body at the very time (six to eighteen months) it is beleaguered by a motor incapacity that renders it more helpless and defenseless than perhaps any other mammal of the same age. The intense motor incapacity and uncoordination, resulting from "prematuration" (or, in evolutionary terms, from neoteny), engulfs the infant in an almost cinematographic world of uncontrollable visions. One of these mystic-like visions is of itself (or the mother) with a coordination and in a place where it does not in fact exist, along the rectilinear mirror plane. This form of mystical identificatory representation with an image or imago Lacan designates as "image-inary." As a fictional form of the I, it is comforting and effects the discrete sense of self from toddler on into adulthood, the sense of self that has been catered to by American ego psychology in contradistinction to the original Freudian insights and painstaking deconstructions of a psyche (psycho-analysis) formerly presumed to be whole. The Lacanian psychoanalytic revamping of the myth of Narcissus suggests that what we perceive to be our body, as the locus of our "self," is in fact plastic, malleable; and indeed, the lability of the imaginary view of self has come to the fore in the first technology-mediated glimpses of a new image of the human body: Earth from space (Sagan 1990b). This rapidly proliferating image, now recognized as our ecological or biospheric home, will, with further population growth, interspecies interdependencies, and optimization of global media, begin to be re-cognized as body.

Already the shift from biosphere-as-home to biosphere-as-body has become apparent in the scientific work of James E. Lovelock, whose Gaia hypothesis, with mythical allusions of its own, has inspired a planetary search for "geophysiological" climatological and biogeochemical mechanisms (Sagan 1990a). Biospheric individuality was already recognized by Julian Huxley, who wrote:

> the whole organic world constitutes a single great individual, vague and badly co-ordinated it is true, but none the less a continuing whole with inter-dependent parts: if some accident were to remove all the green plants, or all the bacteria, the rest of life would be unable to exist. This individuality, however, is an extremely imperfect one—the internal harmony and the subordination of the parts to the whole is almost infinitely less than in the body of a metazoan, and is thus very wasteful; instead of one part distributing its surplus among the other parts and living peaceably itself on what is left, the transference of food from one unit to another is usually attended with the total or partial destruction of one of its units (Huxley 1912).

As positivists, materialists, or physical reductionists in the western scientific tradition, we would like to think that the picture of the body as an adequately closed topological surface is necessary and sufficient prima facie self-evidence—for the self. And so it is within a certain rectilinear closure. However, as we—and even this coauthorial "we" must be put in quotation marks as we ponder the self, the subject, person, etc.—intimated, the egotistic I is clear only in the sense of a fundamentally fictional or topologically displaced mirror image; there is nothing behind the mirror. Emphasizing tactility rather than vision, on a sensual level it is easy to imagine a conception of the human environment as beginning with the fingernails, hair, bones, and other substances no longer considered to be body parts because they are bereft of sensation. Conversely, technological introjection exemplified by devices such as tele-vision (video, movies, etc.) and tele-portation (automobiles, airplanes, and so forth) suggests a topological extension of the human into what formerly would have been considered the environment. Therefore the body, the material or corporeal basis for "self," has no absolute time-independent skin-encapsulated topological fixity. It is a sociolinguistic psychoanalytic evolutionary construct. Mucus, excrement, urine, spittle, corpses, pornography, and other detachments from and marginal representations of the human body call its essential hegemony, its universal nature, into question.

Chastising the Spanish artist for painting unrepresentative cubistic abstractions, a layman withdrew a photograph of his wife from his pocket, and held it up to Picasso with the admonition. "Why can't you paint realistically, like that?" "Is that what your wife really looks like?" Picasso asked. "Yes," replied the man. "Well, she's very small, and quite flat." Our working assumption of what the self is—like the layman's view of what his wife "really looks like"—is based on a model of representation that takes far too much for granted. Representation itself has, in postmodernist philosophy, fallen into disfavor in a manner similar, perhaps, to that in which figurative realistic painting fell into disfavor with the innovation of the camera. This does not mean that the possibilities of representational or propositional truth, of the correspondence theory of reality still so entrenched in science, is necessarily dead; on the other hand, the difficulties posed by the evidence of quantum mechanics, not least of which is the philosophical nonsolution of the Copenhagen interpretation of the structure of the atom, suggest that most scientific models of reality may be neither so enlightened nor *au courant* as they assume. Indeed, what is in question is the very possibility of modeling reality at all.

Psychoanalytically, when we broach the topic of castration, amputation, dismemberment, the infant's polymorphic perverse sensations and perceptions of the body being, as in a picture by Hieronymus Bosch, in bits and pieces, is probably close to the true state of nature, if such a state there be. In other words, the infant's primordial presocialized experience of the world should not be considered inaccurate but rather, precisely because it precedes sociocultural linguistic norms, less prejudiced and potentially more "realistic." And, apart from parturition, there may be a biological basis for these perceptions, which, later in life, are recalled as amputation, castration, dismemberment. Permitting ourselves a wee bit of abstraction here we splice in a couple more comments by Huxley:

> . . . certain bits of organic machinery are of such a nature that it is physically impossible for the animal to live at all if they are seriously tampered with. It is just because our blood-circulation is so swift and efficient and our nervous system so splendidly centralized that damage to heart or brain means almost instant death to us, while a brainless frog will live for long, and a heart-less part of a worm not only will live but regenerate. Thus here again sacrifice is at the root . . . and only by surrendering its powers of regeneration and reconstitution has life been able to achieve high individualities with the materials allotted her. . . . We have seen the totality of living things as a continuous slowly-advancing sheet of protoplasm out of which nature has been ceaselessly trying to carve systems complete and harmonious in themselves, isolable from all other things, and independent. But she has never been

> completely successful: the systems are never quite cut off, for each must take its origin in one or more pieces of previous system; they are never completely harmonious (Huxley 1912).

Given the abiding prevalence of an image-inary or representational world view in Western science, it is impossible to overestimate the theoretical importance of this relatively abstract, nonrepresentational splicing or grafting that crosses cellular, species, and taxonomic boundaries. Light, no less than matter, cannot be understood simply as a collection of particles but must also be comprehended as a wave: with quantum mechanics the Democritean atomistic Newtonian world view has come to a functional end, although the momentum of scientific discourse has prevented it from reckoning with the consequences of this theoretical shipwreck.

Comparable with the end of the Newtonian age in physics, evidence of the dwindling of an atomistic model of organismic identity in the biological realm is reflected by the debate over the essential unit of selection in Darwinian evolution, whether it is really genetic, the gene—*inside* the organism—or the "individual" competing organism—as Darwin stressed—or group levels such as species or multicellular assemblages. Hierarchy theory entertains species and multicellular assemblages—extended phenotypes *outside* the organism and beyond the traditional confines of the self—to be the crucial units of selection (Dawkins 1982). Certainly the paradoxical notion of group selection seems necessary to explain epochal evolutionary transformations such as those from protoctist colonies to the first plants, animals, and fungi.

The minimal autopoietic, or living, system is the membrane-bounded cell. A cell, or any other autopoietic entity of even more complexity, undergoes continual chemical transformations easily recognizable as "being alive." In the process of this ubiquitous metabolism, each living entity is materially contained within at least a membranous boundary of its own making. In addition to the universal plasma membrane of all living cells, other boundaries, for example, skin, theca, or cuticles, may be self-produced. Such borders include the black, smooth skin of humpback whales, the glycocalyx of some amoebae, the hard overwintering thecal coat of hydra eggs, or the waxy cuticle of a cactus. Minimally the autopoietic unit produces the plasma membrane but often cells and organisms make cellulosic walls, coccoliths, or siliceous spines—complex material extensions found just outside, adjacent, or attached to the universally required membrane. All autopoietic entities continually construct, adjust, and reconstruct these dynamic physical structures by which they are bounded.

We recognize autopoietic entities as "individuals," or "individual organisms." A tree, a potted plant, a swimming euglena, and a cat are immediately perceived as single living organisms. Minimally, all such autopoietic entities are comprised by at least one genomic system: a DNA-containing genome (that is, the sum total of all the genes of the organism) and the RNA-driven protein-synthetic, ribosome-studded internal cellular apparatus associated with that genome.

What is the lowest common denominator of individual life? The minimal autopoietic entity, a single genomic system, is a bacterial cell. Bacteria contain chromonemal DNA, that is, DNA uncoated with histone protein, that codes, via RNA, for an accompanying protein synthetic system itself comprised of RNA and protein. This interacting, metabolizing unit of perhaps some 3000 identifiable genes and proteins bounded by dynamically changing membrane makes and is the bacterial genomic system. Live bacterial cells are single genomic entities in this sense. Whereas single-celled bacteria, uninfected with viruses or plasmids, are comprised of single genomic systems, those so infected have supernumerary genomes—both large (chromonemal) and small replicons (viruses, plasmids). Multicellular bacteria, for example, *Polyangium, Fischerella, Arthromitus*—there are myriads of them—comprised of many copies of the same genomic system, are thus polygenomic. Filamentous, tree-shaped, branched, or spherical colonies, such organisms are comprised of homologous genomic systems in direct physical contact with each other. In some cases, like swarms of cyst-forming myxobacteria (for example, *Chondromyces, Myxococcus*), the component genomes sense each other and fuse, forming a larger structure—no membranes are breached. In others, as when the akinetes of a cyanobacterium float away, the genomic systems disperse. Multicellular bacteria—*Stigmatella, Fischerella,* and the like (Figure 5.1)—are polygenomic beings in which each of the comprising genomic systems, each of the cells, has very recent common ancestors.

All organisms of greater morphological complexity than bacteria, that is, nucleated or eukaryotic organisms (whether single or multicellular), are also polygenomic, having selves of multiple origins (Table 5.1). All these "selves," comprised of heterologous (different-sourced) genomic systems, evolved from more than one kind of ancestor. Because the organelles (nucleocytoplasm, mitochondria, plastids, and so forth) of eukaryotic cells had independent origin among the bacteria, any such cell—any

FIGURE 5.1. Multicellular bacteria: examples. **(A)** *Stigonema* sp., a cyanobacterium from the Alps grows in patches like plants. **(B)** *Stigmatella* sp., a heterotrophic soil myxobacterium. The single cells are capable of producing new "treelike structures." **(C)** *Arthromitus*-like gram-positive spore-forming symbiont from the termite gut shows true branching because single cells are capable of growth at three different sites on the surface. **(D,E)** Microbial mat *Gomphosphaeria,* a cyanobacterium from a microbial mat environment, lacks a single cell stage: it forms these colonies, which reproduce by fragmentation of the entire colony into two smaller, roughly equal colonies (the light micrograph taken with fluorescence microscopy on the right indicates the distribution of chlorophyll). **(F)** Unidentified heterotrophic colonial organisms in which the entire colony fragments into two. It is likely that most bacteria in nature are multicellular.

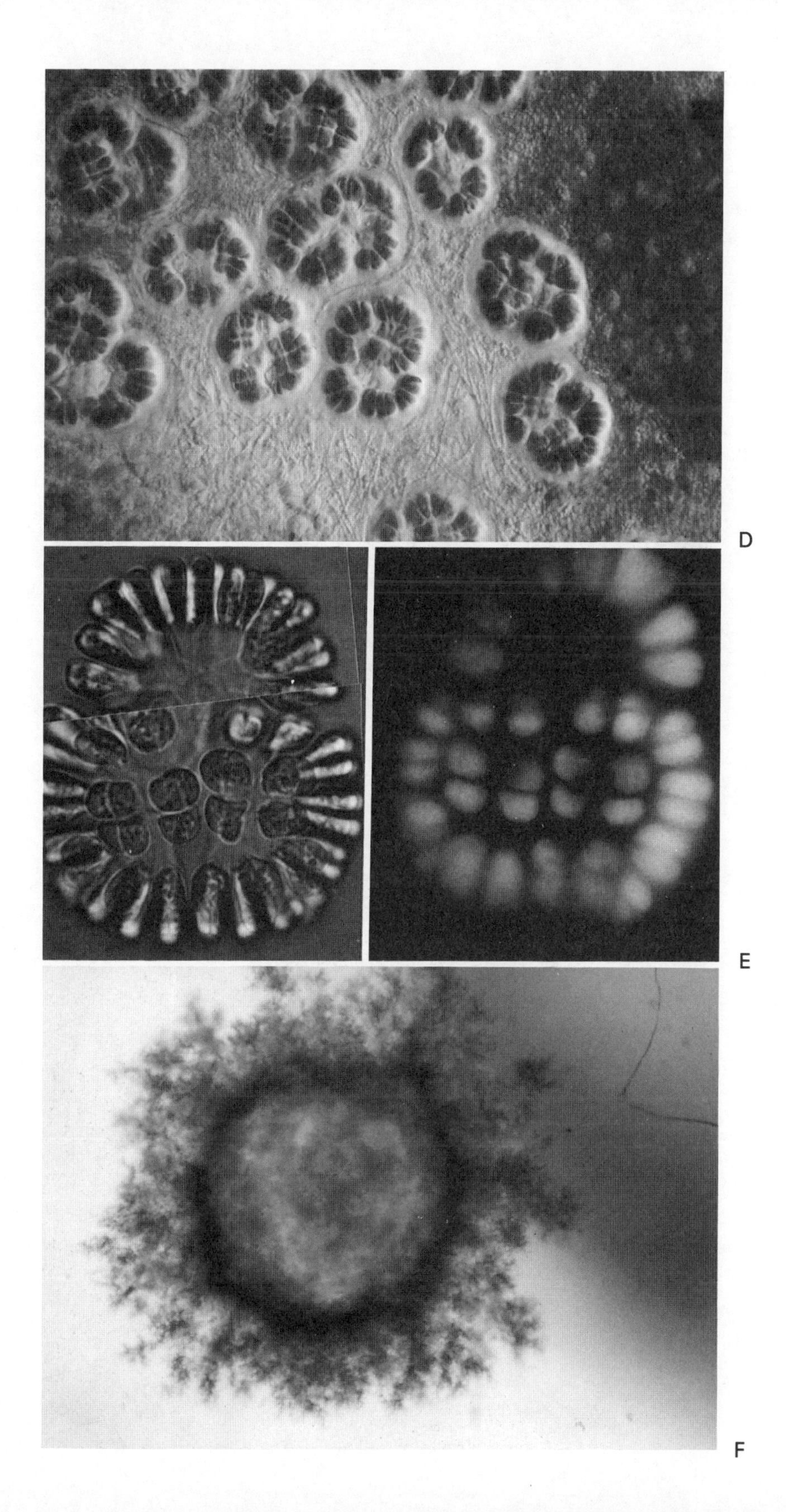
D
E
F

Table 5.1. Multiple Origins of Self in Evolution[a]

Ancestral Bacteria	Extant Organelles or Organs	Hypothetical Minimum Number of Genomes	"Individual" Organisms
Thermoplasma (archaeobacterium)	Nucleocytoplasm	2	Eukaryotes: anaerobic protists
Spirochaeta (eubacterium)	Kinetosomes, centrioles, microtubule organizing centers	2	Eukaryotes: anaerobic protists[b]
Respiring eubacteria	Mitochondria	3	Most heterotrophic eukaryotes (with mitochondria)
Cyanobacteria	Cyanelle	4	*Cyanophora, Glaucocystis, Cyanidium*
Cyanobacteria (*Synechoccocus*)	Rhodoplast	4	Red algae
Chlorobacteria (*Prochloron*)	Chloroplast	4	Green algae, plants
Sulfide-oxidizing eubacterium *Vibrio fischeri*	Thiosome, trophosome	5	Vestiminiferan, tube worms[c]
Carotenoid-producing eubacterium	Red nidemental gland	5	*Loligo* (squid)

[a]For technical details of integrations of genomic systems in endosymbiotic origin of eukaryotic cells see Margulis (1993); for nontechnical, Margulis and Sagan (1996, 1997).
[b]Margulis (1991c).
[c]Vetter (1991).

eukaryotic genomic system—must be comprised of heterologous parts. Each component cell is derived as a chimera; ultimately it emerged from a diversity of bacterial ancestors with only remote common ancestry. In a plant cell, for example, the ancestor of the mitochondria is only remotely related to that of the chloroplast—both descend from gram-negative photosynthetic bacteria with complex respiratory pathways. Neither mitochondria nor plastids are very related to the nucleocytoplasm in which they are embedded. The nucleocytoplasm itself is of archaeobacterial ancestry. Such polygenomic eukaryotic systems are intrinsically and unambiguously chimeric, always enclosed within membranes of course, and often within other self-produced structures that lie external to these membranes. In order to qualify as an autopoietic entity, that is, as an individual organism, any such material-metabolizing entities must be bounded by membranes made by their own metabolism. Biologically, any individual is minimally a metabolic system, made of, in some cases, many genomic entities, hetero- or homogenomic, but all are always bounded by a single, continuous covering. The breaching of the boundary signals disintegration or loss of autopoietic status.

We now see a possible correspondence of the "sense-of-self" to "autopoietic entity" or "live individual." All individuals, all living organisms actively self-maintain. From the early Archean Eon (3500 million years ago) and its bacterial inhabitants through the protists of the Proterozoic Eon (2500 to 520 million years ago), and the fungi, plants, and animals of the Phanerozoic Eon (520 million years ago to the present), the "sense-of-self" seems synonymous with the nature of autopoiesis; boundaries resist breaching while biochemistry acts to maintain integrity. It is the nature of life to interact with the material world to incessantly integrate its components, rejecting, sorting, and discriminating among potential food, waste, or energy sources in ways that maintain organismal integrity.

What is remarkable is the tendency of autopoietic entities to interact with other recognizable autopoietic entities. These interactions may be neutral, as in an amoeba and a pebble; that is, no obvious reaction may occur at all. Two approaching organisms may be indifferent. Alternatively, two heterologous organisms may be destructive—disintegrative—towards each other. One, for example, may produce extracellular enzymes that destroy the other and, relieving it of its autopoiesis, break it down to component metabolic parts. The resulting chemical breakdown products may then be used as food in a trophic relation whereby the still-intact autopoietic being consumes and incorporates the chemical components of its victim. Though relations between organisms may be disintegrative or neutral, those interactions between autopoietic entities that lead beyond destruction to integrative mergers we find to be the most fascinating. Such mergers (fertilization, partner integration in symbiosis) lead to autopoietic entities of still greater complexity. For example, the integration of a fungus attacking an alga for nutrients often—perhaps 25,000 times—has lead to a balance between the disintegrative responses of both fungal and algal partner. Eventually a lichen emerges. A lichen is neither a fungus nor an alga—as a "lichen" it is a composite symbiotic complex that itself is an autopoietic entity at a more complex level of organization. The scholars and botanists are not incorrect in naming the lichen a plant—even though, lacking embryos within maternal tissue, one today would not place lichens within the plant kingdom in any classification scheme. In every level of biological organization from beyond bacteria toward the present, the "sense-of-self" can be inferred from the integrating and discriminating chemical and motility behavior of the components of what we, after the fact, recognize as the individual organism.

An amoebae, *Paratetramitus jugosus,* with a vacuole is shown in Figure 5.2. In the vacuole are two entities. One, interpreted to be a bacterium, is in the process of being broken down, digested, and reutilized as food for the amoeba. Given the terms developed above we can say that the food bacterium, as a disintegrating homologous genomic system, is present in the vacuole. The second structure, a propagule (p), probably a "chromidium," an integrated heterologous genetic system (nucleocytoplasm plus mitochondria) is seen on its way outside the cell. Chromidia are interpreted to be very immature amoebae, that is, stages in the reproduction of these free-living amoebae of the vahlkampfid sort (Dobell 1913; Wheery 1913; Margulis, Enzien, and McKhann 1990). (Vahlkampfids are members of a family—Vahlkampfidae—of small shell-less "monopodial"—or "one-foot" amoebae. They tend to slowly streak forward rather than move simultaneously in many directions or form an exuberance of spines—as other amoebae do.) Thus, at the amoeba level of biological organization, "self" inside the same cell—indeed, inside the same vacuole of the same cell—can already clearly be distinguished from "food." Inspection of the microbiological literature shows, in fact, "sense-of-self" awareness is already present in the virus-infected bacterial world.

Although cell-to-cell mergers are conspicuously lacking in all interacting bacteria, such prokaryotes do accept—take into their membrane-bounded bodies—single genomes in the form of chromonemal DNA: plasmids, viruses, phages. Such DNA is transferred after cell contact directly from a second cell or from the fluid medium. The DNA, from syringe-like bacteriophages—may be forceably ejected through the bacterial membrane. Membranes from more than a single bacterial cell may touch, but they never open to accept another live, bacterial being. The only types of bacteria known to be capable of penetration of the membrane of a second bacterium prey on and destroy that second bacterium. Predatory behavior involving the breaching of membranes, destruction,

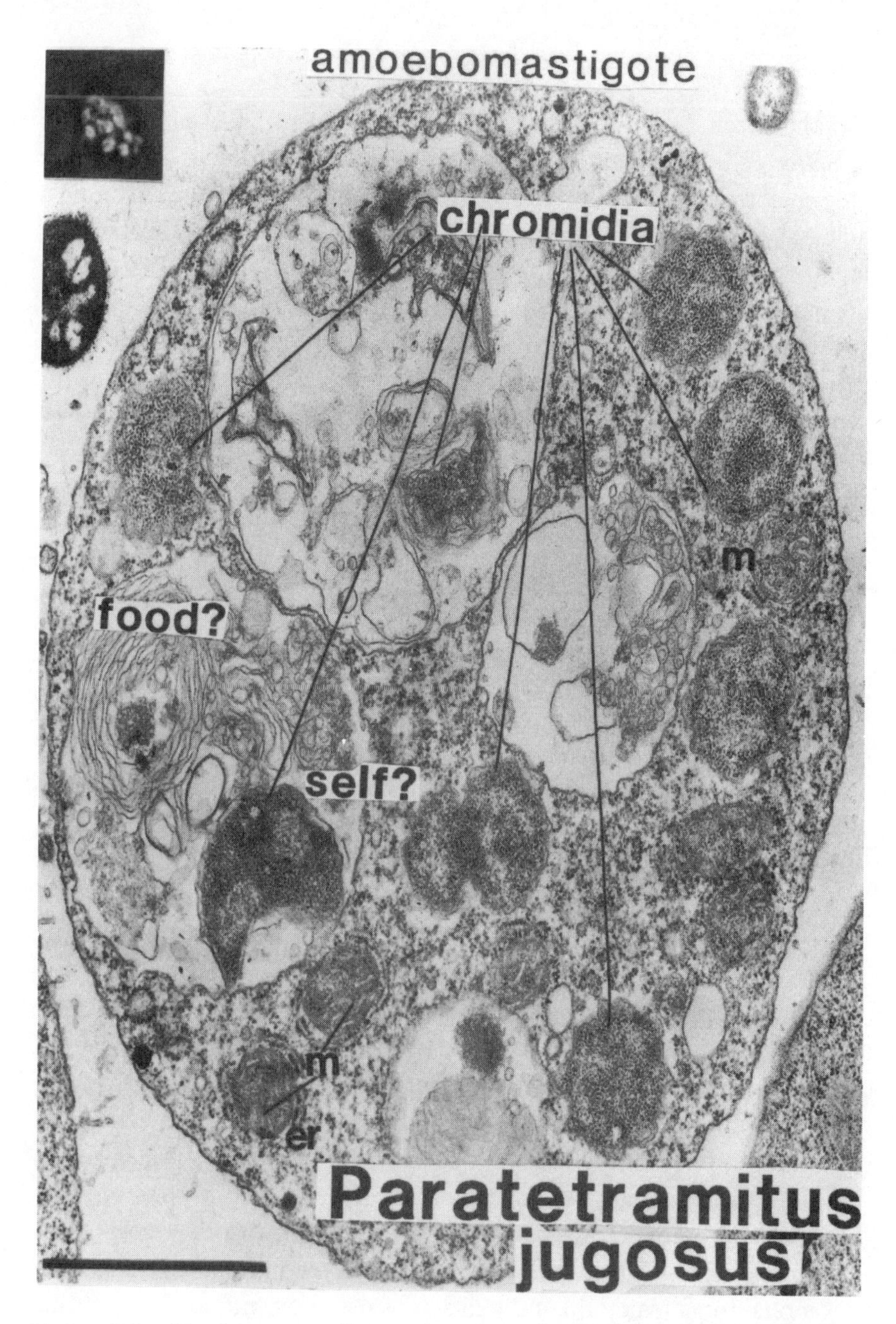

FIGURE 5.2. Food remains of bacteria can be distinguished from structures interpreted to be chromidial propagules (p) in a single vacuole of *Paratetramitus jugosus*, an amoebomastigote taken from a Baja California microbial mat. Perhaps the presence of membrane around the chromidium provides the signal to resist digestion of "self." Electron micrograph in which bar = 1 μM; m = mitochondrion, er = endoplasmic reticulum surrounding the mitochondria. At upper left is a comparable photo of the live organism, apparently releasing chromidia. See Margulis, Enzien, and McKhann, 1990 for details.

and the inevitability of their death is characteristic of *Daptobacter* and *Vampirococcus* attacks on *Chromatium* or *Bdellovibrio* assaults of *Spirillum serpens*, for example (Guerrero et al. 1987), in undestructive encounters only naked DNA slips through the membrane of one bacterial cell to another, changing its genes, with health and survival of the recombinant as the outcome. Because a virus-infected bacterium becomes immune—it resists superinfection by the same sort of virus—there can be little doubt that an integrating sense-of-self already protects uncontrolled loss of autopoiesis—resistance to death—among the world's smallest creatures. Antigens, parts of proteins, appear on the surface of virus-infected bacteria, signalling to the outside world that these bacteria harbor the viral genome. Although other viruses may attach and even enter the already-infected cell, the humble "immune system" of the bacterium refuses to replicate the new virus, which then is lost. Thus signs and signals, self-identification, occur already in prokaryotes, of which the human being represents (if we can still use this word) a kind of massive, three-dimensional pointilist elaboration.

With regard to the later-day three-dimensional pointilist elaboration of the arcane immunity of virus-infected bacteria, we are admonished to ponder the connections. The AIDS-infected human differs little—in principle—from the *E. coli* bacterium infected with lysogenic bacteriophage. The "independence" of the nervous system (mind) from the immune system (body) is severely questioned. Candace Pert defiantly speaks only of bodymind or mindbody. Interviewed by her friend Nancy Griffiths-Marriott, she points to an overemphasis of the blood–brain barrier and the model of the nervous system as a network of penetrating, penile-shaped cells that control the body. Pert emphasizes that monocytes cross that "barrier" within seconds; furthermore, these cells of the immune system transform to become the glial cells of the nervous system. (Glial cells are ten times more abundant than neurons in the mature nervous system). Like gut and brain cells, such monocytes bear neuropeptide receptors—surface proteins—sensitive to the endorphin peptides—natural or endogenous drugs inside the individual—of the neuroimmune system that bring on feelings of elation and ecstacy. Neuropeptides, small communicative molecules, include vasointestinal peptides and endorphins that signal to monocytes. Such protein-like molecules attach to the cell receptors at the surface of gut or brain or monocyte cells at the same place the AIDS virus gets stuck. No, says Pert, there is no mind/body, controller/controlled, male/female, neuron/glial cell dichotomy. Rather there is "mindbody-bodymind," a dynamic system kept informed by devastating news, transforming monocytes, neuropeptide messengers—and hundreds of other integrating mechanisms that confirm the mobile self (Pert and Griffiths-Marriott 1988).

Beginning as latter-day evolutions of bounded endosymbiotic bacterial communities we—as densely packed biomineralizing complexes of eukaryotic cells—should not be too sanguine about the longevity of the modern notion of self. Already in the nineteenth century Samuel Butler clearly and successfully deconstructed personality by parasitizing Charles Darwin's texts. Between the ovum and the octogenarian, held Butler, lie differences greater than those between human beings and other species. What with the vagaries of memory and experience, it is essentially arbitrary to believe that the zygote and the eighty year old are the same person, whereas the father and the son have different selves. Genotypically we may argue with Butler, but to do so phenotypically would be a far more difficult chore. Butler demonstrates the essential arbitrariness of our definitions of organismic identity, of organic integrity and "individuality," even more strikingly by taking the case of a moth. Here we have a being, Butler says, that undergoes radical bodily change between egg and chrysalis, between pupa and winged insect; and yet the only time we say it dies is after the adult moth form stops moving its wings, despite the other radical phenotypic changes during which the genotype has nonetheless been preserved. We might as easily, Butler reminds us, have chosen to consider the transfer from egg to chrysalis or from chrysalis to moth as "death"—and construed the demobilization of the moth as a sloughing-off similar to the shedding of a skin. Indeed, to seriously consider death at all entails a certain ignorance—a certain disregard for the continuity of the "personality" (let us not be too quick to say germ cells, and to invoke the same philosophy of the subject, the self, at a deeper level) despite its radical transformations. So you see that with this figure in which the moth's "self" is held aloft on the tenterhooks of quotation marks "we" have provisionalized identity—not least of all by avoiding the traditional figure of the rectangle that enframes the essay, representing thoughts in an enclosed form that seems to mirror the hegemony of a rigidly structured Platonic body. Topologically the self has no homunucular inner self but comes

6

Power to the Protoctists

Lynn Margulis

Perhaps because we are divided into two sexes, the human tendency to dichotomize—to divide things into either this or that—is very strong. According to traditional systems of classification, anything alive must be either plant or animal. But taxonomy, or placing organisms into categories, is not just an exercise of science—it promotes a frame of mind that pervades our thinking, colors our values, and affects our actions. Furthermore, that frame of mind may persist even when the classification system becomes obsolete. So it is with the plant/animal legacy. If we view microbes (all those organisms invisible to the unaided eye) as mere "germs," hence unworthy of our consideration as part of biodiversity, we slight those organisms that provide our air and fertilize our soil, and we separate essential processes from the web of life. We codify our ignorance and preclude learning to use the recycling and gas production skills of the so-called lower organisms. The old labels impede the spread of knowledge about the mutually dependent diversity of life and its importance to our well-being.

The two-kingdom system—and our position in it—started unravelling with the invention of the microscope in the late 1600s, which enabled the Dutchman Anton van Leeuwenhoek for the first time to see subvisible organisms. Those that swam reminded him of tiny animals, so he named these *animalcules*. Microscopic beings that didn't move or were green he called tiny plants. But, on closer examination, none of the vast world of microorganisms is so easily pigeonholed.

What about an organism such as *Euglena gracilis?* With a microscope, one can see green parts in *Euglena,* which look just like those in the leaves of a plant. Because it photosynthesizes, *Euglena* would seem to be a vegetable. But *Euglena* cells also swim. Each has a single moving appendage closely resembling a human sperm tail. Swimming, a kind of locomotion, traditionally is a defining trait of animals. Botanists claimed *Euglena* was a plant, zoologists classified it as an animal, and potential biology students fled to study more logical fields, such as chemistry.

As observations of the microcosm blossomed, more and more oddballs appeared that further muddied the distinctions between plant and animal: Are malarial parasites animals? Are slime molds not fungi and therefore plants? Aren't diatoms phytoplankton, hence marine plants? Are amoebae single-celled animals? And is dry yeast dead? Or is it an animal, a fungus, or a plant?

The problem lies not with the swimming green *Euglena,* but with our old classification system, which promotes an obsolete view of the world. The two-kingdom system—formalized in the eighteenth century by Carl von Linné (Carolus Linneaus)—developed in a hostile world. Floods, earthquakes, plagues, and pestilences, which humans could neither understand nor master, seemed to have nothing to do with living nature. Little wonder that our ancestors comforted themselves that they were the apex of God's creation, given dominion over nature and set apart from it as unique and independent. Western science then embraced humans' egocentric view of themselves as the pinnacle of a linear evolution from the lower, "primitive" to the highest, "most evolved" forms, us.

The combination of more powerful microscopes, molecular biology, and modern genetics and paleontology has enabled scientists to refine taxonomic distinctions to the level of genes and proteins. These sophisticated methods upset the old biological dichotomy. It is indisputable that all life on Earth today derived from common ancestors; the first to evolve—and the last to be studied in detail—are tiny, oxygen-eschewing bacteria. So significant are bacteria and their evolution that the fundamental division in life forms is not that between plants and animals, but between prokaryotes (bacteria)—organisms composed of small cells with no nuclear membrane surrounding their genes—and eukaryotes (all other life forms, including humans, composed of cells with those nuclear membranes). In the first two billion years of life on Earth, bacteria—the only inhabitants—continuously transformed the Earth's surface and atmosphere, and invented all of life's essential, miniaturized chemical systems. Their ancient biotechnology led to fermentation, photosynthesis, oxygen breathing, and the fixing of atmospheric nitrogen into proteins. It also led to worldwide crises of bacterial population expansion, starvation, and pollution long before the dawn of larger forms of life.

Bacteria survived these crises because of special abilities that eukaryotes lack and that add whole new dimensions to the dynamics of evolution. First, bacteria can routinely transfer their genes to bacteria very different from themselves. The receiving bacterium can use the visiting, accessory DNA (the cell's genetic material) to perform functions that its own genes cannot mandate. Bacteria can exchange genes quickly and reversibly, in part because they live in densely populated communities. Consequently, unlike other life, all the world's bacteria have access to a single gene pool and hence to the adaptive mechanisms of the entire bacterial kingdom. (This extreme genetic fluidity makes the concept of *species* of bacteria meaningless.) The result is a planet made fertile and inhabitable for larger life forms by a worldwide system of communicating, gene-exchanging bacteria.

Bacteria also have a remarkable capacity to combine their bodies with other organisms, forming alliances that may become permanent. Fully 10 percent of our own dry weight consists of bacteria, some of which—such as those microorganisms in our intestines that produce vitamin B_{12}—we cannot live without. Mitochondria live inside our cells but reproduce at different times using different methods from the rest of the host cell. They are descendants of ancient, oxygen-using bacteria. Either engulfed as prey or invading as parasites, these bacteria then took up residence inside foreign cells, forming an uneasy alliance that provided waste disposal and oxygen-derived energy in return for food and shelter. Without mitochondria, the nucleated plant or animal cell cannot breathe and therefore dies.

This symbiogenesis, the merging of organisms into new collectives, is a major source of evolutionary change on Earth. The results of these first mergers were protoctists, our most recent, most important—and most ignored—microbial ancestors. Protoctists invented our kind of digestion, movement, visual, and other sensory systems. They came up with speciation, cannibalism, genes organized on chromosomes, and the ability to make hard parts (like teeth and skeletons). These complex microscopic beings and their descendants even developed the first male and female genders, and our kind of cell-fusing sexuality involving penetration of an egg by a sperm.

Scientists thus have discovered that bacteria not only are the building blocks of life, but also occupy and are indispensable to every other living being on Earth. Without them, we would have no air to breathe, no nitrogen in our food, no soil on which to grow crops. Without microbes, life's essential processes would quickly grind to a halt, and Earth would be as barren as Venus and Mars. Far from leaving microorganisms behind on an evolutionary ladder, we are both surrounded by them and composed of them. The new knowledge of biology, moreover, alters our view of evolution as a chronic, bloody competition among individuals and species. Life did not take over the globe by combat, but by networking. Life forms multiplied and grew more complex by co-opting others, not just by killing them.

Discovering the microcosm within and about us changes—indeed, reverses—the way we look at living things and picture their evolution on the planet. For instance, because all life on Earth evolved from bacteria, it makes more sense now to think of beetles, rose bushes, and baboons as communities of bacteria than it does to think of bacteria as tiny animals or plants. This new world view, in turn, requires a new, more representative labelling system.

But ignorance and resistance have stalled that process. Overhauling the two-kingdom convention—a vast information retrieval system on which biologists depend—would require, for starters, changing how we file and compile bibliographies, how we handle agricultural permits and customs declarations, and how we compute ocean diversity and measure ecological stability. More important, the traditional two-kingdom system and the attitude it embodies endure because shifting from the belief in "man, the highest animal" to a more egalitarian view of the world that respects and empowers *all* life is an enormous step. Acknowledging that our roots are bacteria is humbling and has disturbing implications. Besides impugning human sovereignty over the rest of nature, it challenges our ideas of individuality, uniqueness, and independence. It even violates our view of ourselves as discrete physical beings separated from the rest of nature and—still more unsettling—questions the alleged uniqueness of human intelligent consciousness.

Not surprisingly, the idea of the subvisible microcosm strongly and consistently interacting with us still has not expanded beyond the world of biologists. Only in the 1970s did any scientists begin to take seriously alternatives to a two-kingdom system. Most of humanity—including those who make political decisions about biodiversity—still clings to the two-kingdom view. We still discount the importance of any organism that is neither pet, nor relative, nor food, nor directly useful to us—especially if we can't see it. People ignore the microbially based productivity and waste-recycling capacity of wetlands, for instance, because wetlands seem useless as real estate if not drained.

Biologists have now shifted to new taxonomies. All recognize that live organisms—from multicellular bacteria to marmosets—are co-evolving products of nearly four billion years of evolution. None is more evolved than any other. One scheme reorganizes all life into five kingdoms, listed here in order of their origin:

BACTERIA • PROTOCTISTS • ANIMALS • PLANTS • FUNGI

Because viruses are incapable of any metabolic transformation, including DNA replication outside living cells, they are not alive. Unrelated to each other, they are probably runaway fragments from diverse living cells. The perhaps half-million different kinds of bacteria, fungi, and protoctists are neither animals nor plants. Such former "misfits" as slime molds, yeasts, and *Euglena* have finally found a niche.

Bacteria—the most metabolically diverse and smallest cells on the planet—reproduce by dividing. Without chromosomes (unlike all other life), bacteria have a more informal arrangement of DNA that probably allows more flexible and frequent gene exchange between bacteria but precludes the rigid relation between sex and reproduction found in many animals and plants. Among bacteria's other functions are the ability to digest cellulose in the guts of cows, to color swimming pools blue-green with photosynthesizing cells, and to fix nitrogen in the water and soil. Every spoonful of garden soil contains some 10^{10} bacteria; the total number in anyone's mouth is greater than the number of people who have ever lived. We rely on our personal bacterial populations to help us digest our food and to

FIGURE 6.1A. *Calonympha grassii,* cutaway view of nuclei, kinetosomes and undulipodia. See Chapters 3 and 4. Drawing by K. Delisle.

keep us healthy by restraining the overgrowth of harmful microbes. Babies born without their microbial symbionts must be kept alive in germ-free bubbles at the cost of $100,000/day! (Figure 6.1A).

Protoctists—all the eukaryotes that are neither animals, plants, nor fungi—include ciliates, amoebae, malarial parasites, slime molds, plankton, seaweeds, and single-celled photosynthetic swimming microbes such as *Euglena.* Protoctists are aquatic: some live primarily in the oceans, some primarily in freshwater, some in the watery tissues of other organisms. Some are parasitic. Nearly every animal, fungus, and plant—perhaps every species—has protoctist associates. While most are harmless, protoctists cause many tropical diseases (Chagas disease, giardiasis, malaria, and African sleeping sickness), red tides, and major crop and animal infestations. Because nearly all phytoplankton are protoctists, they also form the basis for the ocean food chain. Protoctists show remarkable variation in cell organization, cell division, nutrition, and life cycles but are far less metabolically diverse than the bacteria (Figure 6.1B).

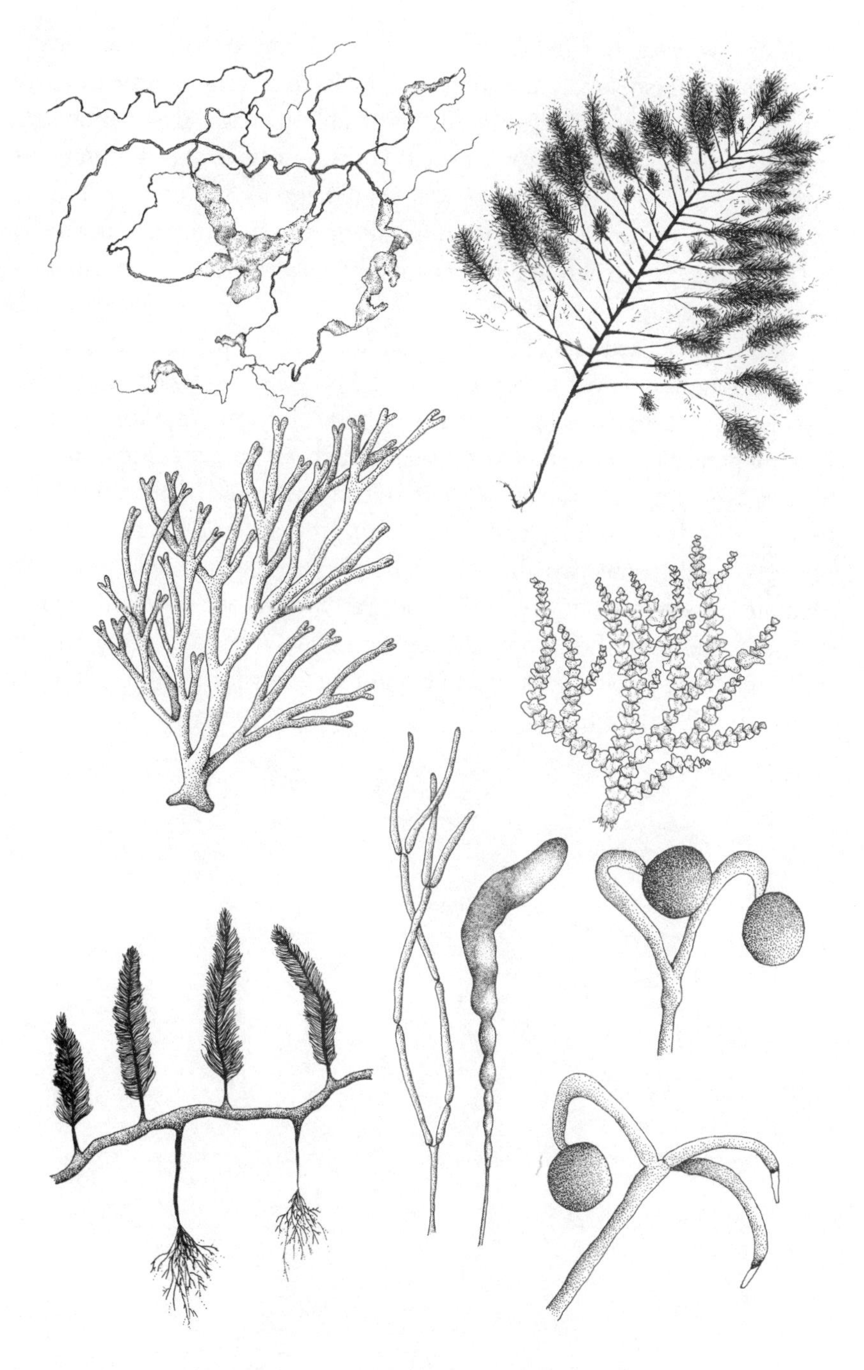

FIGURE 6.1B. Large protoctists: red algae. See Margulis et. al. 1990 or Margulis, McKhann and Olendzenski, 1993 for details. Drawing by Sheila Manion-Artz.

Fungi—yeasts, molds, truffles, puffballs, and mushrooms—are mostly terrestrial organisms that, after the bacteria, were among the first to move onto land. Fungi are tenacious microbes, able to resist desiccation and harsh conditions. Some grow in acid; others survive in environments largely lacking nitrogen, an essential ingredient for all growth. Fungal growth is responsible for rotting fruit, raising and molding bread, smelly feet, ripening cheese, fermenting beer and wine, and producing antibiotics such as penicillin.

Those who speak only for the special interests of human beings fail to see how interdependent life on Earth really is. We cannot view evolutionary history in a balanced manner if we think of it only as a four-billion-year preparation for "higher" organisms, such as humans. Most of life's history has been microbial. We are recombinations of the metabolic processes of bacteria that appeared during the accumulation of atmospheric oxygen some two thousand million years ago. Intellectually we separate ourselves from the rest of life, yet without it we would sink in feces and choke on the carbon dioxide we exhale. Like rats, we have done well separating ourselves from and exploiting other forms of life, but our delusions will not last.

7

From Kefir to Death

Lynn Margulis

It happens to the "individual." Death is the arrest of the self-maintaining processes we call metabolism, the cessation, in a given being, of the incessant chemical reassurance of life. Death, signalling the disintegration and dispersal of the former individual, was not present at the origin of life. Unlike humans, not all organisms age and die at the end of a regular interval. The aging and dying process itself evolved, and we now have an inkling of when and where. Aging and dying first appeared in certain of our microbial ancestors, small swimmers, members of a huge group called protoctists. Some two billion years ago, these ancestors evolved both sex by fertilization and death on cue. Not animals, not plants, not even fungi nor bacteria, protoctists form a diverse—if obscure—group of aquatic beings, most of which can be seen only through a microscope. Familiar protoctists include amoebae, euglenas, ciliates, diatoms, red seaweeds, and all other algae, slime molds, and water molds. Unfamiliar protoctists have strange names: foraminifera, heliozoa, ellobiopsids, and xenophyophores. An estimated two hundred and fifty thousand species exist, most of which have been studied hardly at all.

Death is the loss of the individual's clear boundaries; in death, the self dissolves. But life in a different form goes on—as the fungi and bacteria of decay, or as a child or a grandchild who continues living. The self becomes moribund because of the disintegration of its metabolic processes, but metabolism itself is not lost. Any organism ceases to exist because of circumstances beyond its control: the ambience becomes too hot, too cold, or too dry for too long; a vicious predator attacks or poison gas abounds; food disappears or starvation sets in. The causes of death in photosynthetic bacteria, algae, and plants include too little light, lack of nitrogen, or scarcity of phosphorus. But death also occurs in fine weather independently of direct environmental action. This built-in death—for example, Indian corn stalks that die at the end of the season and healthy elephants that succumb at the end of a century—is programmed. Programmed death is the process by which microscopic protoctists—such as *Plasmodium* (the malarial parasite) or a slime mold mass—dry up and die. Death happens as, say, a butterfly or a lily flower made of many cells matures and then disintegrates in the normal course of development.

Programmed death occurs on many levels. Monthly, the uterine lining of menstruating women sheds as its dead cells (the menstrual blood) flow through the vagina. Each autumn, in deciduous trees and shrubs of the north temperate zone, rows of cells at the base of each leaf stem die. Without the death of this thin layer, cued by the shortening of day length, no leaf would fall. Using genetic-engineering techniques, investigators such as my colleague at the University of Massachusetts, Professor Lawrence Schwartz can put certain "death genes" into laboratory-grown cells that are not programmed to die. The flaskful of potentially immortal cells, on receipt of this DNA, then die so suddenly that the precipitous cessation of their metabolism can be timed to the hour. The control cells that have not received the death genes live indefinitely. Menstrual blood, the dying leaf layer, the rapid self-destruction of the cells that receive the "death genes," and the slower, but more frightening aging of our parents and ourselves are all examples of programmed death.

Unlike animals and plants that grow from embryos and die on schedule, all bacteria, most nucleated microscopic beings, namely, the smaller protoctists and fungi such as molds and yeast, remain eternally young. These inhabitants of the microcosm grow and reproduce without any need for sexual partners. At some point in evolution, meiotic sex—the kind of sex involving genders and fertilization—became correlated with an absolute requirement for programmed death. How did death evolve in these protoctist ancestors?

An elderly man may fertilize a middle-aged woman, but their child is always young. Sperm and egg merge to form the embryo, which becomes the fetus and then the infant. Whether or not the mother is thirteen or forty-three years old, the newborn infant begins life equally young. Programmed death happens to a body and its cells. By contrast, the renewed life of the embryo is the escape from this predictable kind of dying. Each generation restores the *status quo ante,* the microbial form of our ancestors. By a circuitous route, partners that fuse survive, whereas those that never enter sexual liaisons pass away.

Eventually, the ancestral microbes made germ cells that frantically sought and found each other. Fusing, they restored youth. All animals, including people, engage in meiotic sex; all descended from microbes that underwent meiosis (cell divisions that reduce by half chromosome numbers) and sex (fertilization that doubles chromosome numbers).

Bacteria, fungi, and even many protoctists were—and are—reproducing individuals that lack sex lives like ours. They must reproduce without partners, but they never die unless they are killed. The inevitability of cell death and the mortality of the body is the price certain of our protoctist ancestors paid—and we pay still—for the meiotic sex they lack.

Surprisingly, a nutritious and effervescent drink called *kefir,* popular in the Caucausus Mountains of southern Russia and Georgia, informs us about death. Even more remarkably, kefir also illustrates how symbiogenesis—the appearance of new species by symbiosis—works. The word *kefir* (also spelled *kephyr*) applies both to the dairy drink and to the individual curds or grains that ferment milk to make the drink. These grains, like our protoctist ancestors, evolved by symbiosis.

Abe Gomel, the Canadian businessman and owner of Liberté (Liberty) dairy products, manufactures real kefir of the Georgian Caucausus as a small part of his line of products. He and his diligent coworker, Ginette Beauchemin, descend daily to the basement vat room of his factory to inspect the heated growth of the thick, milky substance on its way to becoming commercial kefir. Like all good kefir makers, they know to transfer the most plump and thriving pellets at between nine and ten every morning, weekends included, into the freshest milk. Although nearly everyone who lives in Russia, Poland, or even Scandinavia drinks kefir, this "champagne yogurt" of the Caucasian peoples is still almost unknown in western Europe and the Americas. Abe Gomel and Ginette Beauchemin have been able to train only two other helpers, who must keep constant vigil over the two vats that are always running.

Legend says the prophet Muhammad gave the original kefir pellets to the Orthodox Christian peoples in the Caucausus, Georgia, near Mount Ebrus, with strict orders never to give them away. Nonetheless, secrets of preparation of the possibly life-extending "Muhammad pellets" have of course been shared. A growing kefir curd is an irregular spherical being. Looking like a large curd of cottage cheese, about a centimeter in diameter, individual kefir pellets grow and metabolize milk sugars and proteins to make kefir the dairy drink. When active metabolism that assures individuality ceases, kefir curds dissolve and die without aging. Just as corncobs in a field, active yeast in fermenting vats, or fish eggs in trout hatcheries must be tended, so kefir requires care. Dead corn seeds grow no stalks, dead yeast makes neither bread nor beer, dead fish are not marketable, and in the same way, kefir individuals after dying are not kefir. Comparable with damp but "inactive" yeast or decaying trout eggs, dead kefir curds teem with a kind of life that is something other than kefir: a smelly mush of irrelevant fungi and bacteria thriving and metabolizing, but no longer in integrated fashion, on corpses of what once were live individuals.

Like our protoctist ancestors that evolved from symbioses among bacteria, kefir individuals evolved from the living together of some thirty different microbes, at least eleven of which are known from recent studies (Table 7.1). These specific yeasts and bacteria must reproduce together—by coordinated cell division that never involves fertilization or any other aspect of meiotic sex—to maintain the integrity of the unusual microbial individual that is the kefir curd. Symbiogenesis led to complex individuals that die (like kefir and most protoctists) before sexuality led to organisms that *had* to die (like elephants and us). A kefir individual, like any other, requires behavioral and metabolic reaffirmation.

During the course of brewing the yogurt-like beverage, people inadvertently bred for kefir individuals. In choosing the best "starter" to make the drink, villagers of the Caucasus "naturally selected," which means they encouraged the growth of certain populations and stopped the growth of others. These people inadvertently turned a loose confederation of microbes into well-formed populations of much larger individuals, each capable of death. In trying to satisfy their taste buds and stom-

Table 7.1. Kefir: List of Components, Live Microbes

Each individual (see Figure 7.1) is composed of:
Kingdom Bacteria (Monera)
Streptococcus lactis
Lactobacillus casei, Lactobacillus brevis
Lactobacillus helveticus, Lactobacillus bulgaricus
Leuconostoc mesenteroides
Acetobacter aceti
Kingdom Fungi (yeasts, molds)
Kluyveromyces marxianus, Torulaspora delbrueckii
Candida kefir, Saccharomyces cerevisiae
and at least fifteen other kinds of unknown microbes

achs, kefir-drinking Georgians are unaware that they have created a new form of life.

The minute beings making up live kefir grains can be seen with high-power microscopy (Figure 7.1): specific bacteria and fungi inextricably connected by complex materials, glycoproteins and carbohydrates of their own making—individuals bounded by their own skin—so to speak. In healthy kefir, the bacterial and fungal components are organized into a curd, a covered structure that reproduces as a single entity. As one curd divides to make two, two become four, eight, sixteen, and so on. The reproducing kefir forms the liquid that after a week or so of growth becomes the dairy drink. If the relative quantities of its component microbes are skewed, the individual curd dies and sour mush results.

Kefir microbes are entirely integrated into the new being just as the former symbiotic bacteria that became components of protoctist and animal cells are integrated. As they grow, kefir curds convert milk to the effervescent drink. "Starter," the original Caucasian kefir curds, must be properly tended. Kefir can no more be made by the "right mix" of chemicals or microbes than can oak trees or elephants. Scientists now know, or at least strongly suspect from DNA sequence and other studies, that the oxygen-using parts of nucleated cells evolved from symbioses when certain fermenting larger microbes (thermoplasma-like archaebacteria) teamed up with smaller, oxygen-respiring bacteria.

Mitochondria, which combine oxygen with sugars and other food compounds to generate energy, are found almost universally in the cells of

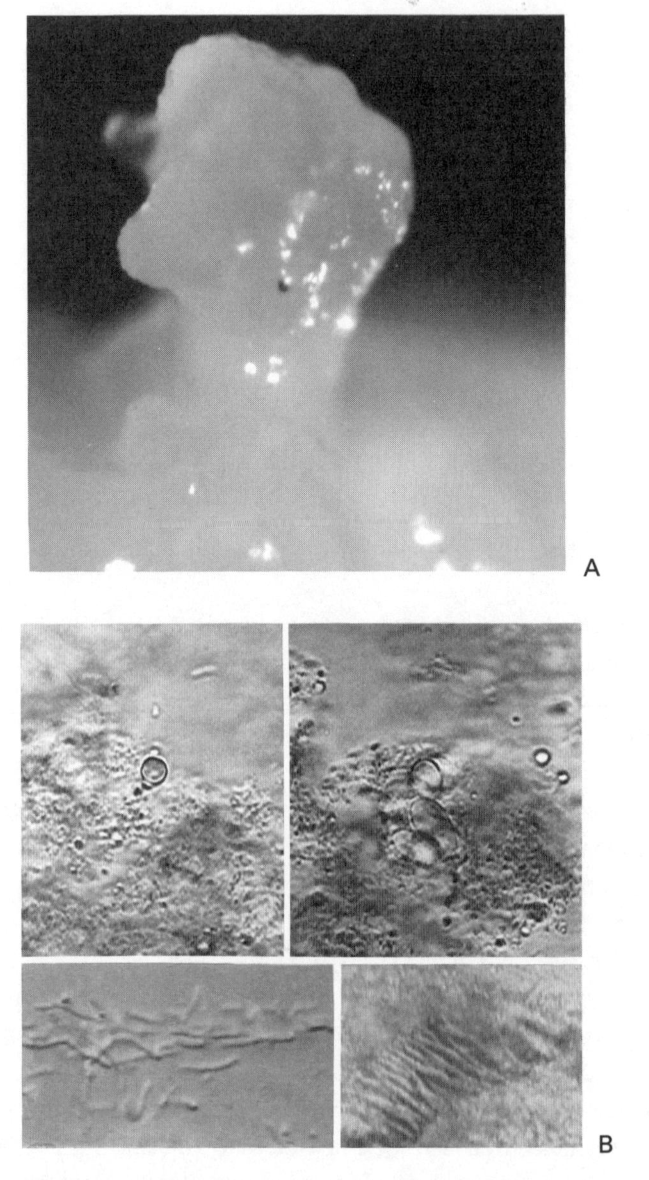

FIGURE 7.1. (A) The complex kefir "individual" live as seen by low-power microscopy (magnification five times). (B) Kefir microbes.

protoctists, fungi, plants, and animals. We, as all mammals, inherit our mitochondria from our mother's egg. Like kefir, we, and all other organisms made of nucleated cells, from amoebae to whales, are not only individuals, we are aggregates. Individuality arises from aggregation, communities whose members fuse and become bounded by materials of their own making. Just as people unconsciously selected the new kefir life-form, so other beings caused evolution of new life, including our ancestors, as microbes—feeding on each other's fats, proteins, carbohydrates, and waste products but only incompletely digesting them—selected each other and eventually coalesced.

Plants come from ancestors that selected but did not entirely digest each other as food. Hungry ancestral swimming cells took up green photosynthetic microbes called cyanobacteria. Some resisted digestion, surviving inside larger cells and continuing to photosynthesize. With integration, green food grew as a part of a new self. The bacterium outside was now an integral part inside the cell. From partly digested cyanobacterium and a hungry translucent swimmer, a new individual, the alga, evolved. From green algal cells (protoctists) came the cells of plants.

Kefir is a sparkling demonstration that the integration processes by which our cells evolved still occurs. Kefir also helps us recognize how the origin of a complex new individual preceded programmed death of the individual on an evolutionary time scale. Kefir instructs us, by its very existence, about how the tastes and choices of one species (ours) influence the evolution of others, the thirty intertwined microbes that became kefir. Although kefir is a complex individual, a product of interacting aggregates of both non-nucleated bacteria and nucleated fungi, it reproduces by direct growth and division. Sex has not evolved in it, and, relative to elephants and corn stalks, both of which develop from sexually produced embryos, kefir grains undergo very little development and display no meiotic sexuality. Yet when mistreated they die and, once dead, like any live individual, never return to life as that same individual.

Knowing that symbionts become new organisms illuminates individuality and death. Individuation, which evolved in the earliest protoctists in a manner similar to the way it did in kefir, preceded meiotic sexuality. Programmed aging and death was a profound evolutionary innovation, limited to the descendants of the sexual protoctists that became animals, fungi, and plants.

The development of death on schedule, the first of the sexually transmitted diseases, evolved along with our peculiar form of sexuality, a process that kefir lacks now and always has done without. The privilege of sexual fusion—the two-parent "fertilization-meiosis" cycle of many protoctists, most fungi, and all plants and animals—is penalized by the imperative of death. Kefir, by not having evolved sex, avoids having to die by programmed death.

8

KINGDOM ANIMALIA: THE ZOOLOGICAL MALAISE FROM A MICROBIAL PERSPECTIVE

LYNN MARGULIS

With respect to the old belief in steady progress nothing could be stranger than the early evolution of life—for nothing much happened for ever so long . . . The oldest fossils are some 3.5 billion years old . . . but multicellular animals appeared just before the Cambrian explosion some 570 million years ago.

STEPHEN J. GOULD
Wonderful Life:
The Burgess Shale and the Nature of History

Autopoiesis and Sexual Cyclicity of Animals

A modern view of the animal kingdom states that all members are diploid organisms displaying gametic meiosis. All develop from an anisogametous fertilization leading to the gamonts (gamete-producing adults) through a blastular embryo. Animals evolved from protoctists: heterotrophs derived from symbiotic associations of bacteria (Margulis 1993). The awareness of animals as co-evolved microbial communities that must undergo sexual cycles of fusion (fertilization) and restoration of the haploid (by meiotic reduction) permits an assessment of the position of the kingdom relative

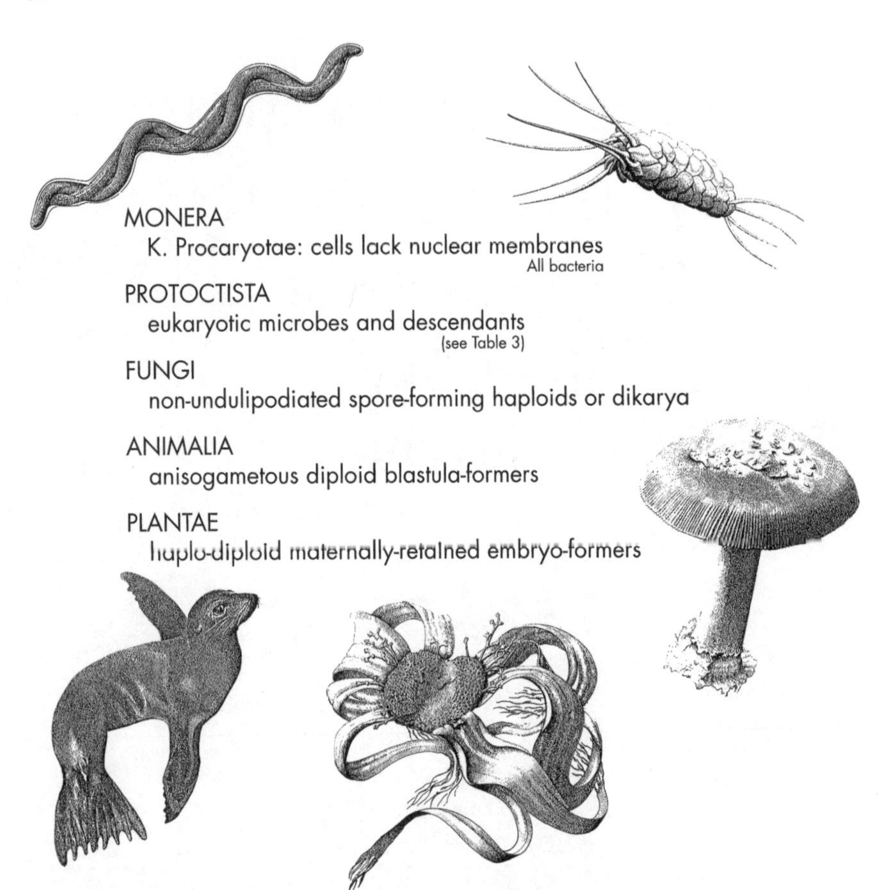

FIGURE 8.1. Five kingdoms: Genetic definitions and examples. Clockwise: *Cristispira, Mallomonas, Amanita, Welwitschia* and a seal. Drawings by Christie Lyons.

to the other four highest taxa: prokaryotes (Starr et al. 1981), protoctists (Margulis et al. 1990), fungi (Kendrick 1997), and plants (Margulis and Schwartz 1997) (Figure 8.1).

Animals and all other organisms of the Earth's biota display self-maintaining properties leading to their characterization as autopoietic entities. This concept, to be discussed later, involves properties listed in Table 8.1. Whereas the smallest recognizable autopoietic entity in today's biota is a tiny bacterial cell, the largest is Gaia, the organismal-environmental regulatory system at the Earth's surface, comprised of more than thirty million extant species (Lovelock 1988).

Table 8.1. Criteria of Autopoiesis

Criteria	Summary	Life today
Identifiable boundary around discrete components	System identity; materially open	Plasma-membrane bounded cells
Entity is a material system in which the component interactions and transformations are determined by component properties	Physicochemical operations	Cell metabolism, including ionic regulation within membrane (K^+, N^+, H^+, OH^-, Cl^-)
Boundary of the entity is determined by relations among its components	Self-maintained boundary	Cell metabolism constructs plasma membrane
Boundary components produced by component interactions and transformations	Components produced, transformed, and organized by system from external sources	External sources of C, H, N, O, P, S, etc., organized by multienzyme gene-directed pathways, resulting in cell maintenance and waste production
System components are produced by component interactions and transformations	System component interactions involve transduction of energy	Photo-/chemo-/auto-/hetero-/litho-organotrophic modes energy/matter transformation (Table 8.5)

[a]After Fleischaker (1988), with permission.

The popular neo-Darwinian view, in which all evolutionary innovation is assumed to be generated by chance accumulation of mutations, I claim, is mechanical, parochial, and repressive. Neo-Darwinism, sociobiology, and other prevailing zoological philosophies tend to ignore immense nonzoological scientific literature; they closely reflect the larger social context of contemporary science that produced them. My hope is that insufficient and inadequate neo-Darwinian analysis soon will be replaced with autopoietic concepts of living beings that include a Gaian view of organism-environmental interactions. Such a trend, returning from neo-Darwinism toward Darwin's original intentions (for example, Darwinism), will reintroduce microbiological and geological (including sedimentological and geochemical) awareness into zoological thinking.

Animal–Plant Dichotomy

Throughout the history of the science of life, zoology, the study of animals, has been contrasted with botany, the study of plants. In the two-kingdom system, animals are organisms that move in search of food, whereas plants, incapable of locomotion, produce their own food and our oxygen by photosynthesis. This charming dichotomy has been totally overturned by a microbial perspective of the animal kingdom. Two other myths of zoology, that animals are independent beings and that an animal body is an individual organism, have also been supplemented. These views of the plant/animal dichotomy, the independence of an animal, and individuality directly derive from cultural concepts; even the term *kingdom* comes from a western European political organization that was prevalent up to the late nineteenth century. My purpose here, which directly contradicts the aforementioned statement by Stephen J. Gould, is to recognize animals embedded in the context of their microbial predecessors. They are not "superior," or "higher" forms of life, to be contrasted with the "lower" animals and "higher" plants. Rather, animals are peculiar, if familiar, descendants of coevolved microbial communities. Rather than force the restrictive "animal/plant" dichotomy from above, my attempt will be recognition of defining characteristics of animals from below, that is, from the subvisible life that preceded animals in the fossil record: prokaryotes and protoctists.

What Is an Animal?

Striving to be man, the worm mounts through all the spires of form.

R.W. EMERSON (1803–1882)

The characters usually associated with animals include the following: the ability to locomote, in particular to swim and run; behavioral traits that involve attraction, repulsion, and recognition of potential food and danger; and multicellularity, that is, the formation of a distinctive describable body comprised of various types of cells. Even though these characters are often taken to define "the animal," it is clear that locomotion, sensory behavior and response, predation, multicellularity, and even programmed death of differentiated cells are features of the microcosm; all are found in the microbial predecessors to animals.

Indeed, the uniqueness of animals comes from their developmental and genetic pathways. Photosynthetic animals are well known (for exam-

Table 8.2. Examples of Multicellularity in Bacteria, Protoctists and Fungi

Kingdom	Phylum	Genus	Comments
Monera	Myxobacteria	*Polyangium*	All aerial "spore-formers"
Protoctista	Ciliophora	*Sorogena*	
Fungi	Ascomycota	*Neurospora*	

ple, *Elysia, Plachobranchus, Convoluta roscoffensis;* Smith and Douglas, 1987). However, no organisms besides animals have blastular anisogamy wherein an egg, fertilized by a sperm, becomes a zygote, which subsequently cleaves, developing into an embryo: a hollow ball of cells, the blastula. All animals but no members of the other four kingdoms develop from blastular embryos. Of course, in the vast majority of animals, the blastula goes on to gastrulate, and internal intestinal and external epidermal systems develop. Furthermore, in animals, the haplophase is limited to single "germ" cells, and gametic meiosis is the rule: meiosis accompanies gamete formation, whereas fertilization leads to the diploid blastula. This developmental pattern unambiguously distinguishes animals from all other forms of life. Locomotion, heterotrophy, and multicellularity are inadequate to distinguish animals: multicellularity is not only present in bacteria but in every major lineage of protoctists (Tables 8.2, 8.3).

Animals evolved from their protoctist ancestors sometime in the late Proterozoic eon, that is, the late Riphean or early Vendian era (McMenamin and McMenamin 1990). Although the identification of the protoctist ancestors from which originated animals is still an unsolved problem, by the time the transition to the first animals occurred, according to Steve Gould, "nothing much had happened for ever so long." As much as I admire the work and lyrical writings of Professor Gould, nothing is more inaccurate than this statement. All major evolutionary innovations had occurred by the time the first animals appeared in the fossil record (Margulis 1993).

Undulipodia and Protoctista

What traits have animals in common with their protoctist ancestors? From which of the four groups of phyla of Table 8.3 did the metazoans evolve? The importance of centrioles, kinetosomes, and axonemes (for example, components of undulipodia) and centriole/kinetosome transformation is crucial to the understanding of the relation of protoctists to animals. The

Table 8.3. Four Groups of Protoctista Phyla

1. No undulipodia; no complex sexual cycles: Rhizopoda, Haplosporida, Myxozoa, Microspora
2. No undulipodia; complex sexual cycles: Acrasea, Dictyostelida, Rhodophyta, Conjugaphyta
3. Reversible formation of undulipodia; no complex sexual cycles: Euglenida, Chlorarachnida, Actinopoda, Hyphochytrids, Plasmodiophoromycota, Labyrinthulids, Xenophyophora, Cryptomonads, Glaucocystids
4. Reversible formation of undulipodia; complex sexual cycles: Chrysophyta, Phaeophyta, Dinomastigota, Apicomplexa, Oomycota, Myxomycota, Bacillarophyta, Chlorophyta, Xanthophyta

undulipodium of eukaryotes (which should never be confused with the flagellum of prokaryotes) is depicted in Figure 8.2.

Cells of animals, of course, can form undulipodia: sperm with undulipodia as tails, sensory cells such as retinal rods, and cones. Oviduct and gill cilia, trachea, and other ciliated surface-epithelium tissue are studded with undulipodia. From these observations we deduce, as did the cytologists and zoologists of the last century, that animals have undulipodiated ancestors.

The protist ancestors of animals cannot have belonged to the phyla lacking undulipodia at all stages (groups I, II), whether or not they have complex sexual cycles. The ancestors of animals, in my opinion, did not come from the sophisticated protists in group IV either, although superficially (because those organisms can reversibly form undulipodia and display complex life cycles), they are the most like animals. The key fact is that no undulipodiated animal cell ever has been observed to undergo mitosis: cell division by mitosis and the presence of an undulipodium for swimming or sensory function are mutually exclusive properties in any given animal cell. Unlike many protoctists—which are composed of cells each capable of cell division in the presence of functional undulipodia—in animals an undulipodiated cell is always precluded from mitosis. Swimming or ciliation and mitotic cell reproduction are mutually exclusive. I interpret this to mean that animals evolved from a lineage of protoctists, probably amoebomastigotes, in which the ancestral cells, once they differentiated undulipodia, were permanently incapable both of cell division by

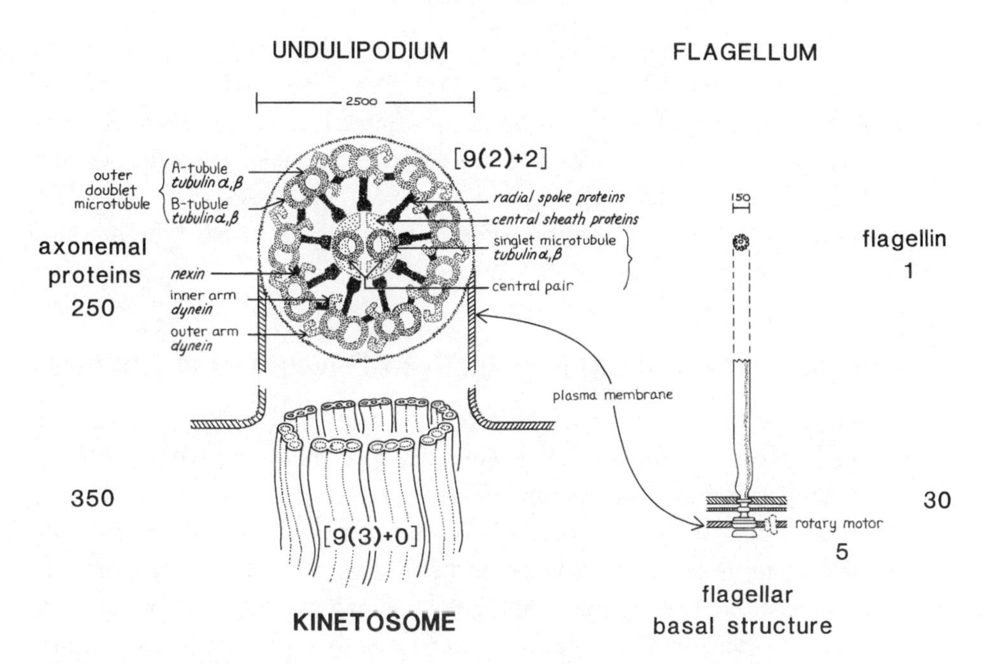

FIGURE 8.2. Undulipodium of nucleated (eukaryotic) cells compared with flagellum of bacteria (prokaryotic cells). Diameters in angstroms, estimated numbers of proteins in each structure given. Drawing by K. Delisle.

mitosis and reproducing these undulipodia. Unlike ciliates, euglenids, chlorophytes, and many other protoctists (groups 3 and 4 in Table 8.3), our animal ancestors never solved the the problem of "How can I divide and at the same time swim by means of undulipodia?" To resolve the locomotion-cell division dilemma, animals evolved a peculiar sort of multicellularity. In the animal ancestor the "two-cell stage" already existed, the separation of *germ plasm* (those failing to differentiate any undulipodia, that is, a cell lacking undulipodia but retaining the capacity for mitotic cell division) and the *soma:* the undulipodiated cell. In the light of the stunning announcement by David Luck and his colleagues of centriolar DNA in *Chlamydomonas* (Hall et al. 1989), I infer that undulipodiated cells of animals either lack altogether or cannot access the six-megabase centriolar DNA required for centriolar reproduction. The organization of the microtubules into functioning components for mitosis and for centriolar reproduction are, in animals, only retained in cells incapable of developing a mature kinetosome–undulipodium complex. Like the kinetosomes of the oral membranellar band of *Stentor* (Younger et al. 1972), those kinetosomes incapable of reproduction (that is, most likely all of them in metazoa) probably lack the requisite kinetosomal/centriolar DNA. Hence undulipodiated animal cells depend on sexual processes of neighboring undifferentiated, nonundulipodiated cells for the retention and reproduction of specific kinetosomal DNA.

Autopoietic Gaia to Replace Neo-Darwinian Mechanics

Autopoiesis, a term invented by Maturana and Varela (1980) and elaborated by other authors (Fleischaker 1988), refers to the living nature of material systems. Well within the materialist view that recognizes the physical-chemical composition of organisms, autopoiesis refers to the self-making and self-maintaining properties of living systems relative to their dead counterparts. Autopoietic, unlike mechanical, systems produce and maintain their own boundaries (plasma membranes, skin, exoskeletons, bark, etc.). Autopoietic systems incessantly modulate their ionic composition and macromolecular sequences (that is, amino acid and nucleotide residues in their proteins and nucleic acids). Some even regulate their internal temperatures. Any autopoietic entity capable of reproduction (that is, cells, organisms, cohesive groups of organisms) is subject to natural selection for the simple reason that all the potential products of reproduction can never survive. The smallest autopoietic systems, spherical and less than a micrometer in diameter, are bacterial cells. (Viruses, plasmids, and other replicons are too simple and small to be autopoietic.) The largest autopoietic system, so far incapable of reproduction, is the modulated surface of the Earth that Lovelock (1988) has named *Gaia.*

For the purposes of this paper, Gaia is defined as the large self-maintaining, self-producing system extending within about 20 kilometers of the surface of the Earth. The Gaia hypothesis states: the surface sediments and troposphere of the Earth are actively regulated by the biota (the sum of the live organisms) with respect to the chemical composition of the reactive elements (for example, H, C, N, O, S), acidity (for example, H^+, OH^-, $CO_3^=$, HCO_3^-), the oxidation-reduction state and the temperature. Gaian regulation, like the physiology of an embryo, is more homeorrhetic than homeostatic in that the internally organized system regulates around moving, rather than fixed-from-the-outside, setpoints. (For example, the values of temperature, atmospheric gas composition, and others around which there is Gaian regulation do change with time.) Gaia, a single enormous system deriving from a 3500 million-year-old common ancestor, is connected through time (by ancestry)—as Darwin realized—and through space (by allelochemics, atmospheric and oceanic currents, and the like, as V.I. Vernadsky [1997] realized). The Gaian system persists in the face of changes (for example, population expansions and extinction, rise and fall of sea level, and so forth). The tendency of all organisms to inject their genes into the next generation is one of these changes.

The validity of the Gaia idea, the self-regulating biosphere, has been forcefully argued by Lovelock (1979, 1988). The Gaia hypothesis has even been called a "grand unified theory" of biology; it has also been recognized as more a point of view than a scientific hypothesis (see p. 185). In autopoietic language, Gaia is the largest unit that displays the properties in Table 8.1. Probably the best way to think about Gaia is to contemplate the assertion that the atmosphere and surface sediments of the Earth are part of the living system. Life does not "adapt" to a passive physicochemical environment as most neo-Darwinians assume; instead, life actively produces and modifies its surroundings. The oxygen we breathe, the humid atmosphere inside of which we live, and the mildly alkaline ocean waters in which the whales are bathed are not determined by a physical universe run by mechanical "laws." In stark contrast to a mechanical, physics-centered world, the metabolizing biosphere is physiologically self-controlled. The breathable oxygen, the humid air, and mildly alkaline oceans result from the growth and metabolism of uncountable and always changing numbers of bacteria, plants, and algae, which produce oxygen using solar energy. Water transportation is driven by the activities of great forests, primarily of neotropical trees; and the neutralization of the acid tendencies of the planet is accomplished by the production of alkaline substances such as ammonia by myriads of organisms, for example, by urination and gas release. Many other examples exist of Gaian Earth-surface regulatory activities (Lovelock 1988; see Part III, p. 125. 261 below).

The Gaian worldview is an autopoietic one; the surface of this planet is alive with a connected megametabolism, which leads to temperature and chemical modulation systems in which humans play a small and epiphenomenal part. (After all, humans as *Homo sapiens sapiens* evolved only some 40,000 years ago, long after the Gaian system, which is over 3000 million years old, was completely in place.)

Neo-Darwinians, who ignore chemical differences between living beings, who never factor autopoiesis into their equations, and who consider organisms as independent entities evolving by accumulation of chance mutations, *must* hate and resist an autopoietic, Gaian worldview.

If we assume that consistency is a scientific virtue, then acceptance of the autopoietic-Gaian worldview requires rejection of philosophical underpinnings of current neo-Darwinism. Neo-Darwinism, from a Gaian perspective, must be dismissed as a minor twentieth-century religious sect within the sprawling religious persuasion of Anglo-Saxon biology. As yet another example of a thought style in the great family of biological-scientific *Weltenschaungen,* past and present, neo-Darwinism (like nineteenth-century German *nature-philosophie* and phrenology) must take its place (like British social Darwinism) as a quaint, but potentially dangerous aberration.

Elsewhere we have argued the advisability of replacing mechanical neo-Darwinian outlooks with Gaian physiological-autopoietic ones. In a review of 150 years of financial support for environmental science, we show that scientists, like everyone else, generate information to bolster the philosophies of those who pay them (see Chapter 16, p. 207). We also detail the neo-Darwinian and other institutionalized resistance to autopoietic-Gaian points of view, showing why there is "big trouble" in biology today (see Chapter 20, p. 265). In what follows I refer to that earlier work, here simply accepting the autopoietic-Gaian concepts as background to analyze the nature of the intrinsic sexuality of animals. The analogy of cyclical meiotic sexuality, a defining feature of animals, with cyclic symbioses is instructive. The explanations for cyclicity in certain symbioses, I believe, aid in the understanding of the maintenance of the sexual cyclicity in animals.

Neo-Darwinian language and conceptual structure itself ensures scientific failure: Major questions posed by zoologists cannot be answered from inside the neo-Darwinian straitjacket. Such questions include, for example, "How do new structures arise in evolution?" "Why, given so much environmental change, is stasis so prevalent in evolution as seen in the fossil record?" "How did one group of organisms or set of macromolecules evolve from another?" The importance of these questions is not at issue; it is just that neo-Darwinians, restricted by their presuppositions, cannot answer them. Molecular evolutionists also face difficulties: they fail to make sense of enormous quantities of data because they lack a framework to functionally distinguish one or another DNA base-pair sequences. Biological education tends to parrot neo-Darwinism, amplifying its deficiencies and leaving a confused public. Those who have seen the neo-

Darwinian exposition on quantitative evolution at the British Museum (Natural History) can easily attest to the validity of this observation.

Origins of novelty and stasis in the history of life, origins and functioning of physiological control systems, and many other questions will remain unanswerable as long as neo-Darwinian strangleholds on the professional zoologist persist. The situation is comparable with the eighteenth century question, "What causes contagious disease?" in which answers were proferred in the total ignorance of the existence and behavior of bacteria, protoctistan parasites, and fungi.

Autopoiesis, Individuals, and Death

Populations are comprised of organisms that are members of the same species, living together at the same time in the same place. Communities, then, are populations of more than a single species living together at the same time and the same place.

Organisms are perceived as "individuals"—plants and fungi are measured on an animal individual standard. All are thought by neo-Darwinians to be reproducing products of chance mutations. But what is an individual?

The term "individual" contains the idea that half (or any other fraction) of whatever is an individual is not itself an individual. Although I know of no formal definition in the zoological literature, I suspect what is meant by practitioners is as follows: an individual is a countable entity that behaves independently, it is an entity that when reduced or divided loses its defining properties. Most biologists, including neo-Darwinians, would probably accept this definition. Yet much neo-Darwinian malaise comes from the critical differences between this "common sense" concept of individual and parallel ideas of autopoietic systems that have not been made explicit. A soccer ball, a book, a pitchfork, a newspaper, and an atomic particle are individuals; yet none is a self-maintaining system produced by its own internal organization. Groups of these individuals are not subject to natural selection—they do not evolve in the Darwinian sense. None is alive; none has autopoietic properties. The ability to count individuals—and hence to practice number games—coupled with the failure to distinguish autopoietic entities from other sorts of countable "individuals," has led academic neo-Darwinian biology down a rosy path from which I see no return in sight.

Neo-Darwinians and Gaia-autopoietic biologists respond differently to scientific "explanations." Some "explanations" are conceived as aspects of the self-maintaining activities of the living, that is, of autopoiesis. An example: Why do Pacific salmon swim upstream to die in the area where they themselves spawned? A neo-Darwinian uses military or economic terms, tending towards an explanation in terms of "reproductive strategies," of offspring outcompeting others with fewer genes in common. In the autopoietic point of view, attention is paid to the chemical components of the fish. For example, that the dead bodies of the upstream adult salmon provide phosphorus for the diatoms that, during the next season, serve as food for salmon fry.

Another example: Why do small quantities (less than or equal to 0.5 ml inocula) of certain bacteria added to fresh growth solution not grow whereas larger ones (greater than or equal to 1.0 ml) grow well? The observation that death of the organisms comprising nearly the entire inoculum provides conditions for growth of the few remaining bacteria is described as pure "altruism" (and thus rejected) by neo-Darwinians. From the chemically self-conscious, autopoietic point of view, it is sufficient to recognize that component lipids and other compounds shed by a large inoculum provide sufficient ambient conditions, probably including food, for the initial growth of at least a few of the bacteria in pure culture. At least ten orders of birds contain species in which parents or nestmates eat their offspring. Cell death, tissue resorption, and cannibalism are common means for the autopoietic imperative of replacement of molecular components.

All organisms require sources of energy, electrons, and certain elements (for example, at least carbon, hydrogen, oxygen, nitrogen, sulfur, and phosphorus) at rates and in ranges determined by the natural history/structure/physiology of the organism itself. Without the continual movement of component matter (that is, metabolism, including nutrient supply, water circulation, and waste removal) comprising the microbe, animal, or plant, integrity is threatened; stress and eventually death ensue. Autopoietic systems are bounded; the minimal boundary is a cell membrane. Autopoietic entities, transforming compounds to ensure self-maintenance, use chemical energy to move electrons, chemical elements, and their compounds.

Some zoological conclusions may be derived. The feeding behavior, social interactions, composition of excretory emissions, respiratory gas concentrations, and most other activities of animals (and all other live beings) are determined directly by the imperatives of the autopoietic system that is the animal in question. The organization of an animal, its population, and community are determined far more by its history than by its present environment (Fleischaker 1988). Each animal's organization is maintained by metabolism (that is, the interacting chemical activity involving several thousand genes, the same number of proteins, four to five classes of RNA, lipids, and other metabolites). Thus, no single gene as a stretch of DNA nor single example of any other class of molecule suffices to define, determine, or represent the entire autopoietic system. The history of a population of organisms cannot, even in principle, be reconstructed by anyone using only a single molecule. No single gene, gene product (such as 16S ribosomal RNA), nor any storage product (such as cellulose or starch) suffices to even approximate the evolutionary history of that organism. Even though very long sequences of nucleic acid storing a great deal of information are far better criteria than single pigments or storage products, no single molecule, in principle, is adequate as a criterion to reconstruct the history of the evolution of a group of living beings.

All animals—or other live organisms—because of their continuous replacement of biochemical components at the expense of matter and energy taken from the environment, have properties of behavior related to their chemical integrity and continuity. Such physicochemical dictates imply that the use of anthropocentric terms such as "selfishness," "altruism," and "group selection" with reference to an organism's behavior or a gene's repeated polynucleotide synthesis is inappropriate labelling. Like all labelling, the naming itself perpetuates ignorance—it leads to dismissal of the phenomenon named from the consciousness of the labeller. Such anthropocentric terminology, so prevalent in neo-Darwinian subfields such as sociobiology (Table 8.4), although appealing to those who seek quantifiable solutions to complex problems, restrict science.

Most aspects of the growth, behavior, and reproduction of animals are dynamically determined by internal organization. Not only chemical exudates and multienzyme pathways, but behaviors (for example, feeding movements, hibernation behavior, and light absorption and emission in luminous fish; McFall-Ngai 1991) tend to be consequences of the autopoietic system itself. Thus, the projection onto nucleotide sequences of DNA, social insects, or hibernating mammals of cost-benefit analyses, altruism, selfishness, or other terms reminiscent of western economic conditions (such as those in Table 8.4) are worse than naive. The use in biology of computer software developed for economic analyses is as pernicious and debilitating as using a rag doll to model a child.

Yet dolls, stars, machines, fashions, and macromolecules do evolve in the sense that they change in describable ways through time. Not

Table 8.4. Neodarwinism: Anthropocentric Terminology

1. Adaptation
2. Altruism, altruistic behavior
3. Cheating, selfish behavior
4. Fitness, inclusive fitness
5. Genetic variation, diversity
6. Genotype, phenotype
7. Group selection
8. Individual
9. Kin selection
10. Levels of selection, units of selection
11. Daughter cells
12. Species, race

autopoietic, these entities do not evolve in the Darwinian sense. All organisms are components of the autopoietic planetary system; all (plants, animals, and microbes) are connected, however circuitously, to all others spatially and by common descent. We are advised to integrate Darwin's original intentions into a modern evolutionary analysis that recognizes autopoiesis as the defining principle of life. Darwin's (1859) fundamental contentions can be reasserted: (1) all organisms are "descended by modification" (they are related by common ancestry) and (2) they are subject to natural selection at all times. Natural selection is a simple consequence of the fact that too many autopoietic entities are potentially self-produced than can possibly survive. "Natural selection" is the inability, in any given case, for the biotic potential to be reached. Biotic potential, the capacity for organisms to self-produce (fission into, hatch, give birth to, etc.) other organisms, is measured by the units: organisms produced per generation (or organisms per unit time). In modern language, we can say that all organisms alive today or in the past share a physical continuity with all others. All organisms are part of a single continuous bounded autopoietic system that has never been breached since the origins of life in the Hadean or Archean eon. While portions of the system (cells, individuals, populations, species) are always losing autopoietic properties, the entire system itself persists. Death must co-occur with life. Failure to retain autopoietic properties is death—and death by loss of components, desiccation, disintegration, and atrophy is intrinsic to the continuity of life. Wounding-scarforming, hair shedding, nail and skin sloughing are examples of programmed cell death characteristic of mammalian autopoiesis.

Darwinians must now ponder autopoiesis. When aspects of autopoiesis are breached (for example, self-maintenance of the organized system), no matter how apparently trivially, first "stress" then moribundity ensues and, with time, survival becomes impossible. Examples of breaching of autopoiesis include failure to deliver sulfur at sufficient rates for the synthesis of the cysteine- and methionine-containing proteins, or rupture of the membranous border of an amoeba. Failure of some autopoietic systems to persist in the face of the survival of others is another way of phrasing Darwin's concept of natural selection.

Failure to achieve biotic potential is an aspect of Darwinian evolution that has accompanied life since its origins. Although the failure to grow and reproduce at maximal capacity (that is, failure of biotic potential to be reached) is equivalent to natural selection, it is not, as Darwin thought it might be, the way in which new species originate.

We can recognize failure of autopoiesis at several levels: parts of cells, cells, populations, or communities. The failure can be random and unpredictable (for example, when lightning strikes a pond and kills all the birds on it or when a fire destroys a wood) or it can be a completely predictable, intrinsic behavior of the autopoietic system (such as flower wilting, metamorphosis in insects, egg cannibalism in birds, or aging in mammals). Failure of any member of a population or species to survive at a given time in a given place is manifested as population or species death, that is, extinction. Natural selection brilliantly explains the "editing" or "correcting" aspects of evolution, but to what do we owe the source of novelty? At least one major source is symbiosis, the protracted association of organisms belonging to more than a single species (Margulis and Fester 1991). Indeed, before the origin of eukaryotes, all of which derived from symbiotic complexes, "species" of organisms probably did not exist (Sonea 1991).

Species as Products of Symbiont Integration

In acceptance of the ideas of Sonea and Panisset (1983), I concur that before eukaryosis (that is, the origins of organisms with membrane-bounded nuclei: protoctista, fungi, plants, and animals), species did not exist. Genetic (sexual) exchange occurred across the Archean biota; distribution of traits in the bacterial world was maintained by the promiscuous sort of bacterial sex because bacteria can "drink up" new genes and alter their metabolism and behavior accordingly. Before protoctists, as Sonea and Panisset (1983) argue, there was no "origin of species" per se: changes in bacterial populations occur continuously as genes on small or large replicons transgress bacterial boundaries. Prior to the earliest protoctists, the Earth was populated only by archaebacteria, cyanobacteria, actinobacteria, and so on. The evolutionary process itself changed with the appearance of the first eukaryotes, some of which later evolved into animals, fungi, and plants. Tens of thousands of species appeared—complex, integrated microbial communities. These supplemented the more loosely organized communities of bacteria.

Bacteria, unlike plants and animals, differ from each other primarily by their metabolic modes: differences in the ways by which they acquire energy, carbon, and electrons for their internal chemical reactions (Table 8.5). The ways in which bacteria rid themselves of waste also differ radically from those of most protoctists, fungi, animals, and plants. Bacterial cells very quickly develop prodigious population numbers—one bacterium can become billions in a day. They just as quickly die. The origin of new varieties of bacteria occurs primarily by acquisition of new fragments of DNA. Bacterial populations are maintained directly by death of their extreme end members.

Bacterial evolution is flexible—in principle, all bacteria can exchange DNA (mate) with any other. Bacteria are easy for us to "engineer" because they have easily made and exchanged genes by themselves for eons. Bacterial evolution, studied in the laboratory, involves acquisition and loss of genes correlated with changes in enzyme reactions (Mortlock 1984). Bacteria evolve as an enormous, worldwide, interacting, loose network (Sonea and Panisset 1983). Imagine if whales and daffodils could mate and form fertile puppies with flowering petals: bacterial modes of evolution differ in principle from those of eukaryotes.

Soon after the first autopoietic system appeared, bacteria of the Archean Earth (3.9 until 2.5×10^9 years ago), the processes of cell division became established—including at first reproduction in the total absence of meiotic sex.

Cyclicity: Sex and Symbiosis

Homogenomous organisms, for discussion purposes, we define as multigenomic eukaryotes that share the same or very similar genomes. They have far more recent common ancestors than heterogenomous ones. These are obviously relative terms: heterogenomous organisms are eukaryotes composed of recently fused but very remote ancestors.

Table 8.5. Modes of Nutrition Known for Life on Earth[a,d]

Energy: Light or Chemical Compounds	Electrons (or Hydrogen Donors)	Carbon Sources	Examples of Organisms, Hydrogen or Electron Donors
Photo- (light)	Litho- (inorganic and C_1 compounds)	Auto-(CO_2)	Prokaryotes: *Chlorobium* H_2S, H_2 *Chromatium*, H_2S, H_2 *Rhodospirillum*, H_2 cyanobacteria, H_2O:H_2S chloroxybacteria, H_2O Protoctista (algae), H_2O Plants, H_2O
		Hetero- $(CH_2O)_n$	None
	Organo- (organic compounds)	Auto-	None
		Hetero-	Prokaryotes: *Chromatium*, org. comp.[e] *Chloroflexa*, org. comp.[e] *Halobacter*, org. comp.[e] *Heliobacterium*, org. comp.[e] *Rhodomicrobium*, C_2, C_3
Chemo- (chemical compounds)	Litho-	Auto-	Prokaryotes: Methanogens, H_2 Hydrogen oxidizers, H_2 Methylotrophs, CH_4, CHOH, etc. Ammonia, nitrite oxidizers, NH_3, NO_2
		Hetero-	Prokaryotes: Manganese oxidizers, Mn^{2+} Iron bacteria, Fe^{2+} Sulfide oxidizers, e.g., *Beggiatoa*
	Organo-	Auto-	Prokaryotes: *Clostridia*, etc., grown on CO_2 as sole source of carbon (H_2, -CH_2)
		Hetero-	Prokaryotes (most) (including nitrate, sulfate, oxygen, and phosphate[b] as terminal electron acceptors) Protoctista[c] (most) Fungi[c] Plants[b] (achlorophyllous) Animals[c]

[a]A list of the sources of energy, electrons, and carbon for metabolism; the name of each mode with examples of growth of organisms to which the names apply is given. Names constructed by addition of suffix "-troph," for example, photolithoautotroph (plants).

[b]Detection of phosphine: Dévai I.

[c]Oxygen as terminal electron acceptor.

[d]Table devised in collaboration with S. Golubic, R. Guerrero, and S. Goodwin

[e]Organic compounds, for example, acetate, proprionate, pyruvate.

All eukaryotes (protoctista, fungi, animals, and plants; Margulis and Schwartz 1988) are complex (multiple heterogenomous) autopoietic entities (Margulis and Bermudes 1985; Margulis 1993; Figure 8.3). By recognizing the autopoietic status of eukaryotes, we can glimpse the origin of a

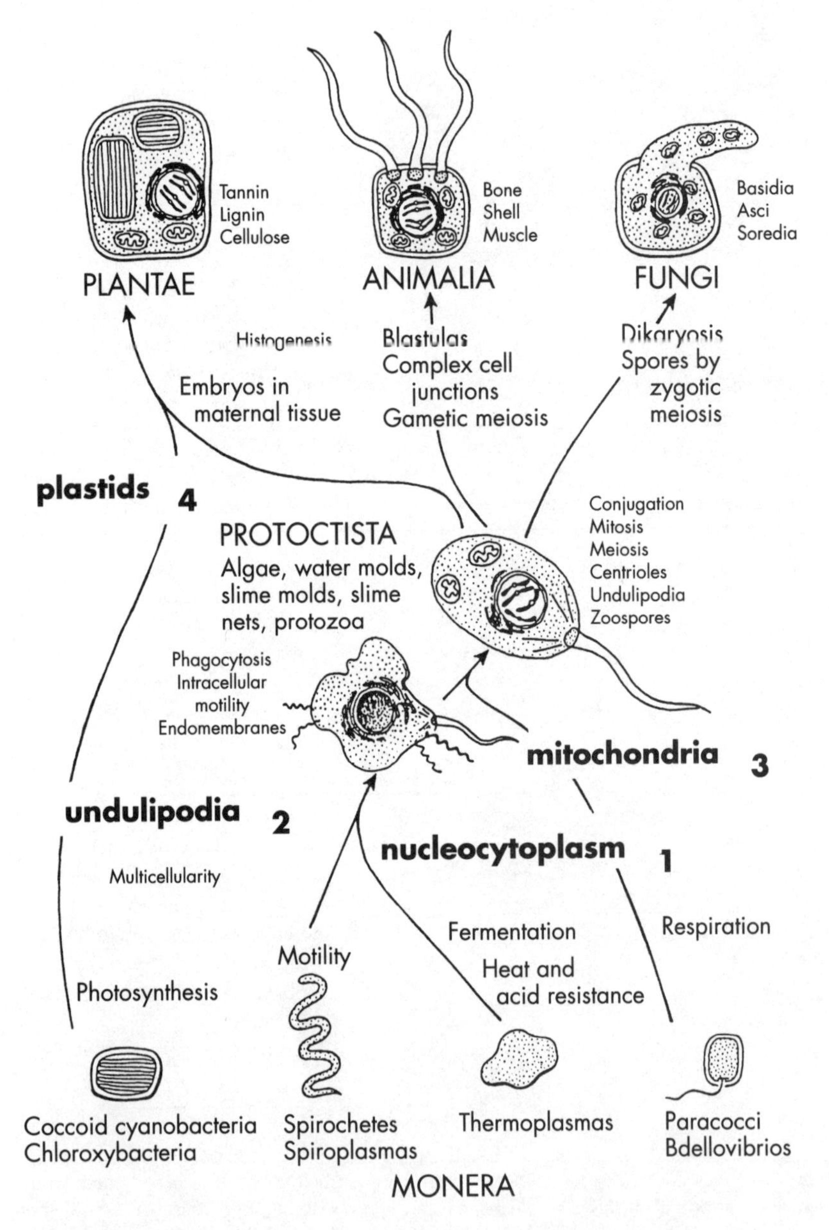

FIGURE 8.3. Symbiogenesis and origin of eukaryotes: The acquisition of mitochondria after undulipodia. Numbers refer to order of acquisition.

fundamental happening in the history of life that began with bacteria. Speciation processes began only after emergence of symbionts as a new type of autopoietic entity. Following the fusion of two or more heterogenomous autopoietic entities that survived to form a new autopoietic entity, the new entity displayed many altered properties relative to its original disparate "parents" that entered the mergers. The process of speciation only began after symbiogenesis (see later).

Many occurences—establishing, maintaining, and dissolving relationships between autopoietic entities—are involved in the formation of new heterogenomous autopoietic entities from existing ones. The process I refer to here can be recognized in the biological literature as the establishment of hereditary symbioses (Bermudes and Margulis 1987). The merged entities are new and different. The crucial question to be posed is if, in an interaction (for example, contact and fusion between two or more autopoietic entities) survival ensues such that the survivors are, or are not, altered after the interactions. Given this analysis, we can describe many familiar biological interactions in these terms.

Fusion and Separation: Autopoiesis and Animal Evolution

Temporary fusion of two (or more) heterogenomous autopoietic entities, including cases in which continuity of one of the entities is either terminated (death) or left unchanged (coexistence), is recognizable as ecological interaction. Many subcategories of temporary fusions of heterogenomic autopoietic entities have been named, especially when they involve the curtailing or cessation of the autopoiesis of one of the entities involved, for example, familiar terms for such processes include carnivory, cropping, feeding, pathogenesis, predation, pruning, and so forth.

Temporary fusion of two (or more homogenomous autopoietic entities, as in animal fertilization and embryogenesis), including cases in which continuity of one of the entities is terminated, has been termed "ontogenesis" (or development). Although all embryological development and maturation involves cell death, cell fusion, and cell reproduction, the analysis of ontogeny from the point of view of autopoietic entities and their termination deserves careful study.

We have asserted that permanent fusion of two (or more) heterogenomous autopoietic entities to form new autopoietic entities is the origin of species or other higher taxa (Margulis and Bermudes 1985,

1993). This concept, known in the Russian and German biological literature as symbiogenesis, has a long tradition (Khakhina 1979; Schwemmler 1989; Margulis 1991b). A major example of cyclical fusion of two or more homogenomous autopoietic entities is meiotic sexuality. This point is central to the thesis of our book on the origins of sex (Margulis and Sagan 1986b). Cyclical fusions of two (or more) heterogenomous autopoietic entities are the cyclical symbioses well known in the symbiotic literature. Hundreds of examples are known, including mycorrhizae (angiosperms with zygomycotes or basidiomycotes), root nodules (angiosperms with rhizobium bacteria), and the fusion of *Tetraselmis* (= *Platymonas*) and *Convoluta* to form the green photosynthetic flatworm *Convoluta roscoffensis.*

Darwinian evolution can be described as change through time of autopoietic entities. Neither the component parts (for example, membranes, macromolecules, plasmids) nor the artifacts of autopoietic entities (for example, shells, machines, trace burrow fossils) evolve by themselves in the Darwinian sense. Like the neo-Darwinians (see Table 8.4), molecular evolutionists too have generated what Fleck (1979) called battle cries. Such terms serve more to identify members in the field than they do to illuminate evolution of life on Earth (Table 8.6). No study that fails to specify the organisms with which it is concerned nor gives the time and place of their existence can, in principle, add knowledge to the evolutionary history of life on Earth. Fascinating and expensive studies on the chemistry of life (such as the papers in Baltscheffsky et al. 1986) cannot intrinsically contribute to unraveling the evolutionary history of life because the chemistry of phylogeny is not distinguished from that of ontogeny. Such chemical observations are as useful for evolutionary studies as unidentified rock samples, out of context of their outcrops, are for stratigraphy. To resolve

Table 8.6. Molecular Evolution: Words Used as Battle Cries

1. Advanced, primitive organisms
2. Archaebacteria, eubacteria, metabacteria, crenarcheota, archae
3. Conserved sequences
4. Eucytes, parkaryotes[a]
5. Higher, lower organisms
6. Molecular homology, convergence, divergence
7. Quickly evolving, slowly evolving molecules
8. Rooted trees

[a]From Lake (1988), with permission.

cognitive dissonance and to ease the pain of disparate scholars (for example, molecular evolutionists and population biologists), zoologists must appropriately regain perspective on the origin and evolution of animals from a rich variety of their pre-Phanerozoic predecessors. Unstated neo-Darwinian and molecular evolutionary assumptions (for example, those of Dawkins 1976; Woese 1987; Hori 1982; Lake 1988) must be returned to consciousness, and mechanistic metaphors replaced with autopoietic Gaian concepts.

9

SPECULATION ON SPECULATION

LYNN MARGULIS

Whereas in science theory is lauded, speculation is ridiculed. A biologist accused in print of "speculation" is branded for the tenure of her career. This biologist finds herself like a ballet dancer imitating a pigeon-toed hunchback: All of the intellectual training to keep my toes turned out emotionally backfires with a request to speculate freely.

In a manuscript lacking data, field and laboratory observations, descriptions of equipment and their correlated methodologies, and deficient in references, I feel a huge restraint as I attempt to slacken the bonds of professionalism and turn my toes in. My well-seasoned inhibitions are nevertheless titillated by the joys of this opportunity to really tell you about the hypothesis that I am always testing, that which I am always questioning: my developing worldview of mind. Asking your patience and indulgence to see beyond the inevitable barriers of language, I can at least try to articulate the unmentioned and hitherto unmentionable.

We have an intuitive grasp of the reality to which these terms refer: perception, awareness, speculation, thought, memory, knowledge, and consciousness. Most of us would claim that these qualities of mind have been listed more or less in evolutionary order. It is obvious that bacteria perceive sugars and algae perceive light. Dogs are aware; whether deciding to chase a ball or not, they seem to be "speculating." Thought and memory are clearly present in nonhuman animals such as *Aplysia,* the huge, shell-less marine snail that can be taught association. *Aplysia,* the sea hare, can be trained to anticipate; it will flee from potential electric shock as soon as a light is flashed. Knowledge, some admit, can be displayed by whales, bears, bats, and other vertebrates, including birds. But conventional wisdom tells us that consciousness is limited to people and our immediate ancestors. Many scientists believe that "mind"—whatever it is—will never be known by any combination of neurophysiology, neuroanatomy, genetics, neuropharmacology, or any other materialistic science. Brain may be knowable by the -ologies, but mind can never be.

I disagree with many versions of this common myth. I believe brain is mind and mind is brain, and that science, broadly conceived, is an effective method for learning about both. The results of the -ologies just listed, as well as of many other sciences, can tell us clearly about ourselves and what is inside our heads. Furthermore, humans have no monopoly whatever on any of these mental processes. As long as we indicate consciousness of *what,* I can point to conscious, actively communicating, pond-water microscopic life (and even extremely unconscious bureaucrats). The processes of perception, awareness, speculation, and the like evolved in the microcosm: the subvisible world of our bacterial ancestors. Movement itself is an ancestral bacterial trait, and thought, I am suggesting, is a kind of cell movement.

We admit that computers have precedents: electricity, electronic circuits, silica semiconductors, screws, nuts, and bolts. The miracle of the computer is the way in which its parts are put together. So, too, human minds have precedents; the uniqueness is in the recombination and interaction of the elements that comprise the mind-brain. My contention is that hundreds of biologists, psychologists, philosophers, and others making inquiries of mind-brain have failed to identify even the analogues of electricity, electronic circuits, silica semiconductors, screws, nuts, and bolts. In the absence of knowing what the parts are and how they came together, we can never know the human mind-brain. Only the very recent history of the human brain is illuminated by comparative studies of amphibian and reptilian brains. The crucial ancient beginnings of the human brain lie in the dancing of bacteria: the intricate mechanisms of cell motility. How do cells locomote? The answer to this puzzle is the beginning of enlightenment for the origins of mind-brain.

I cherish a specific, testable, scientific theory. The means for testing it are biochemical, genetic, and molecular-biological. The facilities for the testing are available in New York City. A conclusive proof would require generosity on the part of at least two hugely successful and highly talented scientists and their laboratory assistants, a Columbia University biochemist and a Rockefeller University geneticist. Charles Cantor, of Columbia University Medical School, has developed new techniques to purify genes (DNA) gently. The purification holds the biological material on agar blocks (a gelatin-like substance) in such a manner that the structures in which the genes reside, the chromosomes, are extracted in their natural long, skinny form. Groups of genes (linkage groups) can be identified. Chromosome counts, difficult to determine microscopically, can be made biochemically.

And David Luck, an active geneticist at The Rockefeller University for over a quarter of a century, has recently discovered a new type of genetic system. He has found a special set of genes that determines the development of structures, bodies called kinetosomes that are present in thousands of very different kinds of motile cells: those of green algae, sperm, ciliates, oviducts, and trachea, for example. These structures, which I think of as assembly systems for nearly universal cell motors, may be determined by a unique set of genes separate from those of the nuclei and other components of cells. These genes, inferred from genetic studies of Luck and his colleagues, may be exactly the spirochetal remnant genes I predicted still must be inside all motile cells that contain kinetosomes (Hall, Ramanis, and Luck 1989).

Although no exorbitant amount of money would be needed, because testing my theory would be limited by the requirement for time and energy of very busy people, it would be expensive. Furthermore, the results of my testing, even if they are ideal, would cure no disease, stop no war, limit no radioactivity, save no tropical forest, and produce no marketable product. At least in the beginning, there would be no immediate profit coming from the work. The very concrete results would simply help us reconstruct the origin of our mind-brains from their bacterial ancestors.

What is the central idea to be tested? I hypothesize that all these phenomena of mind, from perception to consciousness, originated from an unholy microscopic alliance between hungry killer bacteria and their potential archaebacterial victims. The hungry killers were extraordinarily

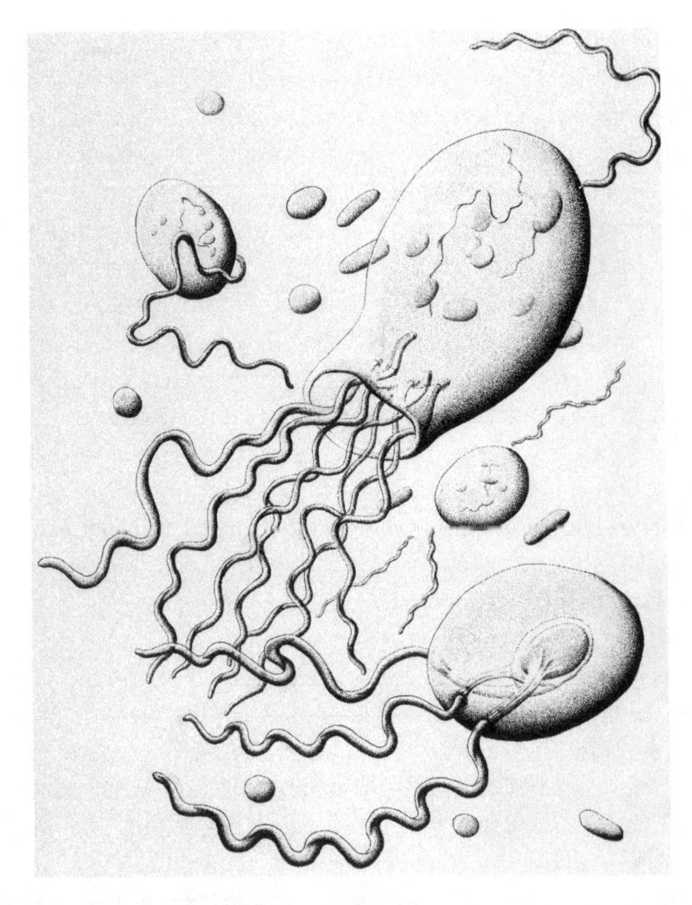

FIGURE 9.1. Spirochetes become undulipodia in the origin of mitosis. Drawing by Christie Lyons.

fast-swimming, skinny bacteria called spirochetes (Figure 9.2). These active bacteria are relatives of the spirochetes of today that are associated with the venereal disease that, in prolonged and serious cases, infects the brain: the treponemes of syphilis. The fatter, slow-moving potential victims, a second kind of bacteria called archaebacteria, were quite different from the spirochetes. By resisting death the archaebacteria incorporated their would-be, fast-moving killers into their bodies. The archaebacteria survived, continuing to be infected by the spirochetes. The odd couple lived together; the archaebacteria were changed, but not killed, by their attackers; the victims did not entirely succumb. (There are precedents for this: plants are green because their intended victims, the chloroplasts that began as oxygen-producing cyanobacteria, resisted death by ingestion.)

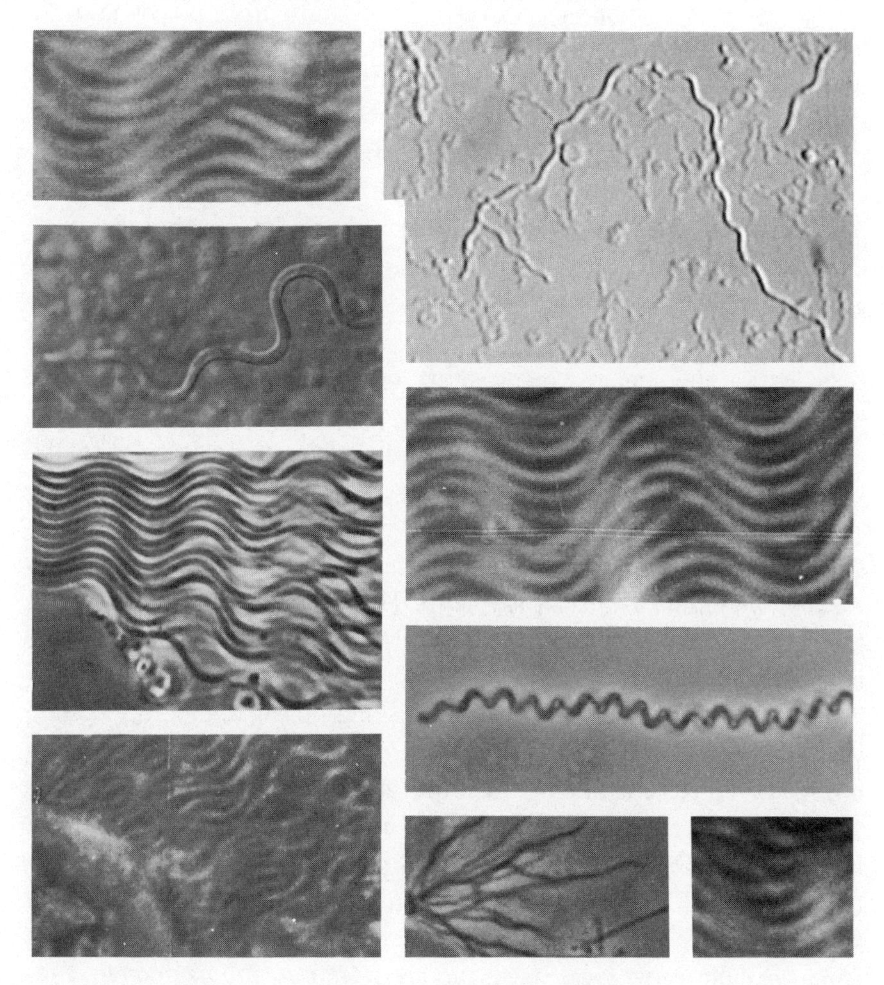

FIGURE 9.2. The antics of spirochetes in nature, photographed live, through a microscope. Whether from the hindguts of termites, the digestive system of clams, the Muddy River at the fens in Boston, Massachusetts, or the salt flats near the delta of the Ebro river in northeast Spain (between Valencia and Barcelona), these microbes carry on their sensuous and social lives.

Our cells, including our nerve cells, may be products of such mergers—the thin, transparent bodies of the spirochete enemies sneakily incorporated inextricably and forever. The wily fast movement, the hunger, the sensory ability of the survivor's enemies, all were put to good use by the evolving partnership. Cultural analogues of such mergers exist: cases in which two very different warring peoples form new identities after the truce; identities, for example, in which unique domesticated plants of one culture become firmly incorporated into that of the second. The presence of Indian corn, tomatoes, and potatoes in Europe is due to the near annihilation of indigenous Native Americans. I see our cell movement, including the movements leading to thought, as the spoils of ancient microbial battles.

My speculations, two thousand million years later, may be the creative outcome of an ancient uneasy peace. If this reckoning is true, then the spirochetal remnants may be struggling to exist in our brains, attempting to swim, grow, feed, connect with their fellows, and reproduce. The interactions between these subvisible actors, now full member-components of our nerve cells, are sensitive to the experience we bring them. Perception, thought, speculation, memory, of course, are all active processes; I speculate that these are the large-scale manifestations of the small-scale community ecology of the former spirochetes and archaebacteria that comprise our brains.

Arcana Naturae Detecta is the name of Anton van Leeuwenhoek's seventeenth-century book revealing the microcosm beneath his single-lens microscope illuminated by a gas lamp. The visible became explicable to him by the machinations of the subvisible. Leeuwenhoek and his followers made clear that "decay," "spoiling," and "rotting food" are all signs of healthy bacterial and fungal growth. In baking, "rising dough" is respiring yeast; in tropical disease, malarial fevers are apicomplexan protists bursting our red blood cells. Fertility is owed in part to semen or "male seed" containing millions of tailed sperm in sugar solution. The disease of Mimi, the heroine of *La Bohème,* is "consumption." From its point of view, "consumption" is the healthy growth of *Mycobacterium* in the warm, moist lungs of the lovely young woman. Speculation, I claim, is the legacy of the itching enmities of unsteady truce. Speculation is the mutual stimulation of the restrained microbial inhabitants that, entirely inside their former archaebacterial enemies, have strongly interacted with them for hundreds of millions of years. Our nerve cells are the outcome of an ancient, nearly immortal marriage of two archenemies who have managed to coexist: the former spirochetes and former archaebacteria that now comprise our brains.

Like animated vermicelli married and in perpetual copulatory stance with their would-be archaebacterial victims, these former free-living bacteria are inextricably united. They probably have been united for more than one thousand million years. The fastidiously described speculation is indistinguishable from the theory. I continually play with an idea: the origin of thought and consciousness is cellular, owing its beginnings to the first courtship between unlikely bacterial bedfellows who became ancestors to our mind-brains.

My goal in the rest of this essay is to explain what I mean and why I make such a bizarre assertion.

What needs to be explained? My basic speculation is that mind-brain processes are nutrition, physiology, sexuality, reproduction, and microbial community ecology of the microbes that comprise us. The microbes are not just metaphors; their remnants inhabit our brain, their needs and habits, histories and health status help determine our behavior. If we feel possessed and of several minds, if we feel overwhelmed by complexity, it is because we are inhabited by and comprised of complexities (Margulis and Sagan 1997).

The detailed consequences of the theory of spirochete origin of microtubules of brain cells do not belong in an essay about speculation for The Reality Club. Indeed, it is unlikely that such a statement would even be considered for publication in the *Proceedings of the National Academy of Sciences*. Rather, I ask only that the unmentionable become discussable over mulled wine and friendship so that the consequences of the hypothesis

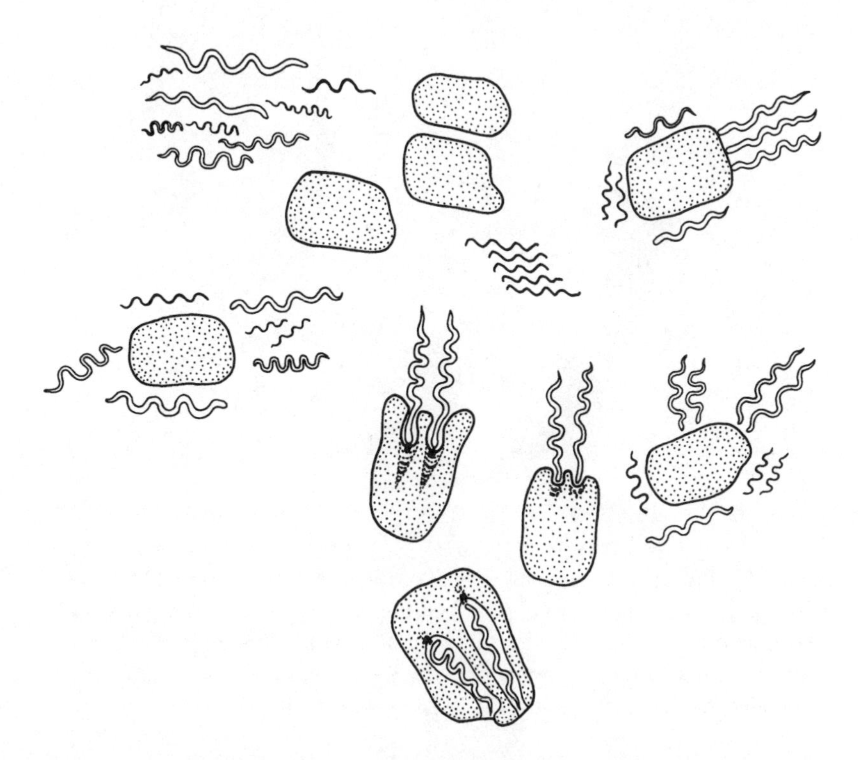

FIGURE 9.3. Spirochetes form many types of relations with other kinds of microbes, shown here as blobs. Drawings by Laszlo Meszoly.

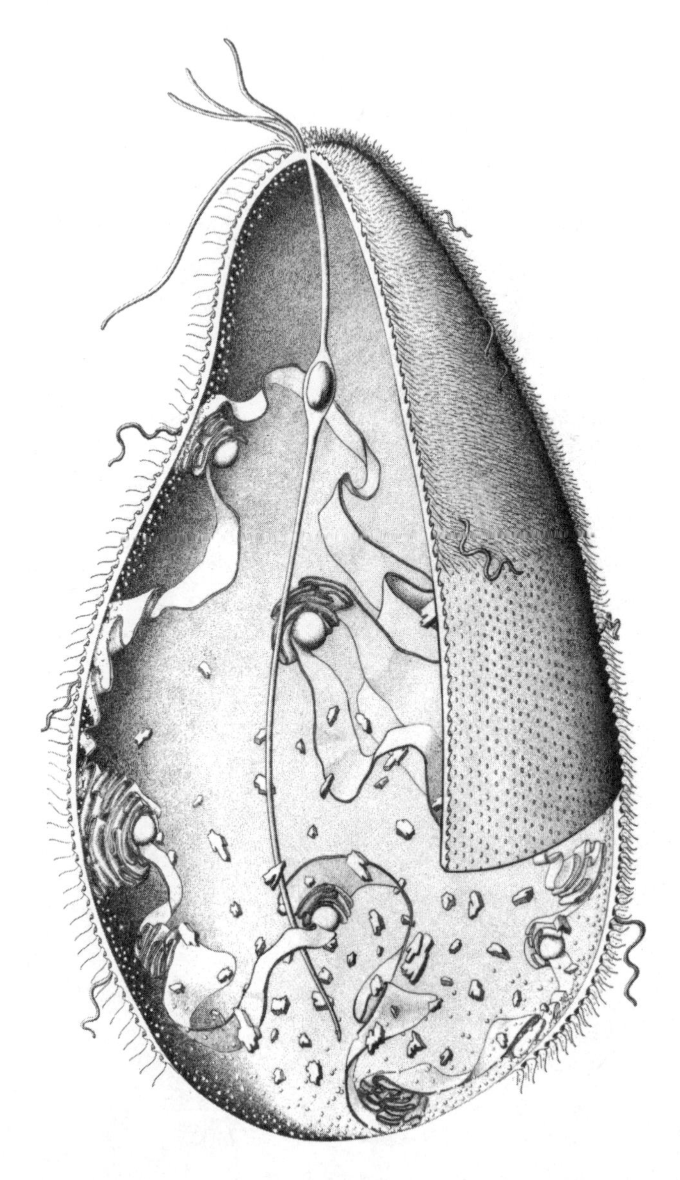

FIGURE 9.4. Two kinds of spirochetes, large unidentified ones and small treponemes (like those associated with syphilis) are permanently attached to *Mixotricha paradoxa,* a "large" wood-eating termite microbe from Australia. Simultaneous movement by hundreds of attached spirochetes make *M. paradoxa* swim forward. Drawing by Christie Lyons, based on detailed micrographs by A.V. Grimstone and the late L.R. Cleveland.

may be speculated upon. Could thought, speculation, and awareness really have evolved from fast-moving bacteria and their interactions, their hungers, their activities, their satiations, their associations with their fellows, both like and unlike, and their waste-removal processes? Is it possible that we are as entirely unaware of the microbial inhabitants that comprise us as a huge ship tossing in the waves is unaware that her responses are determined by the hunger, thirst, and eyesight of the captain at the helm and his communications with the crew? (Figure 9.3, 9.4).

What might be the implications for mind-brains if this bacterial origin of speculation is correct? Let us list a few. They all may be incorrect, but they are all testable within the rigors of the scientific tradition.

1. **Nerve impulses and the firing of nerves.** These become explicable as our motile spirochetes' trying to swim; as Betsey Dyer (biologist at Wheaton College, Norton, Massachusetts) says, captive former spirochetes are spinning their wheels unable to move forward. They have become uncoupled motors going around and around. This quasi-movement is the nerve impulse. It occurs because small, positively charged ions (for example, sodium, potassium, calcium) are accumulated and released across what is now our nerve cell membrane. These ions, their protein and membrane interactions, derive from the membranes of the original spirochetes.

2. **Sweet memories.** Two different kinds of memory systems exist: short term (seconds to minutes) and long term (indefinite). The storage of memories is markedly enhanced by adrenaline and other substances that lead directly to increased availability of sugar to the brain cells (Gold 1987). Sugar, like anything penetrating the blood–brain barrier—that is, entering the brain from the blood—is very carefully monitored and controlled.

Short-term memory arises every time from casual encounters between the sticking-out parts of former spirochetes and their friends. These interactions begin in seconds; it probably takes a few minutes at most while two or more neurons, née spirochetes, interact. The casual encounters occur by small-ion interactions with proteins on the surfaces of what used to be spirochete membranes (now they are our nerve-cell membranes). In brief, short-term memories derive from the physiology of spirochetal remnants in the brain. We know that the pictorial short-term memory, for the recognition of fractal designs, for example, "is coded by temporary activation of an ensemble of neurons in the region of the association cortex that processes visual information" (Miyashita and Chang 1988). Presumably the short-term memory is stored when visual information is processed and not in special compartments for short-term memory. The "temporary activation," if I am correct, will be directly homologous to spirochete behavioral interaction—not analogous to it or to computer software manipulation.

Long-term memory is stable; it depends on new protein synthesis. Long-term memory works because it stores the short term. What were repeated casual encounters between former spirochetal remnants become stabilized attachment sites. "Synapse," if I am correct, is the neurophysiologist's term for the well-developed spirochetal remnant attachment site. In brief, long-term memories derive from the growth of spirochetal remnants, including their attachment sites, in the brain.

Sugar enhances memory processes because it feeds preferentially the spirochetal remnants so that they can interact healthily and form new attachment sites. Sugar has been the food of spirochetes since they squiggled in the mud.

As Edelman (1985) has pointed out, no two monkeys, no two identical twins, are identical at the level of fine structure of their neuronal connections. "There must be a generator of diversity during the development of neural circuits, capable of constructing definite patterns of groups but also generating great individual variation. Variation must occur at the level of cell-to-cell recognition by a molecular process. Second, there must be evidence from group selection and competition in brain maps and reentrant circuits. This must occur not in the circuitry but in the efficacy of preformed connections or synapses" (Edelman 1985). I believe Edelman is discovering the actively growing latter-day populations of microbes that comprise every brain. Edelman's "populations" are nerve cells and their connections. I interpret Edelman's populations literally as remnants of ancestral microbial masses. The spirochetal remnants, either poised or ready to grow, attach and interact depending on how they are treated during a human's crucial stages (fetal development, infancy, and early childhood). Neural Darwinism, differential growth by selection of spirochete associations, determines the way in which the brain develops.

Mental health is, in part, how we feed the normal spirochetal remnants that make up our brain. Learning becomes a function of the number and quality of new connections—attachment sites—that these wily apo-beings forge. The spirochetal remnants grow faster, dissolving temporary points of contact while consolidating firm connections that are our nerve cell endings during our infancy and childhood. More potential changes occur early—in infancy and adolescence—relative to those of adulthood. The growth patterns of nerve cells née spirochetes are sensitive to the food, such as essential fatty acids, that the rest of our body provides for them; experience is always active, always participatory, and, if registered in long-term memory, unforgotten. Our memories are their physical networks. Our crises and climaxes are their "blooms," their population explosions. Senility is spirochetal-remnant atrophy. It is no coincidence that salt ions and psychoactive drugs, including anesthetics, have strong effects on spirochetal movement of the free-living mud-bound cousin spirochetes.

Clearly these enormous contemplative issues cannot be solved here alone by me. All I ask is that we compare consciousness with spirochete microbial ecology. We may be vessels, large ships, unwitting sanctuaries to the thriving communities comprising us. When they are starved, cramped, or stimulated we have inchoate feelings. Perhaps we should get to know ourselves better. We might then recognize our speculations as the dance networks of ancient, restless, tiny beings that connect our parts.

PART III

GAIA

10

The Atmosphere as Circulatory System of the Biosphere—The Gaia Hypothesis

Lynn Margulis
and James E. Lovelock

We would like to discuss the Earth's atmosphere from a new point of view—that it is an integral, regulated, and necessary part of the biosphere. In 1664 Sachs von Lewenheimb, a champion of William Harvey, used the analogy shown in Figure 10.1 to illustrate the concept of the circulation of blood. Apparently the idea that water lost to the heavens is eventually returned to Earth was so acceptable in von Lewenheimb's time that Harvey's theory was strengthened by the analogy (Pagel 1951).*

*Pagel quotes Harvey himself as saying: "I began to think whether there might not be a motion as it were in a circle. Now this I afterwards found to be true; . . . which motion we may be allowed to call circular, in the same way as Aristotle says that the air and the rain emulate the circular motion of the superior bodies; for the moist earth, warmed by the sun evaporates; the vapours drawn upwards are condensed, and descending in the form of rain moisten the earth again; and by this arrangement are generations of living things produced . . . And so in all likelihood, does it come to pass in the body, through the motion of the blood; the various parts are nourished, cherished, quickened by the warmer more perfect vaporous spiritous, and, as I may say, alimentive blood; which, on the contrary, in contact with these parts becomes cooled, coagulated, and, so to speak, effete; whence it returns to its sovereign, the heart, as if to its source, or to the inmost home of the body, there to recover its state of excellence of perfection."

FIGURE 10.1. Frontispiece* to Sachs von Lewenheimb, 1664, Oceanus Macro-Microcosmicus. This illustration stresses the analogies between the circulation of the blood and the circulation of water. According to W. Pagel (1951),

> "The subtitle of the dissertation (which addresses itself to the famous anatomist Thomas Bartholinus) explains that it deals with the analogies between the circular motion of the water from and back to the sea, on the one hand, and that of the blood from and back to the heart, on the other. This motion is "circular," not because it describes the geometrical figure of a circle, but because it reverts to its point of departure. The earth resembles the human body in that, like the latter, it is pervaded by canals and harbours an internal fire. The sea lets water rise by evaporation and return in the form of rain whereby the rivers and subterranean waters are nourished and these finally return the same water to the sea. The latter thereby acts not unlike the heart from which the blood goes out to the organs, starting on its way attenuated by the influx of heat and 'perfected' in the 'workshops' of the organs; finally, after its absorption and assimilation by the organs, its residue is drawn back into the heart in order to be attenuated again—just as the waters are diluted by joining the sea."

*From the original treatise in the Wellcome Library, courtesy of the trustees, with permission.

Three hundred and ten or so years later, with the circulation of blood a universally accepted fact, we find it expedient to revive von Lewenheimb's analogy—this time to illustrate our concept of the atmosphere as circulatory system of the biosphere. This new way of viewing the Earth's atmosphere has been called the Gaia hypothesis (Lovelock 1972). The term "Gaia" is from the Greek for "Mother Earth," and it implies that certain aspects of the Earth's atmosphere—temperature, composition, oxidation reduction state, and acidity—form a homeostatic system, and that these properties are themselves products of evolution (Lovelock and Margulis 1974a,b).

From recent articles and books (Rasool 1974; Kellogg and Schneider 1974) one gets the impression that fluid dynamics, radiation chemistry, and industrial pollution are the major factors determining the properties of the atmosphere. The Gaia hypothesis contends that biological gas exchange processes are also major factors, especially processes involving microorganisms. The human impact on the atmosphere may have been overestimated. Humans are only one of some three million species on Earth, all of which exchange gas and most of which exchange gas with the atmosphere. Humans have been around for only a few million years, while microorganisms have existed for thousands of millions of years. The atmosphere is probably not so much the product of humans as of the several billion smaller organisms living in every pail of rich soil or water.

It seems to us that early twentieth-century nonmicrobiological analysis of the Earth's lower atmosphere will one day be considered as ignorant as early nineteenth century nonmicrobiological analysis of fermentation or disease is today. In an excellent introduction to atmospheric science, Goody and Walker (1972) said, "There is a great difference between research in the laboratory and studies of the Earth and planets. In the laboratory the scientist can perform controlled experiments, each carefully designed to answer questions of his own choosing. Except in minor respects, however, the Earth and planets are too large for controlled experimentation. All we can do is observe what happens naturally in terms of the laws of physics and chemistry."

We agree that the laws of physics and chemistry are basic to the understanding of atmospheric phenomena but insist that the laws of biology must be considered as well. It is our contention that the paucity of overall understanding of certain aspects of the atmosphere, especially composition and temperature, is due to too narrow a paradigm: the idea that the

Table 10.1. Reactive gases in the atmosphere (billions of tons/year)

Gas	Concentration in Parts per Million	How much of the Gas Comes from: Inorganic Sources Volcanic, etc.?	Biological Sources		Residence Time	Where Does the Gas Come From Principally?
			Gaian?	Human?		
Nitrogen (N_2)	790,000	0.001	1	0	1–10 million years	Bacteria from dissolved nitrate in soil
Oxygen (O_2)	210,000	0.00016	110	0	1000 years	Algae and green plants, given off in photosynthesis
Carbon Dioxide (CO_2)	320	0.01	140	16	2–5 years	Respiration, combustion
Methane (CH_4)	1.5	0	2	0	7 years	Fermenting bacteria
Nitrous oxide (N_2O)	0.3	less than 0.01	0.6	0	10 years	Bacteria and fungi
Carbon monoxide (CO)	0.08	less than 0.001	1.5	0.15	Few months	Methane oxidation (methane from bacteria)
Ammonia (NH_3)	0.006	0	1.5	0	Week	Bacteria and fungi
Hydrocarbons $(CH_2)_n$	0.001	0	0.2	0.2	Hours	Green plants, industry
Methyl iodide (CH_3I)	0.000001	0	0.03	0	Hours	Marine algae
Hydrogen (H_2)	0.0000005	0	?	?	2 years	Bacteria, methane oxidation?
Methyl chloride (CH_3Cl)	0.00000000114	0	?	?	?	Algae?

Gaian = nonhuman biological sources.

atmosphere is an inert part of the inorganic environment and therefore amenable to methods of study that involve only physics and chemistry.

In this chapter we explore what is perhaps a more realistic view—that the atmosphere is a nonliving, actively regulated part of the biosphere. In our model atmospheric temperature and composition are regulated with respect to certain biologically critical substances: hydrogen ions, molecular oxygen, nitrogen and its compounds, sulfur and its compounds, and some others, whose abundance and distribution in the atmosphere are presumed to be under biological control. Biological gas exchange processes, thought to be involved in possible control mechanisms, are discussed elsewhere (Margulis and Lovelock 1974). The purpose of this chapter is simply to present our reasons for believing the atmosphere is actively controlled.

Traditional atmospheric studies have left us with some strange anomalies. The atmosphere is an extremely complex blanket of gas in contact with the oceans, lakes, rivers (the hydrosphere), and the rocky lithosphere. It has a mass of about 5.3×10^{21} grams. (The mass of the oceans—the other major fluid on the surface of the Earth—is almost a thousand times heavier, being about 1.4×10^{24} grams.) Because the atmospheric mass corresponds to less than a millionth of the mass of the Earth as a whole, one would expect small changes in the composition of the solid earth to cause large changes in the composition of the atmosphere. Yet even in the face of a large number of potential perturbations, the atmosphere seems to have remained dynamically constant over long periods of time.

Many facts about the atmosphere are known—its composition, its temperature and pressure profiles, certain interactions with incoming solar radiation, and the like (Goody and Walker 1972). Some of these are shown in Tables 10.1 and 10.2. However, as the efficacy of long-range weather forecasting attests, there is no consistent model of the atmosphere that can be used for the purpose of prediction (Kellogg and Schneider 1974). The Earth's atmosphere defies simple description. From the point of view of chemistry, it sustains such remarkable disequilibrium that Sagan (1970) was prompted to remark that given the temperature, pressure, and amount of oxygen in the atmosphere, "one can calculate what the thermodynamic equilibrium abundance of methane ought to be . . . the answer turns out to be less than 1 part in 10^{36}. This then is a discrepancy of at least 30 orders of magnitude and cannot be dismissed lightly."

Table 10.2 shows that given the quantity of oxygen in the atmosphere, not only the major gases such as nitrogen and methane but also the minor

Table 10.2. Composition of the Atmosphere: Gases in Disequilibrium

Gas	Abundance	Flux (moles/yr $\times 10^{13}$)	Disequilibrium Factor	Oxygen Used up in the Oxidation of these Gases (moles/yr $\times 10^{13}$)	Abiological Processes	Source of Gas % Contribution by Biological Process	
						Human	Gaian*
Nitrogen	78%	3.6	10^{10}	11	0.001	0	>99
Methane	1.5 ppm	6.0	10^{30}	12	0	0	100
Hydrogen	0.5 ppm	4.4	10^{30}	2.2	?	0	?
Nitrous oxide	0.3 ppm	1.4	10^{13}	3.5	0.02	0	>99
Carbon monoxide	0.08 ppm	2.7	10^{30}	1.4	0.001	10	90
Ammonia	0.01 ppm	8.8	10^{30}	3.8	0	0	100

*Gaian = nonanthropogenic biological sources; for details see Table 1; ? = some quantities not known; ppm = parts per million.

atmospheric components are far more abundant than they ought to be according to equilibrium chemistry. Even though the minor constituents differ greatly in relative abundance, they sustain very large fluxes—comparable with those of the major constituents. The Earth's atmosphere is certainly not at all what one would expect from a planet interpolated between Mars and Venus. It has too little CO_2, too much oxygen, and is too warm. We believe the Gaia hypothesis provides the new approach that is needed to account for these deviations.

A new framework for scientific thought is justified if it guarantees new observations and experiments. The recognition that blood in mammals circulates in a closed, regulated system gave rise to meaningful scientific questions such as: How is blood pH kept constant? By what mechanism is the temperature of mammalian blood regulated around its set point? What is the purpose of bicarbonate ion in the blood? What is the role of fibrinogen? If the blood were simply an inert environment (as the atmosphere is presently viewed), such questions would seem irrelevant and never be asked at all.

Let us consider another analogy. Bees have been known to regulate hive temperatures during midwinter at about 31°C, approximately 59°C above ambient (Wilson 1970). Under threat of desiccation they also maintain high humidities. While the air in the hive is not alive, it maintains an enormous disequilibrium due to the expenditure of energy by the living insects—ultimately, of course, solar energy. How is the hive temperature maintained? How does the architecture of the hive aid to reduce desiccation? How does the behavior of the worker bees alter temperature? These are all legitimate scientific questions, generated by the circulatory system concept.

The Gaia hypothesis of the atmosphere as a circulatory system raises comparable and useful scientific questions and suggests experiments that based on the old paradigm would never be asked, for example: How is the pH of the atmosphere kept neutral or slightly alkaline? By what mechanism(s) has the mean midlatitude temperature remained constant (not deviated more than 15°C) for the last 1000 million years? Why are 0.5×10^9 tons nitrous oxide (N_2O) released into the atmosphere by organisms? Why is about 2×10^9 tons of biogenic methane pumped into the atmosphere each year (representing nearly 10% of the total terrestrial photosynthate)? What are the absolute limits on the control mechanisms, that is, how much perturbation (emanations of sulfur oxides, chlorinated compounds, and/or carbon monoxide; alterations in solar luminosity; and so forth) can the atmosphere regulatory system tolerate before all its feedback mechanisms fail?

The Gaia approach to atmospheric homeostasis has also led to a number of observations that otherwise would not have been made, for example, an oceanic search was undertaken for volatile compounds containing

Table 10.3. Critical Biological Elements that May Be Naturally Limiting

Major elements	Use in Biological Systems	Possible Form of Fluid Transport
C (carbon)	All organic compounds	CO_2; food; organic compounds in solution; biological volatiles; carbonate, bicarbonate, etc.; usually not limiting
N (nitrogen)	All proteins and nucleic acids	N_2, N_2O, O_3^-, NO_2^- (often limiting)
O, H (oxygen, hydrogen)	H_2O in high concentration for all organisms	Rivers, oceans, lakes
S (sulfur)	Nearly all proteins (cysteine, methionine, etc.); key coenzymes	Dimethyl sulfide; dimethyl sulfoxide, carbonyl sulfide
P (phosphorus)	All nucleic acids; adenosine triphosphate	Unknown (biological volatiles? spores? birds? migrating salmon?)
Na, Ca, Mg, K (sodium, calcium, magnesium potassium)	Membrane and macromolecular function	Usually not limiting, except in certain terrestrial habitats (Botkin et al. 1973)
Trace Elements		
I (iodine)	Limited to certain animals (e.g., thyroxine)	Methyl iodide
Se (selenium)	Enzymes of fermenting bacteria (production of ammonia, hydrogen; animals (Stadtman 1974)	Unknown (dimethyl selenide?)
Mo (molybdenum)	Nitrogen fixation enzymes of bacteria, including cyanobacteria; carbon dioxide reductase (*Clostridium*)	Unknown

elements that are limiting to life on the land, and large quantities of methyl iodide and dimethyl sulfide were in fact observed (Lovelock, Maggs, and Rasmussen 1972).

Given the Gaia hypothesis, one deduces that all the major biological elements (Table 10.3) must either be not limiting to organisms (in the sense that they are always readily available in some useful chemical form) or must be cycled through the fluids on the surface of the Earth in time periods that are short relative to geological processes. (Attempts to identify volatile forms of these elements are in progress.) The cycling times must be short because biological growth is based on continual cell division, which requires the doubling of cell masses in periods of time that are generally less than months, and typically days or hours. On lifeless planets there is no particular reason to expect this phenomenon of atmospheric cycling, nor on the Earth is it expected that gases of elements that do not enter metabolism as either metabolites or poisons will cycle rapidly; for example, based on the Gaia hypothesis, nickel, chromium, strontium, rubidium, lithium, barium, and titanium will not cycle, but cobalt, vanadium, selenium, molybdenum, iodine and magnesium might (Egami 1974). Because biological solutions to problems tend to be varied, redundant, and complex, it is likely that all of the mechanisms of atmospheric homeostasis will involve complex feedback loops [see Margulis and Lovelock (1974) for discussion.] Because, for example, no volatile form of phosphorus has ever been found in the atmosphere, and because this element is present in the nucleic acids of all organisms, we are considering the possibility that the volatile form of phosphorus at present is totally "biological particulate." Figures 10.2 and 10.3 rather fancifully compare the Earth's atmosphere at present with what it might be if life were suddenly wiped out.

Ironically, it is the past history of the Earth, with its extensive sedimentary record (fraught, as it is, with uncertainties in interpretation), that might provide the most convincing proof for the existence of continued biological modulation. If one accepts the current theories of stellar evolution, the Sun, being a typical star of the main sequence, has substantially increased its output of energy since the Earth was formed some 4500 million years ago. Some estimates for the increase in solar luminosity over the past history of the Earth are as much as 100%; most astronomers apparently accept an increase of at least 25% over 4.5 billion years (Oster 1973). Extrapolating from the current atmosphere, given solar radiation output and radiative surface properties of the planet, it can be concluded that until about 2000 million years ago either the atmosphere was different (for example, contained more ammonia) or the Earth was frozen. The

FIGURE 10.2. Earth's atmosphere at present: examples of major volatiles. (Key: the following compounds and spores are depicted. It is left to the reader to identify them. See Gregory, 1973 for many details.) Spores of: ferns, club mosses, zygomycetes, ascomycetes, basidiomycetes, slime molds, bacteria. All contain nucleic acids and other organic phosphates, amino acids and so forth. Animal products: butyl mercaptan, plant products: myoporum, catnip (nepetalactone), eugenol, geraniol, pinene, isothiocyanate (mustard); disparlure; PAN (paroxacetyl nitrate), dimethyl sulfide, dimethyl sulfoxide; gases: nitrogen, oxygen, methane, carbon monoxide, carbon dioxide, ammonia. Painting by Laszlo Meszoly.

most likely hypothesis is that the Earth's atmosphere contained up to about one part in 10^5 ammonia, a good infrared absorber (Sagan and Mullen 1972). Other potential greenhouse gases apparently will not compensate for the expected lowered temperature because they do not have the appropriate absorption spectra or are required in far too large quantity to be considered reasonable (Sagan and Mullen 1972). [There are good arguments for the rapid photodestruction of any atmospheric ammonia (Ferris and Nicodem 1974).] However, it has been argued that ammonia is required for the origin of life (Bada and Miller 1968), and there is good evidence for the presence of fossil microbial life in the earliest sedimentary rocks [3400 million years ago (Barghoorn 1971).] There is no geological evidence that since the beginning of the Earth's stable crust the entire

FIGURE 10.3. The present atmosphere were life deleted. Painting by Laszlo Meszoly.

Earth has ever frozen solid or that the oceans were volatilized, suggesting that the temperature at the surface has always been maintained between the freezing and the boiling points of water. The fossil record suggests that, from an astronomical point of view, conditions have been moderate enough for organisms to tolerate, and the biosphere has been in continuous existence for over 3000 million years (Barghoorn 1971; Cloud 1968). At least during the familiar Phanerozoic (the last 600 million years of Earth history for which an extensive fossil record is available), one can argue on paleontological grounds alone that through every era the Earth has maintained tropical temperatures at some place on the surface and that the composition of the atmosphere, at least with respect to molecular oxygen, could not have deviated markedly. That is, there are no documented cases of any metazoa (animals, out of about 3 million species) that can complete their life cycles in the total absence of O_2 (Augenfeld, 1974, personal communication). All animals are composed of cells that divide by mitosis. The mitotic cell division itself requires O_2 (Amoore 1961). Thus it is highly unlikely that current concentrations of oxygen have fallen much below their present values in some hundreds of millions of years. By implication, oxygen and the gases listed in Table 10.2 have been maintained at stable atmospheric concentrations for time periods that are very long relative to their residence times. (Residence time is the time it takes for the concentration of gas to fall to 1/e or 37% its value; it may be thought of as turnover time). Furthermore, because concentrations of atmospheric oxygen only a few percent higher than ambient lead to spontaneous combustion of organic matter, including grasslands and forests, the most reasonable assumption is that the oxygen value of the atmosphere has remained relatively constant for quite long time periods (Lovelock and Lodge 1972).

How can these observations be consistently reconciled? How can we explain the simultaneous presence of gases that are extremely reactive with each other and unstable with respect to minerals in the crust and at the same time note that their residence times in the atmosphere are very short with respect to sediment forming and mountain building geological processes? In this respect Table 10.3 can be instructive. One can see that even though absolute amounts of the gases vary over about three orders of magnitude, the fluxes are remarkably similar. These gases are produced and removed primarily by nonhuman biological processes (see Table 10.1); (Margulis and Lovelock 1974). While the processes involved in atmospheric production and removal of reactive gases are not primarily dependent on human activity, for the most part they are not based on animal or plant processes either. (See Margulis and Lovelock 1974 for a version of the table that lists these.) It is mainly the prokaryote microorganisms that are involved in gas exchange—the rapidly growing and dividing masters of the microbiological world that make up in chemical complexity and metabolic virtuosity what they lack in advanced morphology. These organisms presumably played a similar role in biogeochemical processes in the past as they do today. There is direct fossil evidence for the continued existence of Precambrian microorganisms (Barghoorn 1971). That they have an ancient history can also be deduced from current studies of their physiology. Among hundreds of species of these prokaryotic microorganisms are many obligate anaerobes, that is, organisms poisoned by oxygen. (All organisms are poisoned by oxygen at concentrations above those to which they are adapted.) Hundreds of others are known that are either microaerophils (adapted to concentrations of oxygen less than ambient) or facultative aerobes (can switch their metabolism from oxygen requiring to oxygen nonrequiring).

As a group, the prokaryotic microbes show evidence that the production and release of molecular oxygen into the atmosphere was an extremely important environmental determinant in the evolution of many genera. Prokaryotic microbes (formerly known as the blue-green algae, cyanobacteria) were almost certainly responsible for the original transition to the oxygen-containing atmosphere about 2000 million years ago (Barghoorn 1971; Cloud 1968).

Figure 10.4 and 10.5 present scenes before and after the transition to oxidizing atmosphere, respectively. Figures 10.6 and 10.7 are reconstructions of anaerobic cycles corresponding to Figures 10.4 and 10.5, respectively. Figure 4 attempts to reconstruct the scene as it might have looked 3400 million years ago, admittedly in a rather geothermal area. Although no free oxygen (above that produced by photochemical processes and hydrogen loss) is available in the atmosphere, the scene is teeming with life—microbial life. For example, entire metabolic processes, as shown in Figure 10.6, are available within the group of anaerobic prokaryotic microbes today. Because at the higher taxonomic levels (kingdoms and phyla) once successful patterns evolve they tend not to become extinct (Simpson 1960), it is likely that ancestors of present-day microbes were available to interact with atmospheric gases very early on the primitive Earth. Certainly life was very advanced metabolically by the time the first

FIGURE 10.4. Scene from a geothermal area in Fig Tree times (about 3400 million years ago). Drawing by Laszlo Meszoly.

FIGURE 10.5. Scene from a geothermal area in Gunflint times (about 2000 million years ago). Drawing by Laszlo Meszoly.

stromatolitic rocks were deposited. With the evolution of oxygen-releasing metabolism by cyanobacteria came the stromatolites. These layered sediments are extremely common, especially in the late Precambrian (Awramik 1973). With the stromatolites come other Precambrian evidence for the transition to the oxidizing atmosphere. By the middle Precambrian, about 2000 million years ago—the time at which the stromatolites and microfossils become increasingly abundant (Barghoorn and Tyler 1965; Schopf 1970)—the scene might have looked like that in Figure 10.5. The metabolic processes accompanying that scene are shown in Figure 10.7. It is obvious that from among metabolic processes in prokaryotic microbes alone there are many that involve the exchange of atmospheric gases. This figure shows how oxygen-handling metabolism was essentially superimposed on an anaerobic world, a concept that is consistent with the observation that reaction with molecular oxygen tends to be the final step in aerobic respiratory processes. All of the processes shown in Figures 10.6 and 10.7 are known from current microorganisms (and, by definition, those that haven't become extinct are evolutionarily successful).

The fossil evidence, taken together, suggests that the Earth's troposphere has maintained remarkable constancy in the face of several enor-

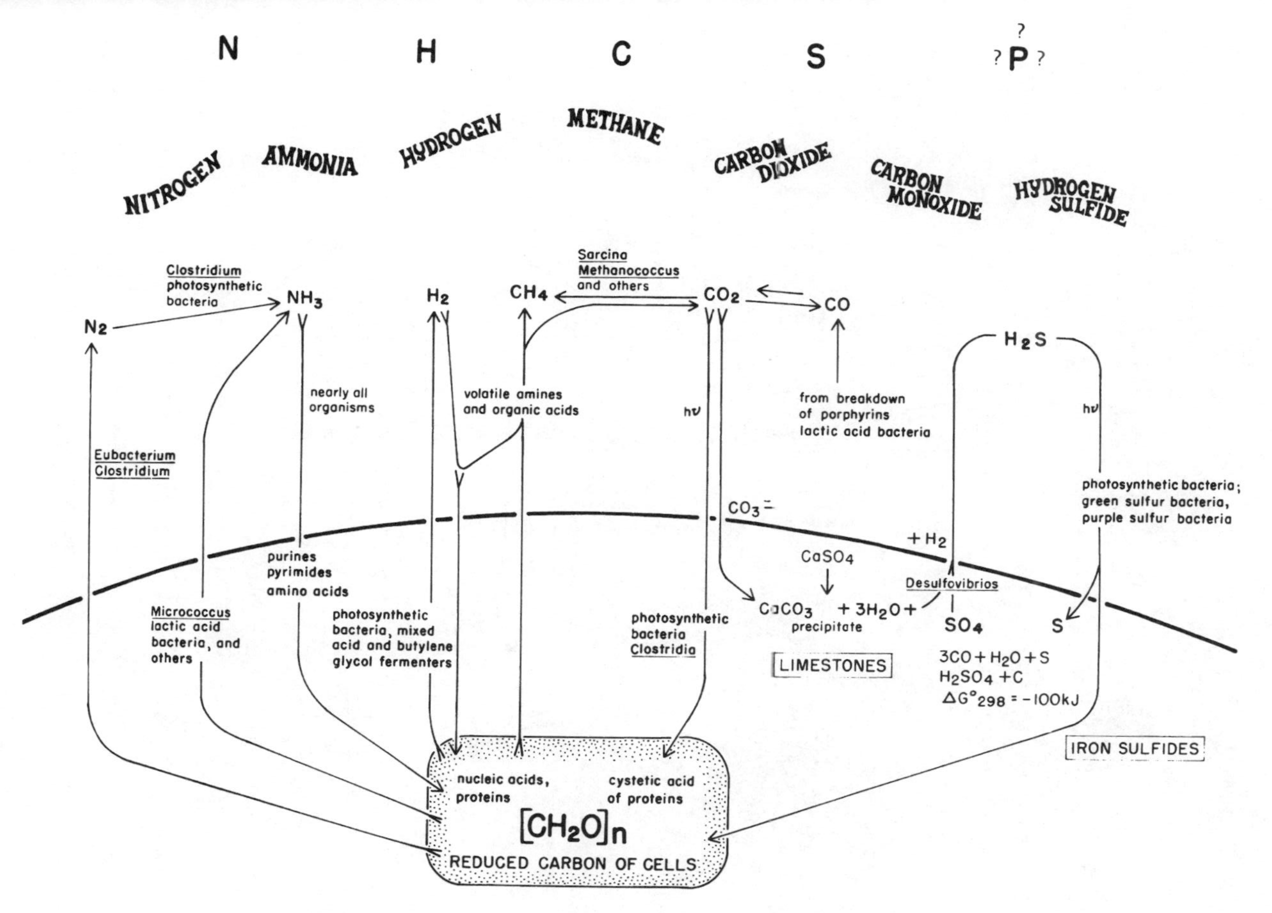

FIGURE 10.6. A reconstruction of possible anaerobic cycles: 3400 million years ago. (Genera of microorganisms catalyzing the reactions are underlined; drawing by Laszlo Meszoly.)

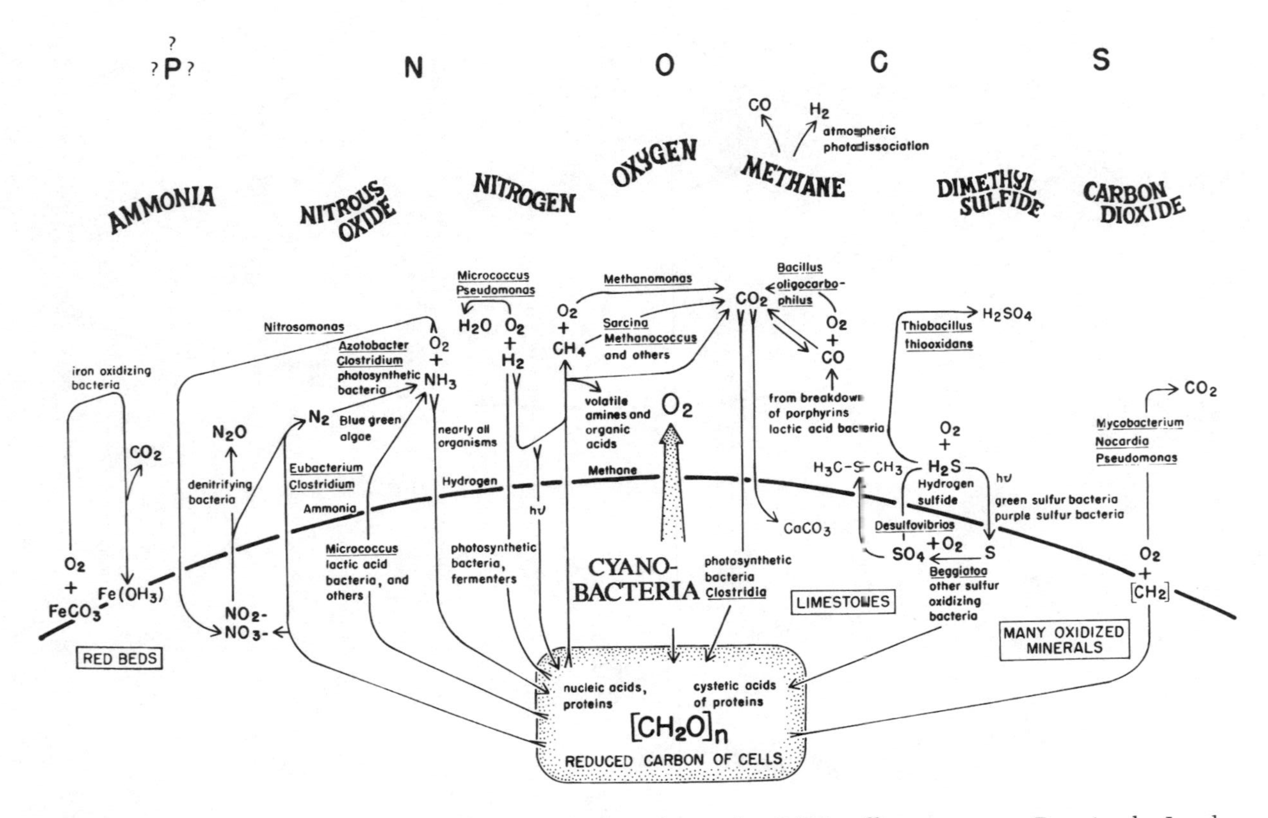

FIGURE 10.7. A reconstruction of possible microbial aerobic cycles: 2000 million years ago. Drawing by Laszlo Meszoly.

mous potential perturbations—at least the increase in solar luminosity and the transition to the oxidizing atmosphere. The Earth atmosphere maintains chemical disequilibria of many orders of magnitude containing rapidly turning over gases produced in prodigious quantities. The temperature and composition seem to be set at values that are optimal for most of the biosphere. Furthermore, the biosphere has many potential methods for altering the temperature and composition of the atmosphere (Margulis and Lovelock 1974). The biosphere has probably had these methods available almost since its inception more than 3000 million years ago. Is it not reasonable to assume that the lower atmosphere is maintained at an optimum by homeostasis and that this maintenance (at the ultimate expense of solar energy, of course) is performed by the party with the vested interest: the biosphere itself?

11

Gaia and Philosophy

Dorion Sagan
and Lynn Margulis

The Gaia hypothesis is a scientific view of life on Earth that represents one aspect of a new biological world view. In philosophical terms this new world view is more Aristotelian than Platonic. It is predicated on the earthly factual, not the ideal abstract, but there are some metaphysical connotations. The new biological world view, and Gaia as a major part of it, embraces the circular logic of life and engineering systems, shunning the Greek-western heritage of final syllogisms.

Gaia is a theory of the atmosphere and surface sediments of the planet Earth taken as a whole. The Gaia hypothesis in its most general form states that the temperature and composition of the Earth's atmosphere are actively regulated by the sum of life on the planet—the biota. This regulation of the Earth's surface by the biota and for the biota has been in continuous existence since the earliest appearance of widespread life. The assurance of continued global habitability according to the Gaian hypothesis is not a matter merely of chance. The Gaian view of the atmosphere is a radical departure from the former scientific concept that life on Earth is surrounded by and adapts to an essentially static environment. That life interacts with and eventually becomes its own environment; that the atmosphere is an extension of the biosphere in nearly the same sense that the human mind is an extension of DNA; that life interacts with and controls physical attributes of the Earth on a global scale—all these things resonate strongly with the ancient magico-religious sentiment that all is one. On a more practical plane, Gaia holds important implications not only for understanding life's past but for engineering its future.

The Gaia hypothesis, presently a concern only for certain interdisciplinarians, may someday provide a basis for a new ecology—and even become a household word. Already it is becoming the basis for a rich new world view. Let us first examine the scientific basis for the hypothesis and then explore some of the metaphysical implications. Innovated by the atmospheric chemist James Lovelock, supported by microbiologist Lynn Margulis, and named by novelist William Golding, the Gaia hypothesis states that the composition of all the reactive gases as well as the temperature of the lower atmosphere have remained relatively constant over eons. (An eon is approximately a billion years.) In spite of many external perturbations from the solar system in the last several eons, the surface of the Earth has remained habitable by many kinds of life. The Gaian idea is that life makes and remakes its own environment to a great extent. Life reacts to global and cosmic crises, such as increasing radiation from the sun or the appearance for the first time of oxygen in the atmosphere, and dynamically responds to ensure its own preservation such that the crises are endured or negated. Both scientifically and philosophically, the Gaia hypothesis provides a clear and important theoretical window for what Lovelock (1979) calls "a new look at life on earth."

Astronomers generally agree that the Sun's total luminosity (output of energy as light) has increased during the past four billion years. They infer from this that the mean temperature of the surface of the Earth ought to have risen correspondingly. But there is evidence from the fossil record of life that the Earth's temperature has remained relatively stable (Margulis and Lovelock 1974). The Gaia hypothesis recognizes this stability as a property of life on the Earth's surface. We shall see how the hypothesis explains the regulation of temperature as one of many factors whose modulation may be attributed to Gaia. The temperature of the lower atmosphere is steered by life within bounds set by physical factors. With a simple model that applies cybernetic concepts to the growth, behavior, and diversity of populations of living organisms, Lovelock has most recently shown how, in principle, the intrinsic properties of life lead to active regulation of Earth's surface temperature. There is nothing mystical in the process at all. By examining in some detail the life of a mythical world containing only daisies (about which, more later), even skeptical readers can be convinced that it is theoretically possible for living, growing, responding communities of organisms to exert control over factors concerning their own survival. No unknown conscious forces need be invoked; temperature regulation becomes a consequence of the well-known properties of life's responsiveness and growth. In fact, perhaps the most striking philosophical conclusion is that the cybernetic control of the Earth's surface by unintelligent organisms calls into question the alleged uniqueness of human intelligent consciousness.

In exploring the regulatory properties of living beings, it seems most likely that atmospheric regulation can be attributed to the combined metabolic and growth activities of organisms, especially of microbes. Microbes (or microorganisms) are those living beings seen only with a mi croscope. They display impressive capabilities for transforming the nitrogen-, sulfur-, and carbon-containing gases of the atmosphere (Margulis and Lovelock 1974). Animals and plants, on the other hand, show few such abilities. All or nearly all chemical transformations present in animals and plants were already widespread in microbes before animals and plants evolved. Until the development of Lovelock's Daisy World, the discussion of control of atmospheric methane (a gas that indirectly affects temperature and is produced only by certain microbes, known as methanogenic bacteria) has provided the most detailed exposition of the maintenance of atmospheric temperature stability (Watson, Lovelock, and Margulis 1978). The concentration of water vapor in the air correlates with certain climatic features, including the temperature at the Earth's surface. The details of the relationship between temperature and forest trees, determining the production and transport of huge quantities of water in a process called evapotranspiration, was recently presented by meteorologists in a quantitative model (Shukla and Mintz 1982). Although these scientists did not discuss their work in a Gaian context, they have inadvertently provided a further Gaian example. Indeed, as Hutchinson originally recognized when he described the geological consequences of feces and, as the new ecology book by Botkin and Keller (1982) shows, many observations concerning the effects of the biota in maintaining the environment can be reinterpreted in a Gaian context (Hutchinson 1954).

How can the gas composition and temperature of the atmosphere be actively regulated by organisms? Although willing to believe that atmospheric methane is of biological origin and that the process of evapotranspiration moves enormous quantities of water from the soil through trees into the atmosphere, several critics have rejected the Gaia hypothesis as such because they fail to see how the temperature and gas composition of an entire planetary surface could be regulated for several billion years by an evolving biota that lacks foresight or planning of any kind (Doolittle 1981; Garrels, Lerman, and MacKenzie 1981).

Primarily in response to these critics, Dr. Lovelock and his former graduate student Dr. Andrew Watson formulated a general model of temperature modulation by the biota, to which they pleasantly refer as "Daisy World." Daisy World uses surface temperature rather than gas composition to demonstrate the possible kinds of regulating mechanisms that are consistent with how populations of organisms behave. Daisy World exemplifies the kind of Gaian mechanisms we would expect to find, based as it is on an analogy between cybernetic systems and the growth properties of organisms. In an admittedly simplified fashion, it shows that temperature regulation can emerge as a logical consequence of life's well-known properties. These include potential for exponential growth, and growth rates varying with temperature such that the highest rate occurs at the optimal temperature for each population, decreasing around the optimum until growth is limited by extreme upper and lower temperatures. We will describe the Daisy World in detail shortly.

Some such model, explaining the regulation of surface temperature, is required to explain several observations. For example, the oldest rocks not metamorphosed by high temperatures and pressures, both from the Swaziland System of southern Africa (Margulis 1982; Schopf 1983; Walter 1976) and from the Warrawoona Formation of western Australia (Awramik, Schopf, and Walter 1983), contain evidence of early life. Both sedimentary sequences are over three billion years old. From three billion years ago until the present, we have a continuous record of life on Earth, implying that the mean surface temperature has reached neither the boiling nor the freezing point of water. Given that an ice age involves less than a 10°C drop in mean midlatitude temperature and that even ice ages are relatively rare in the fossil record, the mean temperature at the surface of the Earth probably has stayed well within the range of 5° to 25°C during at least the last three billion years. Solar luminosity during the last four billion years is thought by many astronomers to have increased by at least

10 percent (Newman 1980). Thus life on Earth seems to have acted as a global thermostat. Any current estimate for the increase of solar luminosity, which varies from less than 30 to more than 70 percent (Newman 1980), does not alter the outcome of Daisy World's conclusions. A relative increase of solar luminosity from values of 0.6 to 2.2 (its present value is 1.0) is consistent with Daisy World assumptions because a range of values has been plotted by Lovelock and his collaborator Watson.

Cybernetic systems, as is well known to science and engineering, are steered. They actively maintain specified variables at a constant in spite of perturbing influences. Such systems are said to be homeostatic if their variables, such as temperature, direction travelled, pressure, light intensity, and so forth, are regulated around a fixed set point. Examples of such set points might be 22°C for a room thermostat or 40 percent relative humidity for a room humidifier. If the set point itself is not constant but changes with time, it is called an operating point. Systems with operating points rather than set points are said to be homeorrhetic rather than homeostatic. Gaian regulatory systems, such as the embryological ones described by C.H. Waddington (1976), are more properly described as homeorrhetic rather than homeostatic. Fascinatingly enough, both homeorrhetic and homeostatic systems defy the most basic statutes of western syllogistic thought, although not thought itself, because most people do not think syllogistically but in an associative fashion. For instance, if a person—surely a homeorrhetic entity—is hungry, he or she will eat. Thereupon hunger ceases. Put syllogistically, the sense of such a series becomes nullified: I am hungry; therefore I eat; therefore I am not hungry. The thesis leads to an antithesis without ever being synthetically resolved. This circular, tautological mode of operations is characteristic of cybernetic systems, including, of course, all organisms and organismic combinations. It is consonant with the emotive poetic power of contradictory statements, dichotomous personalities, and oxymoronic lyrics, such as references to a midnight sun.

Even minimal cybernetic systems have certain defining properties: a sensor, an input, a gain (the amount of amplification in the system), and an output. In order to achieve stability or to increase complexity, the output is compared with the set or operating point so that errors are corrected. Error correction means that the output must in some way feed back to the sensor so that the new input can compensate for the change in output. Positive or negative feedback, usually both, are involved in error correction. A first attempt to apply this sort of cybernetic analysis to the

Gaia hypothesis involved development of the Daisy World mathematical model, first by Lovelock (1983a) and later by Watson and Lovelock together (Watson and Lovelock 1983; Lovelock 1983b). We turn now to the description of the model.

The Daisy World model is used to demonstrate how planetary surface temperature might be regulated. It makes simple assumptions: the world's surface harbors a population of living organisms consisting only of dark and light daisies. These organisms always breed true. Each light daisy produces only light offspring daisies, and each dark daisy produces only its kind. Totally black daisies absorb all of the light coming on them from the sun, and totally white daisies reflect all of the light. The best temperatures for growth for both dark and light daisies are considered to be the same: no growth below 5°, increasing growth as a function of temperature to an optimum at 20°C, and decreasing growth rate above the optimum to 40°C, at which temperature all growth ceases.

At lower temperatures darker daisies are assumed to absorb more heat, and thus to grow more rapidly in their local area than lighter daisies. At higher temperatures lighter daisies reflect and thus lose more heat, leading to a greater rate of growth in their local area. The details have been published in technical journals (Lovelock 1983a,b; Watson and Lovelock 1983) and have recently been explained in a more popular way by us in the British magazine *The Ecologist: Journal of the Post-Industrial Age* (Sagan and Margulis 1983). In summary, the graphs generated by models using these assumptions show that dark and light daisy life can, because of growth and interaction with light, influence the temperature of the planet's surface on a global scale. What is remarkable about the various forms of Lovelock and Watson's model is that the amplification properties of the rapid growth of organisms (here daisies) under changing temperatures are enough in themselves to provide the beginning of a mechanism for global thermal homeorrhesis, a phenomenon that some would rather see credited only to a mysterious life force. In general, in these models an increase in diversity of organisms, such as a greater difference between the light and darkness of the daisies, leads to an increase in regulatory ability as well as an increase in total population size.

Daisy World is only a mathematical model. Even with its oversimplification, however, the Daisy World model shows quite clearly that temperature homeorrhesis of the biosphere is not something that is too mysterious to have a mechanism. By implication it suggests that other observed anomalies, such as the near-constant salinity of the oceans over vast periods of time and the coexistence of chemically reactive gases in the atmosphere, may have solutions that actively involve life forms. The radical insight delivered by Daisy World is that global homeorrhesis is in principle possible without the introduction of any but well-known tenets of biology. The Gaian system does not have to plan in advance or be foresighted in any way in order to show homeorrhetic tendencies. A biological system acting cybernetically gives the impression of teleology. If only the results and not the feedback processes were stated, it would look as if the organisms had conspired to ensure their own survival.

The Gaia hypothesis says, in essence, that the entire Earth functions as a massive machine or responsive organism. While many ancient and folk beliefs have often expressed similar sentiments, Lovelock's modern formulation is alluring because it is a modern amalgam of information derived from several different scientific disciplines. Perhaps the strongest single body of evidence for Gaia comes not from the evidence of thermal regulation that is modeled in Daisy World but from Lovelock's own field, atmospheric chemistry.*

From a chemical point of view, the atmosphere of the Earth is anomalous. Not only major gases, such as nitrogen, but also minor gases, such as methane, ammonia, and carbon dioxide, are present at levels many orders of magnitude greater than they should be on a planet with 20 percent free oxygen in its atmosphere. It was this persistent overabundance of gases that react with oxygen, persisting in the presence of oxygen, that initially convinced Lovelock when he worked at NASA in the late 60s and early 70s that it was not necessary for the Viking spacecraft to go to Mars to see if life was there. Lovelock felt he could tell simply from the Martian atmosphere, an atmosphere consistent with the dicta of equilibrium chemistry, that life did not exist there (Lovelock and Margulis 1976). The Earth's atmosphere, in fact, is not at all what one would expect from a simple interpolation of the atmospheres of our neighboring planets, Mars and Venus. Mars and Venus have mostly carbon dioxide in their atmosphere and nearly no free oxygen, while on Earth the major atmospheric component is nitrogen and breathable oxygen comprises a good one fifth of the air.

*Incidentally, Lovelock is an inventor as well as a scientist. He devised the electron capture device, a sensor for gas chromatographs that detects freon and other halogenated compounds in concentrations of far less than one part per million in the air. Indeed, it was Lovelock's invention and observations that in large part sparked off ecological worries of ozone depletion, ultraviolet light–induced cancers, and general atmospheric catastrophe.

Lovelock has compared the Earth's atmosphere with life to the way the atmosphere would be without any life on Earth. A lifeless Earth would be cold, engulfed in carbon dioxide, and lacking in breathable oxygen. In a chemically stable system we would expect nitrogen and oxygen to react and form large quantities of poisonous nitrogen oxides as well as the soluble nitrate ion. The fact that gases unstable in each other's presence, such as oxygen, nitrogen, hydrogen, and methane, are maintained on Earth in huge quantities should persuade all rational thinkers to reexamine the scientific status quo taught in textbooks of a largely passive atmosphere that just happens, on chemical grounds, to contain violently reactive gases in an appropriate concentration for most of life.

In the Gaian theory of the atmosphere, life continually synthesizes and removes the gases necessary for its own survival. Life controls the composition of the reactive atmospheric gases. Mars and Venus, and the hypothetical dead Earth devoid of life, all have chemically stable atmospheres composed of over 95 percent carbon dioxide. Earth as we live on it, however, has only 0.03 percent of this stable gas in its atmosphere. The anomaly is largely due to one facet of Gaia's operations, namely, the process of photosynthesis. Bacteria, algae, and plants continuously remove carbon dioxide from the air via photosynthesis and incorporate the carbon from the gas into solid structures such as limestone reefs and eventually animal shells. Much of the carbon in the air as carbon dioxide becomes incorporated into organisms that are eventually buried. The bodies of deceased photosynthetic microbes and plants, as well as of all other living forms that consume photosynthetic organisms, are buried in soil in the form of carbon compounds of various kinds. By using solar energy to turn carbon dioxide into calcium carbonates or organic compounds of living organisms, and then dying, plants, photosynthetic bacteria, and algae have trapped and buried the once-atmospheric carbon dioxide, which geochemists agree was the major gas in the Earth's early atmosphere. If not for life, and Gaia's cyclical *modus operandi,* our Earth's atmosphere would be more like those of Venus and Mars. Carbon dioxide would be its major gas even now.

Microbes, the first forms of life to evolve, seem in fact to be at the very center of the Gaian phenomenon. Photosynthetic bacteria were burying carbon and releasing waste oxygen millions of years before the development of plants and animals. Methanogens and some sulfur-transforming bacteria, which do not tolerate any free oxygen, have been involved with the Gaian regulation of atmospheric gases from the very beginning. From a Gaian point of view animals, all of which are covered with and invaded by gas-exchanging microbes, may be simply a convenient way to distribute these microbes more numerously and evenly over the surface of the globe. Animals and even plants are latecomers to the Gaian scene. The earliest communities of organisms that removed atmospheric carbon dioxide on a large scale must have been microbes. In fact, we have a direct record of their activities in the form of fossils. These members of the ancient microbial world constructed complex microbial mats, some of which were preserved as stromatolites, layered rocks whose genesis both now and billions of years ago is due to microbial activities. Although such carbon-dioxide–removing communities of microbes still flourish today, they have been supplemented and camouflaged by more conspicuous communities of organisms such as forests and coral reefs.

To maintain temperature and gas composition at livable values, microbial life reacts to threats in a controlled, seemingly purposeful manner. Gas composition and temperature must have been stable over long periods of time. For instance, if atmospheric oxygen were to decrease only a few percentage points, all animal life dependent on higher concentrations would perish. On the other hand, as Andrew Watson et al. showed, increases in the level of atmospheric oxygen would lead to dangerous forest fires (Watson, Lovelock, and Margulis 1978). Small increases of oxygen would lead to forest fires even in soggy rain forests due to ignition by lightning. Thus the quantity of oxygen in the atmosphere must have remained relatively constant since the time that air-breathing animals have been living in forests—which has been over 300 million years. Just as bees and termites control the temperature and humidity of the air in their hives and nests, so the biota somehow controls the concentration of oxygen and other gases in the Earth's atmosphere.

It is this "somehow" that worries and infuriates some of the more traditional Darwinian biologists. The most serious general problems confronting widespread acceptance of the Gaia hypothesis are the perceived implications of foreknowledge and planning in Gaia's purported abilities to react to impending crisis and to ward off ecological doom. How can the struggling mass of genes inside the cells of organisms at the Earth's surface know, ask these biologists, how to regulate macroconditions like global gas composition and temperature? The molecular biologist W. Ford Doolittle, for example, a man who because of his work is perhaps predisposed toward viewing evolution at smaller rather than larger levels, sees the Gaia hypothesis as untenable, a motherly theory of nature without a mechanism (1981).

Another scientist, the Oxford University evolutionist Richard Dawkins, is even more forceful in his rejection of the theory. Likening it to the BBC Theorem (a pejorative reference to the television documentary notion of nature as wonderful balance and harmony), Dawkins has extreme difficulty in imagining a realistic situation in which the Gaian mechanism for the perpetuation of life as a planetary phenomenon could ever have evolved. Dawkins, author of *The Selfish Gene,* can only conceive of the evolution of planetary homeorrhesis in relation to interplanetary selection: "The universe would have to be full of dead planets whose homeostatic regulation systems had failed, with, dotted around, a handful of successful, well-regulated planets of which Earth is one" (Dawkins 1982).

These sound like forceful arguments, yet if the critics of Gaia cannot accept the notion of a planet as an amorphic, but viable, biological entity, they must have equal if not greater cause to dismiss the origin of life. Surely at one point in the history of the Earth, a single homeostatic bacterial cell existed that did not have to struggle with other cells in order to survive, because there were no other cells. The genesis of the first cell can no more be explained from a strict Darwinian standpoint of competition among selfish individuals than can the present regulation of the atmosphere. While the first cell and the present planet may both be correctly seen as individuals, they are equally alone, and as such they both fall outside the province of modern population genetics.

Nonetheless, Lovelock, a sensitive man with a deep sense of intellectual mischief, has answered his critics with one of their own favorite weapons: mathematical model making in the form of the aforementioned Daisy World (Watson and Lovelock 1983; Lovelock 1983b). Not believing that the Earth's temperature and gases can be regulated with machine-like precision for billions of years, because organisms cannot possibly plan ahead, Lovelock's critics reject his personification of the planet into a conscious female entity named Gaia. Originally lacking an explicit mechanism and falling outside the major Darwinian paradigm of selfish individualism, it was and still sometimes is difficult for trained evolutionists to refrain from regarding Gaia as the latest deification of Earth by nature nuts. How can an entangled mass of disjointed struggling microbes, they ask, effect global concert of any kind, let alone to such an extent that we are permitted to think about the Earth as a single organism? The answer, of course, is the kind of analysis explored in Daisy World, and one still waits to see how those who accuse Lovelock of conscious mysticism and pop ecology will respond to it in all its mathematical intricacy.

Perhaps the greatest psychological stumbling block in the way of widespread scholarly acceptance of Gaia is the implicit shadow of doubt it throws over the concept of the uniqueness of humanity in nature. Gaia denies the sanctity of human attributes. If intricate planning, for instance, can be mimicked by cunning arrays of subvisible entities, what is so special about *Homo sapiens* and our most prized congenital possession, the human intellect? The Gaian answer to this is probably that nothing is so very special about the human species or mind. Indeed, recent research points suggestively to the possibility that the physical attributes and capacities of the brain may be a special case of symbiosis among modified bacteria (Margulis and Sagan, 1997).

In real life, as opposed to Daisy World, microbes, not daisies, play the crucial role in the continual production and control of rare and reactive compounds. Microbial growth is also responsible, possibly through the production of heat-retaining gases as well as the changing colored surfaces, for the continuing thermostasis of the Earth. Evolutionarily, microbes were responsible for the establishment of the Gaian system. Insofar as larger forms of animal and plant life are essentially collections of interacting microbes, Gaia may be thought of as still primarily a microbial phenomenon (Kaveski, Mehos, and Margulis 1983). We human beings, made of microbes, are part of Gaia no less than our bones, made from the calcium from our cells, are part of ourselves.

In his recent article on classical views of Gaia, J.D. Hughes quoted the ancient Greek work *Economics* by Xenophon: "Earth is a goddess and teaches justice to those who can learn, for the better she is served, the more good things she gives in return." In the classical view, that is, of the Greek Gaia or Earth Goddess and the Latin Tellus, the Earth is a vast living organism. The homeric hymn sings:

> Gaia, mother of all, I sing oldest of gods.
> Firm of foundation, who feeds all creatures living
> on earth.
> As many as move on the radiant land and swim in the sea
> And fly through the air—all these does she feed with her
> bounty.
> Mistress, from you come our fine children and bountiful
> harvests.
> Yours is the power to give mortals life and to take it away.
>
> J. Donald Hughes
> "Gaia: An Ancient View of Our Planet"

Although Gaia is reappearing in modern dress, the modern scientific formulation of the Gaian idea is quite different from the ancient one. Gaia is not the nurturing mother or fertility doll of the human race. Rather, human beings, in spite of our raging anthropocentrism, are relegated to a tiny and unessential part of the Gaian system. People, like *Brontosaurus* and grasslands, are merely one of the many weedy components of an enormous living system dominated by microbes. Gaia has antecedents not only among the classical poets but even among scientists, most notably in the work of the Russian V.I. Vernadsky (1863–1945) (Lapo 1982). But Lovelock's Gaia hypothesis is a modern piece of science: it is subject to observational and experimental verification and modification.

There is something fresh, new, and yet mythologically appealing about Gaia, however. A scientific theory of an Earth that in some sense feels and responds is welcome. The Gaian blending of organisms and environment into one, wherein the atmosphere is an extension of the biosphere, is a modern rationalist formulation of an ancient intuitive sentiment. One implication is that there may be a strong biogeological precedent for the time-honored political and mystical goal of peaceful coexistence and world unity.

Contrary to possible first impressions, however, the Gaia hypothesis, especially in the hands of its innovator, does not protect all the moral sanctions of popular ecology. Lovelock himself is no admirer of most environmentalists. He expresses nothing but disdain for those technological critics he characterizes as misanthropes or Luddites, people who are "more concerned with destructive action than with constructive thought" (Lovelock 1979, p. 95). He claims, "If by pollution we mean the dumping of waste matter there is indeed ample evidence that pollution is as natural to Gaia as is breathing to ourselves and most other animals" (Lovelock, 1979, p. 95). We breathe oxygen, originally and essentially a microbial waste product. Lovelock believes that biological toxins are in the main more dangerous than technological ones, and he adds sardonically that they would probably be sold in health food stores if not for their toxicity. Yet there is no clear division between the technological and the biological. In the end, all technological toxins are natural, biological byproducts that, though via human beings, are elements in the Gaian system. Similarly, legislation and lobbying attempts, such as the recent furor in the United States over the mismanagement of the Environmental Protection Agency, are nothing more or less than part of Gaian feedback cycles.

Ecologically speaking, the Gaia hypothesis hardly reserves a special place in the pantheon of life for human beings. Recently evolved, and therefore immature in a fundamental Gaian sense, human beings have only recently been integrated into the global biological scene. Our relationship with Gaia is still superficial. On the other hand, our ultimate potential as a nervous early warning system for Gaia remains unsurpassed. Deflecting oncoming asteroids into space and spearheading the colonization of life on other planets represent additions to the Gaian repertoire that our species must initiate. On the one hand, Gaia was an early and crucial development in the history of life's evolutionary past. Without the Gaian environmental modulating system, life probably would not have persisted. Now, only by comprehending the intricacies of Gaia can we hope to discover how the biota has created and regulated the surface environment of the planet for the last 3 billion years. On the other hand, the full scientific exploration of Gaian control mechanisms is probably the surest single road leading to the successful implementation of self-supporting living habitats in space. If we are ever to engineer large space stations that replenish their own vital supplies, then we must study the natural technology of Gaia. Still more ambitiously, the terraformation of another planet, for example, Mars, so that it can actually support human beings living out in the open, is a gigantic task and one that becomes thinkable only from the Gaian perspective.

In terms of the metaphysics of inner space, acceptance of the Gaian view leads almost precipitously to a change in philosophical perspective. As just one example, human artifacts, such as machines, pollution, and even works of art, are no longer seen as separate from the feedback processes of nature. Recovering from Copernican insult and Darwinian injury, anthropocentrism has been dealt yet another reeling blow by Gaia. This blow, however, should not send us into new depths of disillusion or existential despair. Quite the opposite: we should rejoice in the new truths of our essential belonging, our relative unimportance, and our complete dependence upon a biosphere that has always had a life entirely its own.

12

The Global Sulfur Cycle and *Emiliania huxleyi*

Dorion Sagan

Certain elements are planetary lifeblood. Like blood, they flow through the biosphere in limited supply. The carbon, sulfur, nitrogen, phosphorus, oxygen, and hydrogen that make up all organisms on Earth are not infinite. They must be continually redistributed, or cycled. Unlike an animal, the Earth has no heart pushing this global flow in a simple beat. Instead, the planet lives on a complex of different forces all pulsing to a syncopated rhythm. These forces include wind and ocean currents, the erosion and production of geological formations, and the motion of living organisms.

Although it has recently become more feasible, tracking global element cycles is still a Herculean task. But the National Aeronautics and Space Administration (NASA), which is used to studying planets as whole entities, is turning its resources toward Earth. Every other year since 1980 a NASA-supported group called Planetary Biology and Microbial Ecology (PBME) brings academics, researchers, and space scientists together to discuss the connections between life and the elements it needs to sustain itself. In 1980, the group looked at many elements. In 1982, the focus was carbon. In 1984, PBME-NASA tried to determine sulfur's elusive path through the "veins" of the world. Nitrogen will be the next mystery element.

Many of the major transformations that keep elements accessible to life transpire in hot springs, salt flats, and deeply textured sediments called microbial mats. To investigate these environments, PBME participants met in San Jose, California and explored the San Francisco Baylands, Alum Rock Wildlife Refuge, and Big Soda Lake in the two-casino town of Fallon, Nevada.

Here scientists tried to piece together the puzzle of sulfur-using microbes and the global sulfur cycle. The program's long-term goal is to blend space technology and microbiology and to come up with a map, as it were, of global metabolism. But in the short term, the scientists must trek amid a stench resembling rotten eggs and cabbages, braving pools of mud and suspiciously colored gunk.

An analogy for the collective work of PBME-NASA is the early anatomical studies of the Renaissance artist and scientist Leonardo da Vinci. PBME is also on the vanguard of exploration, uncovering the mechanics of the biosphere. But whereas Da Vinci cut open bodies and looked inside them to be able to draw and abstract about the human body, today's interdisciplinarians—environmentalists, petroleum geologists, microbial ecologists, soil scientists, oceanographers, and atmospheric scientists—study small samples of the biosphere with the constant knowledge that the larger system they are part of can be viewed at large, imaged in near entirety from space. (Figure 12.1).

Just as metabolism is the complex of chemical activities that maintains the structure of organisms and their component cells, so the metabolic activities of all organisms sharing the Earth are so intimately linked that they form a sort of giant metabolism. Sulfur, part of this Earth-wide metabolism, is found in the proteins of all organisms and is therefore required for all growth.

The element exists in both hydrogen-rich forms and in highly oxidized forms. Chemical reactions, from oxidized to hydrogen-rich compounds and vice versa, yield energy. Life mediates sulfur and other elements through such chemical reactions, building up cell material or releasing energy for physiological processes.

Many bacteria, such as *Desulfovibrio, Desulfuromonas,* and *Desulfotomaculum,* turn oxidized sulfates (SO_4^{-2}) and sulfur into hydrogen-rich sulfides. Sulfides, often in the form of gaseous hydrogen sulfide (H_2S), are

FIGURE 12.1. Coccolithophorid bloom seen from space. The white is the land of the northwest coast of Scotland and its islands. The coccolithophorids produce chlorophyll, which accounts for the dark green of the sea; they are also a major producer of dimethyl sulfide, a gas extremely important in the global sulfur cycle. This image helps us see how a phenomenon on the microorganism level could be discovered by planetary observations from space. Scientists have only recently realized that the dimethyl sulfide so important to the global sulfur cycle comes largely from these microorganisms. Courtesy of Patrick Holligan.

then used as an energy source for other bacteria, such as *Beggiatoa*. *Beggiatoa* need oxygen to get energy from oxidizing sulfide, but sulfide can even be oxidized under conditions where there is no gaseous oxygen by bacteria such as *Chromatium*, which use the oxygen in their cells to effect the transformation in reactions that may have originated on the primordial Earth.

Microbes are key to the concept of element circulation, and they can be important in depositing major sulfur-containing minerals, such as the gypsum ($CaSO_4 \cdot 2H_2O$) found in salt flats. As William Holser, of the University of Oregon, told the PBME group, even pyrite (FeS_2), the familiar iron sulfide mineral known as fool's gold, ultimately depends on bacterial alteration of marine sulfate for its formation in sediments. If such mineral deposits depend on and, in a real sense, are part of life, then why are they considered static, inanimate, and nonliving? In fact, it may be better to look at such deposits as part of a global skeleton or storage system, one that is drawn upon by life in the way a pregnant woman draws upon the calcium of her bones to feed her fetus.

Until the last few years, almost nobody thought there was much sulfur in the atmosphere, except for the oxidized sulfur compounds from coal mining and the like. But atmospheric dimethyl sulfide, $(CH_3)_2S$, a recent focus of attention, exemplifies a change of perception in interdisciplinary global studies toward seeing life and the environment, biology, and geochemistry as inextricably bound.

Sulfur in the Air

Dimethyl sulfide, for example, which makes the sea smell like the sea, was caught 10 years ago carrying huge amounts of sulfur from the ocean to the atmosphere. These sulfurous migrations, like most chemistry on Earth, are largely dependent on life.

Meinrat Andreae, at the Department of Oceanography at Florida State University, discovered a correlation between the population density of marine algae such as *Phaeocystis* and *Emiliania* and the buildup of dimethyl sulfide (see Figures 12.1 and 12.2). Some of this gas, which brings so much sulfur up from seawater into the air, is produced by *Phaeocystis poucheti*. This obscure microbe apparently uses the precursor to atmospheric dimethyl sulfide as an osmolyte, as a compound that regulates intracellular salt concentration. For oceanic plankton exposed to the vicissitudes of changing salt concentrations, osmolytes are hot commodities.

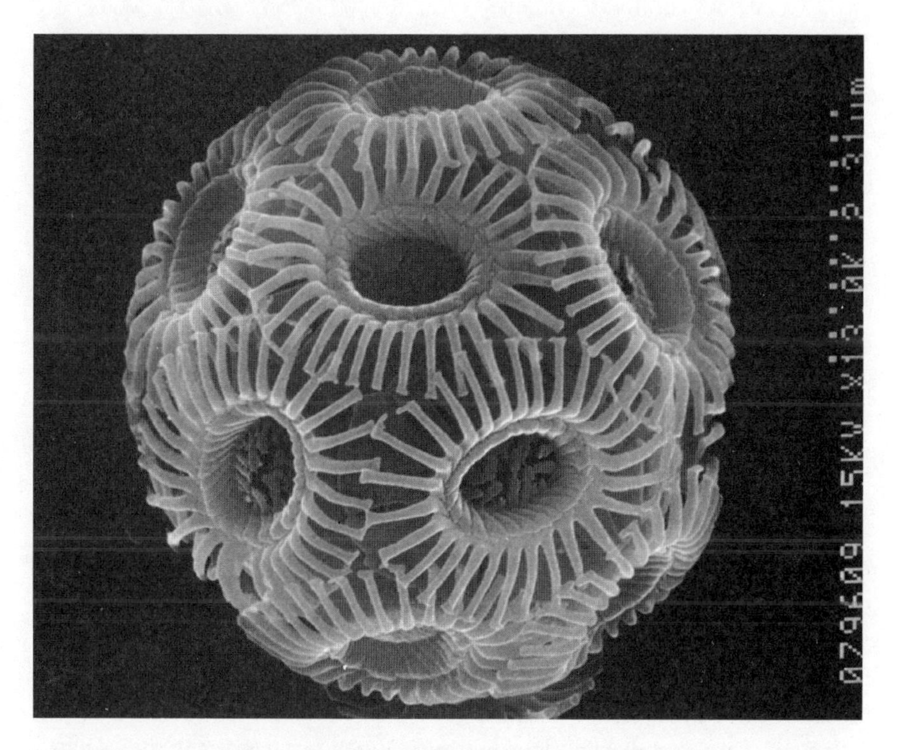

FIGURE 12.2. *Emiliania huxleyi,* a coccolithophorid. The "buttons" (coccoliths) in their immense quantities lead to the dark green color visible off the west of the Hebrides islands. Scanning electron micrograph by Susimo Honjo.

Osmolytes can also be based on nitrogen compounds, but sulfur osmolytes are probably common in ocean-faring organisms, as well as being major sources of atmospheric sulfur gases.

Not all atmospheric sulfur gases are produced by microbes, of course. As New York City commuters from northern New Jersey know only too well, the activities of people also make significant contributions to the sulfur cycle. All factories and automobiles emit at least some sulfur dioxide (SO_2) when sulfur-bearing fossil fuels such as gasoline, coal, and oil are burned. Catalyzed by light, sulfur dioxide and oxygen react in the atmosphere to form sulfur trioxide (SO_3). Sulfur trioxide combines in water to make sulfate droplets that become the sulfuric acid (H_2SO_4) that, swept by winds from such places as the heavily industrialized Ohio Valley, falls as acid rain in New York and New England.

But Andreae said that nonhuman biological processes emit sulfur gases at rates at least comparable with the sulfur dioxide flux from fossil fuel burning. In 1985 the amount of sulfur dioxide given off from the biota to the air was, he said, on the order of a hundred trillion grams. By far the most important processes of the biogenic (natural as opposed to industrial) release of sulfur gases to the atmosphere is the chemical transformation of ocean sulfate into other forms of sulfur compounds by bacteria.

The incorporation of sulfate and organic sulfur compounds by algae and plants is a second immensely important sulfur transformation that occurs on a planetary scale. Indeed, James Lovelock, a British atmospheric chemist, suggests that the quantity of such sulfur compounds—those produced by organisms other than humans and released into the atmosphere—may in fact be far greater than those produced by factories, power stations, and automobiles.

As part of conventional oceanography, environmental sulfur dioxide readings have traditionally been taken at sea. Andreae, Lovelock, and others feel, however, that estimates of sulfur gas production over land are probably wildly inaccurate, leading researchers to overestimate the volume of sulfur produced by industry.

Part of the problem of determining the sulfur cycle is the difficulty of measurements: Sulfur gases can vary by several orders of magnitude over a period of hours at one spot on the coast. Most of the acid rain precursors have been measured on land in the context of some specific, local pollution problem rather than in the context of a total understanding of Earth's atmosphere.

In the Rain

Robert Fuller, of the Department of Civil Engineering at Syracuse University, reminds us that acid rain is only one in a suite of factors determining the acidity of lake water. A lake is frequently a small part of a much larger watershed, where water interacts with vegetation, soil, and the underlying rocks. Watershed characteristics, such as the presence of coniferous vegetation, high levels of soil organic carbon, shallow soils, an inability to adsorb and immobilize sulfate, and low levels of exchangeable and weatherable basic cations, are all factors that can predispose an ecosystem to transfer atmospheric acidity to surface waters. The alkalinity of the rock bed is involved, as well. As an example, neighboring lakes receiving acid rain in upstate New York have been found to have significantly different acidities. But these lakes, beneath the same sky, receive the same amounts of sulfuric acid in their rain.

These observations don't excuse the high sulfur emissions by industry. But they do show that the measured acidity in a lake does not depend only on the quantity of acid in the rain. Most of the furor about high levels of atmospheric sulfur and acid lakes comes from foresters, farmers, and fly-casters. Lakes have even been declared dead because of their relatively high concentrations of sulfuric acid. But not only trees, fish, and forest mammals are affected by acid rain.

In acidified lakes, as in sulfide-rich waters, there are many organisms that positively thrive. Indeed, unusually lush algal and bacterial growth may even identify a lake's acidity. Animals may flourish in high-acid lakes too: While trout are decimated or even totally killed off in very acidic lakes, causing indisputable economic hardship to people who depend on fishing, certain species of crayfish crawl about and reproduce to high population densities unperturbed. The types of bacteria that form coatings and mats, especially along the bottom of acid-rich lakes, are organisms with multibillion-year histories. These prolific microbes must have been involved in the formation of the earliest sulfur cycles.

PBME participants believe the major environmental sulfur transformations are fundamentally biochemical processes that evolved inside bacterial cells. Bacteria coevolved with the earliest biosphere, their remains existing as fossils in some of the oldest unmetamorphosed rocks. Although evidence for sulfur reduction—bacterial conversion of sulfate into sulfur and sulfide—appears in the fossil record only after the appearance of photosynthesis, there is some consensus that sulfate-reducing bacteria evolved before and paved the way for the development of photosynthesis.

A Free Lunch

Early in the history of life, fermenting bacteria partook of the free lunch of energy-rich chemicals left over from the production of the so-called prebiotic soup. Yet soon after, suggests Lynn Margulis, of Boston University and codirector of PBME-NASA, they evolved a more efficient way of deriving energy.

By diverting high-energy electron carriers away from the process of fermentation, some kinds of anaerobic bacteria evolved the ability to breathe the common oceanic ion, sulfate. The ability to breathe sulfate and to use it instead of prebiotically produced complex organic sulfur compounds, such as the amino acids methionine or cysteine, gave such early anaerobic bacteria an evolutionary advantage. The more complete oxidation of organic matter provided them with additional energy.

To reduce carbon dioxide from the air into the hydrogen-rich carbon compounds of cells, microbes needed a source of electrons. An excellent early source of electrons was gaseous hydrogen, which was far more plentiful in the early solar system. As time went on the Sun's high-energy radiation and the Earth's weak gravitational field caused hydrogen to escape into space. Most early hydrogen was eventually lost from the Earth's atmosphere, but hydrogen sulfide, a gas emitted from the Earth's interior through hydrothermal vents, volcanoes, and sulfur hot springs, was still plentiful. Bacteria grappled with this for their electrons instead.

Today the green and purple sulfur bacteria still use hydrogen sulfide as their electron donor in photosynthesis. When cyanobacteria (blue-green algae) began using the hydrogen of water as an electron donor, the global sulfur cycle, along with the other major chemical cycles of the biosphere, changed forever. The use of water rather than hydrogen sulfide led to new waste products.

In the early days photosynthesis was largely dependent on a steady source of hydrogen sulfide, and the gas was converted into yellow sulfur deposits on the ground or into globules in the water that were later oxidized to make ocean sulfate. But now, as water replaced hydrogen sulfide as the largest reserve of electrons for photosynthesis, oxygen began to build up in the air. As the oxygen-producing cyanobacteria spread, the entire planet underwent dramatic oxidation. By 1800 million years ago, during the Proterozoic eon, hydrogen-rich iron, uranium, and sulfur-bearing minerals at the Earth's outer crust practically disappeared. They were replaced by oxygen-rich forms. But the biochemical legacy of the early hydrogen-rich environment was simultaneously preserved in the form of life, making the Earth an astronomical oddity.

Because of life's oxygen waste, Earth underwent many new energizing and energy-releasing reactions, which in turn were exploited by life. The transition to an oxygenic biosphere had many literally Earth-changing consequences, among which was the banishment of some bacteria, those that had previously flourished at the surface, to a new subsurface realm of marine muds and warm geysers. To this day such oxygen-shunning bacteria make up the lower layers of the flat purple and green communities known as microbial mats.

Yehuda Cohen, of the Hebrew University in Jerusalem, introduced the use of microelectrodes as a means of measuring minute concentrations of oxygen, hydrogen, and sulfide in microbial mats. The new technique, first applied to microbial ecology by N.P. Revsbech, of Aarhus University in Denmark, allows detailed vigils over chemical transformations at the Earth's surface. Microelectrode work ("physiology") coupled with ultrastructural study ("anatomy") show that the sedimentary layers of organisms that form these microbial mats are distinct in much the same way that skin, fat, and muscle tissue are composed of differentiated flattened masses of animal cells.

Certain chemical conditions, oxygen and sulfide concentrations, and levels of light penetration typify each layer, but differences in these variables can cause major changes in community interaction, and changes in community interaction in turn can feed back into changes in the variables. Cohen's team examined community relations among microbes in the salt flats near Leslie Salt Co., in Newark, California. They looked at the surface and subsurface microbes in the sulfur springs of Alum Rock State Park too. The tiny millimeter-thick region in both of these locations that separates cyanobacteria from the sulfur bacteria rises slightly during the night and descends correspondingly during the day. At night, there is no photosynthesis to produce the oxygen lethal to sulfur users, and so the microelectrodes detect increased levels of hydrogen sulfide closer to the surface. Like the chest of a sleeper, the chemical boundary moves. Each day the hydrogen-sulfide/oxygen interface rises; each night it falls.

Some bacteria living in this zone are very versatile, for they must be able to cope with potentially poisonous concentrations of both hydrogen sulfide and oxygen. *Oscillatoria limnetica,* for example, uses either hydrogen sulfide or hydrogen from water during photosynthesis. The cosmopolitan microbe *Microcoleus chthonoplastes* has a chameleon physiology. This organism, recognizable because it looks like microscopic bundles of insulated wire (Figure 12.3), sometimes lives like an ancient bacterium, never producing any oxygen. Other times it performs the oxygen-producing photosynthesis typical of plants, but under concentrations of sulfide that would poison plants, animals, algae, and even other bacteria. It seems plausible that such versatility comes from a time when the gas mixture of the Earth's atmosphere was changing from an oxygen-poor to an oxygen-rich one.

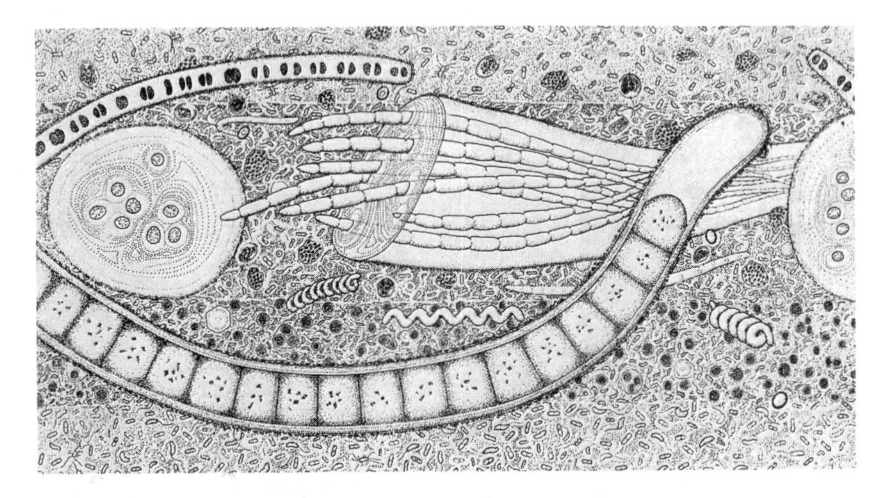

FIGURE 12.3 The inside of a microbial mat. Drawing by Christie Lyons.

Changing Neighborhoods

The daily movement of the boundary layer between oxygen and sulfide may at times not reflect changes in the composition of communities of organisms so much as flexibility in the metabolism of those organisms. The surfaces of marshes, salt ponds, and muds bombarded by light from above and permeated with gas-containing fluids from below present a vast array of energy sources and opportunities. Those organisms able to vary their metabolic repertoire, to complement or enhance the metabolism of others, or that are just generally at home in the melee of deposition and gas exchange around the surface zone of sunlight grow like weeds. And they make the greatest contributions to the sulfur cycle.

To follow globally roaming elements whose territory is the entire surface of the globe is not simple. Sulfur, like any element important to life, has multiple guises and creates a web of activity crossing subtly between animate and inanimate realms. The marriage of microbial and planetary studies is an ambitious new enterprise. It may, like Da Vinci, be ahead of its time.

The late Robert M. Garrels, of the University of South Florida, a PBME participant and expert on element cycling, waxed ironic over global metabolism. Although Garrels takes the idea of a giant circulatory and living system seriously, as shown by his remark that "The Earth's surface environment can be regarded as a dynamic system protected against perturbations by effective feedback mechanisms," he also has a warning. "We all build more and more complicated geochemical models until no one understands anyone else's model. The only thing we do know is that our own is wrong."

But should we then give up trying to understand the global cycling of elements so important to life on Earth? Not necessarily. He explains, "The chief purpose of our models is not to be right or wrong but to give us a place to store our data."

While NASA's life sciences program has been expanding in recent years to include Earth as a planet to be viewed from space and compared with its lifeless neighbors Mars and Venus, microbiology, geology, and chemistry have simultaneously become more circumscribed and circumspect in their university settings. This peculiarity of scientific history has led to an academic struggle, a hybrid sometimes called microbiogeochemistry. We will have to wait to see where this chimeric discipline leads. We still do not know whether it will ever be able to discover the metabolic workings of the Earth or to plot the movement of the elements as gracefully as Da Vinci drew a man. Yet microbiogeochemistry (perhaps better called geophysiology) could be on the verge of a new renaissance.

13

Descartes, Dualism, and Beyond

Dorion Sagan,
Lynn Margulis,
and Ricardo Guerrero

The brilliant French Catholic mathematician, René Descartes (1596–1650), inaugurated the mechanistic dichotomy with his declaration of a universal split between *res extensa,* the determined material reality of nature, and *res cogitans,* the free-thinking reality of people and God. Only humans, Descartes argued, partake of God to the extent that they have souls. Animals, though they seem to feel pain, are in fact soul-less machines: "We are so accustomed to persuade ourselves that the brute beasts feel as we do that it is difficult for us to rid ourselves of this opinion. But if we were as accustomed to seeing automata which imitate perfectly all those of our actions which they can imitate, and to taking them for automata only, we should have no doubt at all that the irrational animals are automatons" (Jonas 1966).

Although Descartes' presentation of the universe as a vast mechanism led to an expansion of scientific investigation, the acceptance of the Cartesian mechanistic universe also had negative implications. On the authority of Descartes, live animals were nailed to boards without remorse to illustrate the facts of anatomy and physiology. Rationalized as unfeeling and inanimate, nature, in the wake of Descartes, was analyzed without fear of trespass. Nature, including the mechanical, automata-like "lower" life-forms, could now be experimented on with impunity. In short, Descartes' philosophy provided a formal justification—a Cartesian license—to investigate virtually everything in an effort to discover the mechanism by which God had "built" the phenomenal world.

By splitting reality into human consciousness and an unfeeling, objective exterior, or in his terms *extensive,* world that could be measured mathematically, Descartes paved the way for a scientific investigation of nature constructed according to the mathematical laws of God. "God sets up laws in nature just as a king sets up laws in his kingdom," he wrote (Berman 1989). The Cartesian license separated matter from form, body from soul, outward spatially extended nature from inner awareness. Matter, body, and nature could—unlike thought or feeling—be measured, compared, and thus ultimately understood by mathematical laws.

This Cartesian license permitted the human intellect, through science, to enter a thousand different realms, from the gigantic to the subvisible. The once divine was now open to scientific exploration. Optical instruments focused on snowflakes and peppercorns or pointed at the pockmarked whiteness of the side-lit moon. Atoms were investigated by chemical combination and physical acceleration. X-rays imaged bones. Radioactive elements clocked the internal metabolism of the human body. Eventually aeronautical engineers even appropriated the seemingly God-given power to fly.

Investigation of the formerly divine realm yielded impressive scientific results. Scientists, perusing nature and not books, returned the Bible and the classics to their dusty shelves. There is a biographical anecdote, perhaps apochryphal, that when Descartes was asked in his urban domicile about the location of his library, he pointed to a dissected calf he had been examining and said, "on top of those books." Scientists began to study the world, "written," as Galileo had put it even prior to Descartes, "in a great book which is always open before our eyes" (Jacob 1973). Galileo had paid dearly for his inquisitive temperament. As quantitative mechanicist, measurer of falling bodies, discoverer of the moons of Jupiter and the rotation of the Earth (Simmons 1996), it was Galileo who had cleared the trail for curious successors such as Descartes, Newton, and the "Prince of Astronomy," William Herschel (1738–1822), who confirmed the Milky Way is a spiral-shaped object formed by distribution of its component stars.†

A defier of potent philosophers and Christian theologians, Galileo provoked the ire of Church authorities. He was, at age 58, brought before the Inquisition and charged with heresy. Galileo recanted his earlier claims that were so at variance with official Church doctrine. He "admitted" that Earth is at the center of the universe. Warned against further heresy, Galileo, who became a prisoner in his own country home, was condemned to three years of weekly psalm recitations. Indeed, his thoughts were censured for nearly 200 years until 1838 Galileo's immensely popular masterpiece, *Dialogue of the Two Chief World Systems,* was banned. With horror, Pope Urban VIII recognized himself in Galileo's imagined character "Simplicio." Correctly believing that he had been mocked, it was Urban who began the censorship.

If Galileo had worked under the Cartesian license he would have fared better. When in 1633 the devout Descartes learned of Galileo's condemnation, he abandoned work on a manuscript that supported a heliocentric rather than an Earth-centered world. Impelled to integrate science into religion, Descartes gave great impetus to modern practices of investigation by doubting everything but the existence of his own doubting mind. Bodies, he held, were clocklike mechanisms, created by a Creator. The body is connected to the mind, he wrote, via the pineal gland, a pea-sized structure at the base of the brain, known at that time in the seventeenth century only in humans. The pineal acted, Descartes suggested, as a valve through which God was connected to the free human soul.

The Cartesian license still rallies scientists to study a universe wide open for investigation. But the "fine print"—to extend the metaphor—of this great card of admission into once-forbidden realms ironically vouchsafes the same repressive, religion-based legacy it was designed to combat. Generating the mechanistic body is the conscious human mind in its deistic incarnation as the mind of God. This vitalistic residue of primordial consciousness remains the ghost within the machine of would-be, wholly materialistic modern science. The Cartesian license still contains in its metaphorical fine print the following assumption: the universe is mechanical and is set up according to immutable laws by God. But neither the

†A large sign saying "Study Nature not Books" and attributed to Louis Agassiz decorates the library at MBL, Woods Hole, Massachusetts.

human exception to the predetermined laws of nature nor the metaphysical assumption of divine mechanism is science. Cartesian philosophy is more imbued with the historical presuppositions of western European culture than the pure objectivity it tauts.

Ultimately, we suggest, the Cartesian license proves to be a kind of forgery. After three centuries of implicit renewal, the permit is still valid, even though the fine print, worn off or ignored, is barely visible. Yet the fine print exempting humans and making machinate the "objective world" is no more peripheral to the Cartesian license than is the Surgeon General's warning on a box of cigarettes. The raison d'être, the rational basis that authorized scientists to follow the spirit of Descartes to proceed with their work and to receive the blessings of society, including the Church, are already implicit in Descartes' license. For many centuries the Judeo-Christian religions had placed "man," man as "made in God's image," high on the ladder of being. People, in the cultural mind of the literate world, are situated perhaps a little lower than the angels but certainly above all the rest of life.

While Descartes cogitated, Europe remained under the rule of royalty. The King and the Lord, representing the power and order of God, reigned supreme. But licensed Cartesians—medical men, explorers, alchemists—soon entered the realms into which they were formerly forbidden to enter for fear of transgressing the sacred.

Scientific revelation of mechanism, part of the new audacity of inquiry, helped unsettle European monarchy. If the universe, made by God, is a giant automaton that works itself, why should people obey any king or lord whose power, God given in the feudal system of medieval Christianity, no longer derived from heavenly decree? Many began to take seriously what they took to be the implications of liberating free inquiry. High-born Frenchman, Donatien Alphonse François Sade, as the infamous Marquis de Sade, for example, keenly wrote about and lived his conviction that the religious basis for morality had vanished. If Nature were a self-perpetuating machine and no longer a purveyor of divine authority, then why did the outrageous acts that he performed matter at all? All was, at best, the morally neutral turning of wheels in a vast, more lifelike than living, automatic mechanism (Klossowski 1991).

In 1776 the British colonists in North America broke free from transatlantic rule. Independence from the burdens of taxes and royalty was proclaimed. In 1789 the French Revolution deposed the king and stripped the lords and ladies of their powers. Irreverent Voltaire claimed that if God did not exist it would be necessary to invent Him. A century later the German philologist and nihilistic aesthetician, Friedrich Nietzsche, declared outright that God is dead. He defined philosophy as the unfettered love of knowledge and the philosopher as he before whom everyone quivers. "Philosophy," he wrote, is "a terrible explosive in the presence of which everything is in danger" (Nietzsche in Wakeford and Walter 1995, pp. 19–37).

England, too, was struck by the revolutionary spirit of the late eighteenth and early nineteenth centuries. Expansionist and socially moderate, however, the English, retaining their king and queen, perceived themselves a bastion of order in a world gone mad.

The Cartesian influence was profound. By the late nineteenth century, western thought suffered a metaphysical reversal. The diminution of importance of the God-given human body and mind was more and more supported by the expanding, skeptical scientific worldview. Our prescientific ancestors tended to consider the universe and everything that moved to be alive. Beings were exempted from life only when they stopped moving, only when the spirit left them by the natural magic trick of death. But now things had changed: in the new, scientific-mechanistic world of Galileo, Descartes, and Newton, the universe and all the beings in it were inanimate (Simmons 1996). The scientific puzzle moved from the mystery of death in a live cosmos to that of life in a dead one.

Inanimate matter had been rendered soul-less and dead by the mechanists. Even animate matter was soul-less and dead in the minds of strict Cartesians, who, with time, began losing their sway. But the universe is neither the dead mausoleum investigated by the Cartesian license nor an enchanted fairyland of invisible spirits.

We all, as citizens, scientists, scholars, or simply curious readers, are interested in life because we admire it from the inside. We feel life is something more than purely mechanical, and yet its freedom, if it exists, seems dubious to credit to a divine God. We do react to stimuli but we also seem to be able to think, to act, to choose. We seem far more than either Cartesian automata or entirely predictable Newtonian machines. Perhaps we are neither. But if we are more than Cartesian automata so, après Darwin, must be the rest of life. Otherwise we risk a great inconsistency.

This dualistic cultural inheritance presents a continuing challenge to science. Given the limited legacy of Cartesian dualism (mind/body, spirit/matter, life/nonlife), it may not be surprising that two of the most profound twentieth century rethinkers of life and its context share a biospheric perspective yet have diametrically opposed views. Russian scientist Vladimir Ivanovich Vernadsky (1863–1945) described organisms as he described minerals—calling them "living matter," whereas our friend and colleague, English scientist, James E. Lovelock (b. 1920), has problematized the Earth's surface in such a way that the entire biosphere, including rocks and air, may be regarded as alive.

Vernadsky portrayed living matter as a geological force—indeed, the greatest of all geological forces. Life moves and transforms matter across oceans and continents. Life, as flying phosphorus-rich seagulls, racing schools of mackerel, and sediment-churning polychaete worms, traverses the near-Earth environment, chemically transforming our planet's surface. Life, at the expense of the Sun's energy, has been largely responsible for the great differences between the third planet and her Solar System neighbors, specifically, the unusual oxygen-rich and carbon dioxide–poor atmosphere of the Earth relative to those of Venus and Mars.

In a tradition begun by Christian Gotfried Ehrenberg (1795–1876), Alexander von Humboldt (1769–1859), and other serious explorers before him, Vernadsky described what he called the "everywhereness of life." Living matter, he noticed, almost totally penetrated into, and consequently became involved in, superficially "inanimate" processes of weathering, water flow and wind circulation. While his contemporaries spoke of the animal, vegetable, and mineral kingdoms, Vernadsky analyzed the Earth's phenomena without labeling and classifying them into these categories. He eschewed preconceived notions of what was and was not alive. Perceiving life not as some abstract entity, with its philosophical, historical, and religious connotations, he referred only to "living matter." This freed him to combine as needed mineralogy, geology, and biology into a new discipline.

Impressed by the movement of machines in the World War I, what struck Vernadsky most was that the material of Earth's crust is packaged into myriad moving beings whose reproduction and growth depend on solar energy while they build and break down matter. Life, he saw, was a global phenomenon. Humans, for example, are accelerators of life's tendency to redistribute and concentrate the chemical elements of the Earth: iron, aluminum, oxygen, hydrogen, nitrogen, carbon, sulfur, phosphorus. Many other elements of Earth's crust are rapidly altered and mobilized by living beings, especially the two-legged, upright wanderers of our own species. People, he explained, have an amazing propensity to dig into, build up, move around, and in countless other ways alter the chemistry of the Earth's surface. We, in Vernadsky's view, represent a new phase in biogeochemical evolution (Lapo 1988).

Vernadsky contrasted gravity, which pulls material vertically toward the center of Earth, with life—growing, running, swimming, and flying against the gravitational force. Life, challenging gravity, moves matter horizontally across the surface. Vernadsky detailed the structure and distribution of aluminosilicates in the Earth's crust and was the first to recognize the importance of heat released from radioactivity to geological change.

But even a resolute materialist like Vernadsky found a place for mind. In Vernadsky's view a special thinking layer of organized matter, growing and changing the Earth's surface, is associated with humans and technology. To describe it he adopted the term nöosphere, from Greek *nöos,* mind. The term nöosphere itself was introduced by Edouard Le Roy, of the Collège de France. Vernadsky met Le Roy along with Pierre Teilhard de Chardin, the French paleontologist and Jesuit priest whose writings would later bring the idea of nöosphere—a conscious layer of life—to a wider audience in Paris in intellectual discussions in the 1920s. Teilhard's and Vernadsky's use of the term nöosphere, like their slants on evolution in general, differed. For Teilhard the nöosphere was the "human" planetary layer forming "outside and above the biosphere," while for Vernadsky the nöosphere referred to humanity and technology as an accelerating, yet integral, part of the planetary biosphere (Grinevald, 1988; Sagan 1990a, pp. 37–38).

Vernadsky distinguished himself from other theorizers by his staunch refusal to erect a special category for life. Life was far less a thing with properties than a happening, a process. Living beings in Vernadsky's writings are moving, chemically curious, but predictable forms of the common fluid, the liquid mineral H_2O we call water. Animated water, life in all its wetness, displays a power of movement exceeding that of limestone, silicate, and even air. It shapes Earth's surface. Emphasizing the continuity of watery life and rocks, such as that evident in coal or fossil limestone reefs, Vernadsky developed the idea, later elaborated by Lapo (1988), that apparently inert geological strata are "traces of bygone biospheres."

Vernadsky and Lovelock, global scientists both but from distinct vantage points, have articulated ways in which life is far more than a Cartesian automaton, or any other sort of machine. The worldviews of both, complementary and complex, were constructed from the usual scientific observations of minutiae. Many eluded them both in spite of their keen powers of observation and sharply focused careers.

Consider this: when offered a variety of food stuffs, bacteria, ciliates, mastigotes, and other swimming microbes make selections—they choose. Squirming forward on retractable pseudopods, *Amoeba proteus* finds *Tetrahymena* delectable but avoids *Copromonas*. *Paramecium* prefers to feed on small ciliates, but if starved for these and other protists, it reluctantly sweeps aeromonads and other bacteria into its cell mouth.

Although "merely" protoctists, foraminifera ("forams" for short) are one of the most diverse groups of fossil-forming small organisms. An astounding variety of magnificent shells are made by these complex single-celled beings, some forty thousand different species of which have evolved in the last 520 million years. Forams outside their shells resemble amoebas with a network of long, thin fusing and branching pseudopods. In certain forams, those called agglutinators, the shells are formed from handy starting materials from the seashore environment. Sand, chalk, sponge spicules, even other foram shells are patched together (agglutinated) to make the coverings. To appropriate their cell-shell homes, these forams place available particles from their surroundings together with an organic cement. Experiments have shown, however, that when presented with a hodgepodge of different particles, foraminifera make distinct "choices" based on shape and size—selecting, for example, small black over larger red glass beads (Lapo 1988). Some will bridle at the term "choice"; however, there seems to be no reliable criterion for distinguishing between the preferential activities of these beings and ourselves. Without brains or hands, these protists pick the building materials from which to construct their body-homes.

Smaller still, and far simpler in cell organization, chemotactic bacteria can sense chemical differences. These little bodies, just two microns (two millionths of a meter) long, swim toward sugar and away from acid. A chemotactic bacterium, without a nose, of course, can "smell" a difference in chemical concentration that is a mere one part in ten thousand more concentrated at one end of its body than at the other.

Biochemist and former editor-in-chief of the leading scientific journal, *Science Magazine,* Daniel Koshland, expressed the spiritual tendencies of the colon bacterium, *E. coli,* as follows:

> "Choice," "discrimination," "memory," "learning," "instinct," "judgment," and "adaptation" are words we normally identify with higher neural processes. Yet, in a sense, a bacterium can be said to have each of these properties . . . it would be unwise to conclude that the analogies are only semantic since there seem to be underlying relationships in molecular mechanism and biological function. For example, learning in . . . [animal] species involves long-term events and complex interactions, but certainly induced enzyme formation must be considered one of the more likely molecular devices for fixing some neuronal connections and eliminating others. The difference between instinct and learning then becomes a matter of time scale, not of principle. (Koshland 1992)

Many organisms too small to be seen without a microscope sense and avoid heat, move toward or away from light. Certain bacteria even detect magnetic fields. Some harbor magnets aligned in rows along the length of their tiny, rod-shaped bodies (Madigan et al. 1996). That bacteria are simply machines, with no sensation or consciousness, seems no more likely than Descartes' claim that dogs suffer no pain. We reject the idea that microbes act without any feeling. Although possible, the idea is ultimately solipsistic. (Solipsism is the idea that everything in the world, including other people, is the projection of one's own imagination.) Cells, alive, act as if they have feelings. Indigestible mold spores and certain bacteria are rejected by protists. Others are greedily ingested. At even the most primordial level, living seems to entail sensation, choosing, mind.

For nineteenth century men of science it was natural and expedient in the Cartesian tradition to invoke physical mechanisms to explain life. Life, as Newton's matter, consists of material bits that predictably respond to forces and obey natural laws. Like well-made clockwork, the world's mechanism was manufactured by the transcendent God, the creative God that constructed magnificent mathematical laws and then withdrew from his perfect and knowable creation.

Life, though, was not created in six days. Ushered in by the shocking contribution of Charles Darwin was the new view of evolution. God, if He existed, was Newton's God. No active interloper in human details, He was a geometer god who made the laws. Beneath the new mathematical god was the ancient residuum of the idea of a more active god.

The earlier view of life, the idea that life itself was evolving but only partially mechanical, was championed by Samuel Butler (1835–1902) an English novelist, painter, musician, and essayist who Gregory Bateson called "Darwin's most able critic" (Bateson, 1928). Butler took issue with the overly mechanistic views of Darwin. He suggested no grand design in nature, but recognized the continuity of life, to which he attributed millions of little purposes. Each purpose or objective was attributable to the cell or organism in its habitat.

To Newtonians, Darwinians, and others in the direct lineage of Descartes, choice or "free will" had been banished from a mechanistic universe. For Descartes, God, of course, has consciousness and people do as well, but only insofar as they communicated with God. When Darwin's painstaking work led to the conclusion that, like nonhuman life, people too had evolved (by the "mechanism" of natural selection), the consciousness that definitively separates Man from the Other suddenly became redundant. Butler, who argued against the special status of cogitating man, brought consciousness back into the discussion. He claimed that life is exuberant matter that chooses now and has chosen in the past. Over the eons choices made by some life-forms have produced more and different organisms, including the colonies of cells that stick together and become human individuals. Butler rejects a perfect immovable mathematical God; his deity is imperfect and dispersed. The properties of life, for Butler, lie in all life. "God" and life are one.

Butler's view that rejects any single, universal omnipotent architect appeals to us. Life is too shoddy a production, both physically and morally, to have been designed by some austere flawless Master. And yet life is more impressive and less predictable than any object whose nature can be accounted for solely by "forces" acting on it deterministically. Butler's godlike qualities of life on Earth include neither omniscience nor omnipotence. Perhaps, though, an argument could be made for the omnipresence of Earthly life.

In the form of myriad cells, from luminescent bacterium to lily-hopping frog, life is virtually everywhere on our third planet. All life is connected through Darwinian time and Vernadskian space. Evolution places us all in the stark but fascinating context of the cosmos. Although mystical powers may determine this cosmos, their existence is impossible to prove. The cosmos, more dazzling than any god of any particular religion, is enough for us. Life is existence's celebration. The features of purpose and determination that our culture tends to ascribe uniquely to people inhere intact in all of life. From life's minimal state as a tiny walled bacterial cell to its huge presence as a calf-nursing elephant or a montane rain forest, its exuberance, its sensible and sentient features, apply to all of its forms.

Butler's theory intrigues us. We agree that mind and body are not separate but part of the unified, functioning whole. Life, sensitive from its onset, has been capable of choice, of decision, of sensing and thinking from the beginning. Such "thoughts," both vague and clear, are physical. They are in the cells of our bodies and in those of other animals.

In comprehending these sentences, certain ink squiggles trigger associations, the electrochemical connections of the brain cells. Glucose is chemically altered by reaction of its components with oxygen, and its breakdown products, water and carbon dioxide, enter tiny blood vessels. Sodium and calcium ions, pumped out, traffic across a neuron's membranes. As you remember, nerve cells bolster their connections, new cell adhesion proteins form, and heat dissipates. Thought, like life, is matter and energy in flux; the body is its complement. Thinking and being are aspects of the same physical organization and its action.

If one accepts the fundamental continuity between body and mind, thought is essentially like all other physiology and behavior. Thinking, like excreting and ingesting, results from lively interactions of a being's chemistry. Even microbial "thinking" derives from cell hunger, movement, growth, association, programmed death, satisfaction, and other intrinsica of all life. Restrained but healthy former microbes find alliances to construct and behaviors to practice. If what is called "thought" results from such cell interactions, then perhaps communicating organisms, each themself thinking, can lead to a process greater than individual thought. This may be implicit in the Vernadskian notion of the nöosphere.

Two modern neuroscientists, Gerald Edelman (Scripps Institute, La Jolla, California) and William Calvin (University of Washington Medical School, Seattle), have each proffered concepts of mind. From Edelman's work and fertile imagination comes the phrase "neural Darwinism." Our brains, both would agree, become minds as they develop by rules of natural selection (Edelman, 1985). This concept ultimately may provide a physiological basis for Butler's insights. In the developing brain of a mammalian fetus, some 10^{12} neurons each become connected with one another in 10^4 ways. These cell-to-cell adhesions at the surface membranes of nerve cells are called synaptic densities. As brains mature, over 90 percent of their cells die. By programmed death and predictable protein synthesis, connections selectively atrophy or hypertrophy. Neural selection against possibilities, always dynamic, leads to choice and learning, as the remaining neuron interactions strengthen. Cell adhesion molecules synthesize and some new synaptic densities form and strengthen as nerve cells selectively adhere and as practice turns to habit. Selection is *against* most nerve cells and their connections, but it is nevertheless *for* a precious few of them. Of course, new work may reveal the physical basis of thought and imagination, but little doubt exists that selective cell death in a vast field of proliferating biochemical possibilities may apply to developing minds in the same manner it does to evolutionary change (Edelman, 1985).

Perhaps Descartes did not dare admit celebratory sensuality of life's exuberance. He negated that the will to live and grow emanating from all live beings, human and nonhuman, is declared by their simple presence. He ignored the existence of nonhuman sensuality. His legacy of denial has led to mechanistic unstated assumptions. Nearly all our scientific colleagues still seek "mechanisms" to "explain" living matter, and they expect laws to emerge amenable to mathematical analysis. We demur; we should shed Descartes' legacy that surrounds us still and replace it with a deeper understanding of life's sentience. In Butler's terms, it is time to put the life back into biology.

It will cost our culture until we recover our senses (Abram 1996) and return to the awareness that we must fully reject Cartesian anthropocentrism. We are interconnected not only to other people but to all other living beings on this planet's surface. The received view is that air travel, telephone lines, internet computer hookups, waterways, and fax machines connect only people. In fact, they connect, through us and others, all life. This incorrect view, symptomatic of residual Cartesian anthropocentrism, is biologically naive. Such rapidly communicating methods link not only us but our planetmates as well. For inhabitants of the urban ecosystem the connections are obvious, whether or not we are conscious of others—cockroaches, sparrows, tomato plants, pigeons, and pubic lice—they clearly enjoy habitat expansion as we "develop" the Earth for more people.

In retrospect, the Cartesian denial is exposed: we see Descartes' strategy as a Christian relic based on philosophical preconception rather than attentive observation. At this late date in our western heritage, we can shed our Cartesian mechanistic legacy at no risk to our scientific credibility. Consistency precludes Cartesianism. Either we are like other live organisms in that both we and they exert choices, or both we and they are mechanistic, deterministic beings whose choosing behavior is essentially illusory. The middle ground is philosophical quicksand. The great majority of the inhabitants of this third planet in our solar system are not humans nor have they ever been human. Indeed, scientists and others who continue to ignore the members of 10 to 30 million species, the other sentient beings, do so to their own great loss. Our planetmates whose existence Descartes and so many of his modern-day successors deny are communicants of the nonhuman splendor that, if we let them, can infuse our lives with joy and meaning.

14

What Narcissus Saw: The Oceanic "Eye"

Dorion Sagan

Sense-knowledge is the way the palm knows the elephant in the total pitch-dark. A palm can't know the whole animal at once. The Ocean has an eye. The foam-bubbles of phenomena see differently. We bump against each other, asleep in the bottom of our bodies' boats. We should try to wake up and look with the clear Eye of the water we float upon.[1]

Rumi (1207 to 1273)

Certain ideas take root in the psyches of their believers, coloring all their perceptions. Kierkegaard noticed that the less support an idea has, the more fervently it must be believed in, so that a totally preposterous idea requires absolute unflinching faith. This perverse balance helps account for the wide variety of beliefs—some "self-evident," others dogmatic—to which people attribute certainty. Abstract and profound ideas, like drawings with an unfinished quality, may contain a certain open-endedness that makes them appeal to many different people. As a virus reproduces itself by infiltrating the cell, so some notions would appear to latch onto the human imagination by being suggestive, self-contradictory, or symbolic. The great ideas leave an empty space in which believers recognize themselves. Fascinated with their own reflection, intrigued by the way a notion speaks directly to their own experience, the converted then proselytize to others on behalf of the idea and its amazing truth. Yet in reality they may be just passing a mirror and saying, "Look."

Whether true or not, subscription to certain philosophical notions puts hinges in the mind with which we can swing open the doors of perception. You may believe (with the Buddhist) that time, space, and individuality are illusions (perpetrated by samsara, the merry-go-round of regeneration). You may believe, as Nietzsche did, that everything you do will recur in the future an infinite number of times—or, conversely, like novelist Milan Kundera, that each act in the play of reality comes only once (floating away into "the unbearable lightness of being"). For the Nietzschean, each thought can have an immense significance: it will be repeated throughout eternity. For Nietzsche the thought of the eternal recurrence of the same raises the stakes of being, because any crisis or pain must be dealt with not just here and now, but forever. For the Kunderan, however, events and thoughts may have no special significance, and may appear meaningless, arbitrary, and random, slipping into the future never to return. Because Nietzsche's idea of the eternal recurrence and Kundera's notion of the lightness of being are diametrically opposed, they cannot both be continuously entertained. Yet each dramatically colors the perception of the true believer.

Again, if you hold that your life has been preordained by God, or that interacting waves and particles whose antecedents were present at the origin of the universe determine your every thought and action, you may be inclined to act less responsibly—and more nihilistically—than if you believe you have perfect freedom of choice. Nietzsche sought to prove his doctrine of cosmic rerun with reference to thermodynamics. Using the example of a Christian belief in eternal damnation, he indicated that an idea need not be true to exert a tremendous effect. The truth or falsity is not a prerequisite for ideational power, the ability of an idea to transform a consciousness. Whether there is heaven and hell or starry void, free will or predestination, reality recurring forever or never, there will be believers. The human mind abhors uncertainty; in the absence of tutelage, whatever philosophy is current will rush in to fill its vacuum. (The disturbances generated by French philosopher Jacques Derrida's tortuous prose result precisely from his "deconstructive" ploy of making scintillating suggestions but anticipating and defusing all would-be conclusions.[2])

People ascribe certainty to their beliefs, reality to their perceptions. From an evolutionary epistemological approach, existence is hindered, discourse impeded by the playful suspension of disbelief. So belief returns. Sheer survival requires that we arrive at and act upon conclusions, no matter how shoddily they are based. Doubt is a stranger to the human heart: to love or live we must believe—in something.

Let us explore now the perceptual implications of one powerfully riveting idea—that the Earth is alive. This is one of those doors that, swung open, reveals a changed world. Many in the past have believed that the whole universe is alive. A corollary of this is that the Earth's surface—our planet with its atmosphere, oceans, and lands—forms a giant global meta-organism. We can say that the Earth is alive. But what does that mean?

Imagine a child of a present or future culture inculcated from childhood to believe that the planetary surface formed a real extension of his or her person, a child whose language implicitly reinforced this connection to such a point that to him or her it would not even seem to be a connection but rather an equation. Such a person would make sense differently. Were nature not a dead mechanism but an immense "exoskeleton" (as the more limited exoskeleton or protective shell of a lobster is not only its house but part of its body), he would be less concerned by what we could not explain. And his perception of the organic would be altered. The arrangement of objects in his home, offhand comments by strangers, walks in the woods, cinema, and vivid dreams would all be linked to the organization of a living organism whose fullness of activity would be beyond his powers of comprehension. His ego no longer encapsulated by skin, he would experience the seas, sands, wind, and soil as numb parts of a body—just as feet and fingers, which, while open to tactile sensation, were yet incapable of speech and sight. The mountains between earth and air would seem to him anatomically placed, as "our" skeleton is between "our" bone marrow and flesh. Putting ourselves in his shoes, landscapes, from jungles and glaciers to deserts and glens, become body parts in a new anatomy, even if, from the limited perspective of that body's minute and only partially sentient parts, the global or geoanatomy remains largely unintelligible. The incomplete sensations of the planetary surface as a live body is no more a metaphor for ignorance than the idea of a skin-encapsulated anatomy. Take an ant crossing a bare human foot. Does it perceive it is touching a life-form? With this scale of differences, would it be able to distinguish a toenail from a rock or shell? Or, what can a bacterium living in the human gut conclude about the life-form that feeds it? Likewise, if we in our daily activities were meandering about upon the surfaces of a giant being, it need not be immediately apparent. Indeed, if one (not a positivist) believes in the necessity of metaphor as a system of explanation to "make known" our ignorance, then the image of a live planetary surface may itself—like Democritus theory of atoms—be enough to launch an entire new epoch of scientific research and individual action. Though inevitably we would reach the borders upon which such a program would be based, it is possible to imagine language itself embedding the structure of such an altered state of affairs and "making it real." The blue Earth itself would color all our perceptions.

Imagine someone from this culture picnicking. She believes her environment—and not just individual plants, animals, fungi, and microbes—to be part of her self. The grass on which she sits is a patch of tissue lining the inside of the superorganism of which she forms a part. The bark at her back, the dragonflies, birds, clouds, the moist air, and ants tickling her foot—all these sensations represent from her point of view not "her," but that which from our point of view we pedantically term the self-perception at one site of a modulated environment. Like the ants, "she" senses what is beyond "her." When "she" pulls her T-shirt over "her" knees, this is no longer human, but one locus of sensation within the kaleidoscopic entrails of a planet-sized photosynthesizing being.

The physiology is vast. The prostaglandins in people's bodies have many functions, ranging from ensuring the secretion of a protective stomach coating that prevents digestive acids from acting on the walls of the stomach to causing uterine contractions when ejaculated along with sperm in the male semen. So, too—looking at it now from an artificial position outside the physiology—the whole woman, by what she says, makes, and does, performs multiple functions within the global anatomy. A hormone is a biochemical produced in one part of the body that is transported through the circulatory system and causes biological reactions. The pituitary gland, at the base of the human brain, for example, stimulates sex hormones in the ovary and testes, causing pubic hair to grow.

As an animal in the Earth breathes, it affects the entire system. Water and atmosphere act as veins conveying matter and information within the geoanatomy. Indeed, the environment is so "metabolic" that minor actions may be amplified until they have major effects, while seemingly major effects may be diminished or negated. The ground is a live repository for metabolisms like the rings of tissue left in the wake of a growing tree. Treelike, the Earth grows, leaving behind it archaeological and paleobiological rings. The "woman" herself is part of a currently active geological stratum; and, far from dead, the air around the body that we, from habit, distinguish as human is fluid and thriving, part of an external circulatory system exploited by life as a whole.

To one raised to believe in the textbook notion of a static geology to which biology adapts, the young woman seems to be eating alone, surrounded only by plant life. Yet from "her" perspective the environment around "her" pulses with communicative life; "she" is at a busy intersection in the heart of nature. A part of nature, "she" is not simply "human" but an action within the self-sensing system of a transhuman being. (Indeed, language's personal pronouns falsify; they do not do justice to "her" but make "us" see "her" as a "thing" in a way that is in fact alien to "her." The self-extension to the environment has altered everything.)

Bearing in mind the idiosyncrasies of "her" perception, let us return to more ordinary language on the condition that without quotations "she" and "her" are still recollected as being imprisoned in such jail bar–like quotes by our more fractured word-biased views. With that said, there are things that to her would seem bona fide but remain quite mysterious from our worldview. Being called by a long-lost friend during the very moment she was thinking of him would not necessarily strike her as being what Carl Jung termed synchronicity—coincidences with such deep significance that one concludes they are more than mere coincidences. For her, strange coincidences come from her ignorance of the huge physiological system of which she forms a small part. Rain forests and seaside sludge she *sees* as vital organs, as inextricable to the biosphere as a brain or heart is to an animal; humans, however, she may construe as lucky beneficiaries of the establishment of superorganism, fluff like fur or skin that can be sloughed off without incurring major harm to the planetary entity as a whole.

Her uncle, a "geophysician," tells her that humanity has caused in the biosphere a physiological disturbance. "The Earth," he says (in so many words), "is oscillating between ice ages and interglacials; it has the global equivalent of malarial chills and fevers . . . Oil in the ground has become a gas in the atmosphere . . . tall tropical forests are being flattened into cattle . . . our vital organs are plugged with asphalt." Deserts, he tells her, are appearing like blotches on the fair face of nature. "But," he tells his niece, "we don't even know if these 'symptoms' are indicative of transformative growth—in which case we are experiencing normal 'growing pains'—or debilitating disease. Perhaps it is both, as in pregnancy, which, if encountered by a being as minute in relation to a pregnant woman as we are in relation to the Earth, might be misdiagnosed as the most bloated and dangerous of tumors."

Let us adopt the mask of metaphysical realism for a moment, and peer through the empty spaces, the (w)holes, which are all it has in the way of eyes. The example of the physical appearance of the Earth's altering due to the popularity of an idea—whether true or not—is an indication of what can happen when philosophy meets technology. But all this speaking of the Earth as if it had a "face" and a "fever"—as if it were some sort of comprehensible living entity—begs the question: Is the Earth really alive? And, if it is an "organism," what kind of organism is it? Can it think? Certainly the biosphere cannot be an animal, but only animal-like. And if the Earth does not resemble any other organism we know, have we reason to call it an organism at all?[3]

Scientific evidence for the idea that the Earth is alive abounds. The scientific formulation of the ancient idea goes by the name of the Gaia hypothesis. The brainchild of British atmospheric scientist James Lovelock, the Gaia hypothesis proposes that the properties of the atmosphere, sediments, and oceans are controlled "by and for" the biota, the sum of living beings. In its most elegant and attackable form, the hypothesis lends credence to the idea that the Earth—the global biota in its terrestrial environment—is a giant organism. Lovelock's Wiltshire neighbor, the novelist William Golding, suggested the name "Gaia" after the ancient Greek goddess of the Earth.

In part because Gaia resonates with a prescientific animism and can bring about a radically different way of perceiving reality, it has been the object of academic dismissal, suspicion, and, now, close scientific scrutiny. The British evolutionary biologist Richard Dawkins (a metaphysical realist if ever there was one) rejects Gaia, arguing that because there is no evidence for other planets with which the Earth has competed, natural selection could never have produced a superorganism. Without planetary competition, how could homeostatic or self-regulating properties on a global scale arise?

Nevertheless, evidence for organismlike monitoring of the planetary environment does exist. Reactive gases coexist in the atmosphere at levels totally unpredictable from physics and chemistry alone. Marine salinity and alkalinity levels seem actively maintained. Fossil evidence of liquid water and astronomic theory combine to reveal a picture in which the global mean temperature has remained at about 22°C (room temperature) for the last three billion years; and this constancy has occurred despite an increase in solar luminosity estimated to be about 40 percent. Under the Gaia hypothesis such anomalies are explained because the planetary environment has long ago been brought under control, modulated automatically or autonomically by the global aggregate of life-forms chemically altering one another and their habitats. All these anomalies suggest that life keeps planetary house, that the "inanimate" parts of the biosphere are in fact detachable parts of the biota's wide and protean body.

It was during the NASA Viking mission to Mars, with its quest to find life there, that Lovelock first thought to use his telescope, "like a microscope," pointing it toward the laboratory of the skies. As a thought experiment, he examined the red planet from Earth for signs of life. His discovery of an unremarkable absence of reactive gases produced under the control of life led him to conclude before the spacecraft landed that Mars was uninhabited (which, however, it became as soon as the outer human or Earthly limb of the sensing spacecraft landed).

To answer Dawkins and others, who required a mechanism of how a self-regulating biosphere could arise in the absence of other competing biospheres, Lovelock and his associate, Andrew Watson, developed computer models that simulated the ecology of a planet containing only light and dark daisies. These models show that neither populations of planets nor foreknowledge on the part of organisms is necessary to stabilize environmental factors on a global scale. Individual organisms grow selfishly when they can bring their environment under control merely by their activities, their continuous metabolic existence. In the model, Daisy World cools itself off despite the increasing brightness of a nearby sun. The cooling comes naturally as clumps of black-and-white daisies absorb and reflect heat as they grow within certain temperature levels that would be normal for them in any field.

On Earth, temperature modulation may be accomplished, at least in part, by coccolithophores, a form of marine plankton invisible to the naked eye but shockingly apparent in satellite images of the northeastern Atlantic Ocean. These tiny beings produce carbonate skeletons as well as a gas called dimethyl sulfide. The gas, pungently redolent of the sea itself, reacts with the air to produce sulfate particles that serve as nuclei for the formation of raindrops within marine stratus clouds. The plankton, then, by growing more vigorously in warmer weather, may enhance cloud cover over major sections of the Atlantic Ocean. But the enhanced density of the clouds leads to more reflection of solar radiation back into space so that the same plankton growing in warm weather cool the planet. In these sorts of ways, the subvisible but remotely sensible beings may be part of a global system of temperature control similar to the thermoregulation of a mammalian body. Without attributing consciousness or personifying them as minute members of some global board of climate control, the organisms may be seen to act together as part of a system of thermoregulation like the one that in us stabilizes our body temperature at approximately 98.6°F. Locally acting organisms apparently can affect the entire planetary environment in a way that builds up organism-like organization.

In a way it is not so surprising that individual action leads to the appearance or, indeed, the actuality of global controls. Academically, the disinclination to accept the possibility that the Earth regulates itself in the manner of a giant living being seems to have less to do with physical and chemical evidence—which lends itself to such interpretation—than it does with the status of modern evolutionary theory. Darwin considered the individual animal to be the unit of selection, but in the modern synthesis of neo-Darwinian theory, natural selection is seen as operating on genes as much as individuals, and evolution is mathematicized as the change in frequency of genes in populations consisting of individual animals. So, too, altruism in sociobiology is often seen as the tendency of genes to preserve themselves in their own and other gene-made organisms; biologists tend to dismiss the idea that groups above the level of the individual can be selected for, because they are not cohesive enough as units to die out or differentially reproduce. Evolutionary biologists lump arguments for selection of populations of organisms with the archaic oversimplification "for the good of the species"; they then perfunctorily dismiss such arguments as misguided, if not altogether disproved. Yet, as elegant as the mathematics combining Mendelian genetics and Darwinian theory sometimes may be, sociobiologists have a deep conceptual problem on their hands with their insistence that natural selection never works at a level above genes and the individual.

First of all, it is not clear what sociobiologists think an *individual* is; they fail to analyze or define this term, assuming that it is self-evident because of a parochial focus on the animal kingdom. The problem is that certain microscopic entities, cells called protists, which in the form of colonies must have given rise to the ancestors of all modern plants, animals, fungi (as well as "protoctists"—algae, slime molds, protozoans, and the like), did, and still do, assume the form of individuals. How, then, can evolution not work at a level above that of the individual if the very first animals were themselves multicellular collections—populations—of once-independent heterogeneous cells?

The animal body itself has evolved as a unit from a morass of individuals working simultaneously at different levels of integration. Sociobiologists and neo-Darwinian theorists disdain "group selection" because they don't have strong enough cases for its existence in populations of animals. But it may well be that, due to their large size and late appearance on the evolutionary stage, animals have not yet achieved the high level of group consolidation found in microbes. No matter how elegant the mathematics, dismissing "group selection" as an evolutionary mechanism requires dismissal of individual animals also, for the body of the academic itself provides a counterexample to the thesis that natural selection (if it "works" at all) never works on "groups." A person is a composite of cells.

Part of the problem here is the restrictive focus on animal evolution when animals themselves are the result of multigenome colonial evolution and represent only a special intermediary level of individuality midway between microbes and multianimal communities. But cells, animal species, and the biosphere all evolve concurrently. The first plants and animals began as amorphous groups of cells, later evolving into discretely organized and individuated communities of interacting cells. The evolution of individual cells led to the group of cells we recognize as the animal body. Groups of animals such as insect societies and planetary human culture begin to reach superorganism-like levels of identity and organization. The human body is itself a group that has differentially reproduced compared with other, more loosely connected collections of cells. That cells of human lung tissue can be grown in the laboratory long after the victim from which they were taken has died of cancer shows that the cells in our body are tightly regimented into tissue groups but still retain the tendency for independent propagation.

To be consistent, mainstream biology should explain how something called "natural selection" cannot be "acting" on groups of organisms if the animal "individual" is in a very deep sense also a "group" of organisms, namely, cells with their proposed histories and origins. Here we can accept, for the sake of argument, that several hundred million years ago multicellular assemblages began to evolve into the animal lineage. These groups left more offspring than their free-living unicellular relatives. Their very bodies contained the principle of social altruism, in which some cells specialized and curtailed their "selfish" tendency toward indefinite propagation for the "benefit" of the group to which they belong.

Working within the framework of evolutionary theory, we must accept the argument that "group selection" exists in the origin of animals—therefore, we must (again, within this framework) concede that evolution favors populations of individuals that act together to re-create individuality at ever higher levels. This somewhat freaky assertion calls into question the very usefulness of trying to isolate the units of natural selection: because of the articulation or community relations of living things, the differential reproduction of units at one level translates into the differential reproduction of units at a higher, more inclusive level. I anticipate that the mathematical theory of fractals, in which the same features are present in interlocking geometrical figures at various scales of analysis, may be useful in illustrating the principle of emergent identity in the series cell, multicellular organism, superorganismic society. In principle, the "animal-like" nature of the Earth can be considered fractally as resulting from the Malthusian dynamics of cells reproducing within a limited space.

If this essay's evolutionary understanding (qualified by placement under the rubric "metaphysical realism") is "right," it may be that the Earth itself represents the most dramatic example of emergent identity. As in Lovelock's Daisy World, the properties of global regulation on Earth result from the metabolic activities of the organisms that comprise our biosphere; on a less inclusive scale, "global" human consciousness and unconscious physiological control mechanisms can be traced to the synergistic effects of billions of former microbes acting locally to comprise the human body and its central nervous system. As an individual, the human body has evolved in isolation from other organisms, whereas the biosphere as a whole does not even have as clear a physical boundary separating it from the abiological cosmic environment, let alone from other organisms. In this sense, the biosphere is much less an individual than an animal. But the lack of biospheric individuality may be as artifactual as it is temporary. A superorganism as large as the Earth has not had the chance to evolve distinctive characters in isolation. Moreover, even if it were far more complex (anatomically, physiologically, and "psychologically") than a mammal, we may have difficulty understanding it precisely because of that complexity. In short, because the Earth is so huge, the Gaian organism may not be as apparent—or as consolidated—as a single animal. Over time, however, the Gaian superorganism can be expected to consolidate and become increasingly apparent; it may in the next centuries even become "obvious" to the majority of human beings.

Russell L. Schweickart, a NASA astronaut from 1963 to 1979, is an adviser on Biosphere II—a private-capital project to build a multimillion-cubic-foot biosphere near Tucson, Arizona, for about the price of a modern skyscraper. "The grand concept," he said recently at a meeting of those working on the project, "of birth from planet Earth into the cosmos—when, 1993, 1994, 2010, 2050, whenever—is a calling of the highest order. I want to pay a lot of respect to everyone associated with that grand vision for their courage to move ahead with this in the face of the unknowns which make the lunar landing look like a child's play toy. There were a lot of complexities there, but we were dealing with resistors, transistors, and optical systems which were very well understood. Now we're wrestling with the real question: that natural process of reproduction of this grand organism called Gaia. And that's what all the practice has been about." Many astronauts space-walking or gazing at the Earth report on the tremendous transformative power of the experience. That looking at the Earth from space could so totally change a person's consciousness suggests that the experience has not yet fully registered upon the body politic. People such as Schweickart who have seen the Earth from "outside" in space may be more prepared to accept the unorthodox idea that the biosphere is not only a living entity, but about to reproduce—as many individuals—and, indeed, many cellular groups arranged into individuals—have done "before."[4]

However, at the Cathedral of St. John the Divine in New York City in 1987, the thoughtful plant geneticist Wes Jackson protested the idea of Gaia on the basis of Gaia's "infertility." Jackson claimed there is no way the Earth could be an organism because all known organisms, from microscopic amoebae to whales, reproduce. Because the Earth has no "kids," it cannot be a real organism. It is only a metaphor, he said—and it may even be a bad one. According to Jackson, we do not even know what the Earth is. ("What is God?" he asked provocatively, suggesting the questions were similar.)

In a way I do agree with Jackson. The Earth seems indefatigable in its capacity to make us wonder about its true nature. Yet I had become convinced that the Earth is, in a sense, reproducing before ever hearing Jackson raise this counter-Gaian argument. The reason for my conviction that the biosphere is on the verge of reproduction has to do with two things: (1) The growing number of scientists and engineers involved in designing, for a variety of reasons, closed or self-sufficient ecosystems in which people or aggregates of life can live; and (2) my assumption that humanity is not special but part of nature. For, if we are part of the Earth, so is our technology, and it is through technology that controlled environments bearing plants, human beings, animals, and microbes will soon be built in preparation for space travel and colonization. In space these dwellings will have to be sealed in glass and metal or other materials so that life will be protected inside them. Such material isolation gives the recycling systems discrete physical boundaries—one of the best indications of true biological "individuality." Thus, the bordered living assemblages necessary for long-term space travel and planetary settlement by their very nature bear a resemblance to biological individuals at a new, higher scale of analysis. They look startlingly like tiny immature "earths"—the biospheric offspring Jackson claims must exist for the Earth to be a true organism.

We can trace a progression in size in these human-made containers of recycling life. Clair Folsome of the University of Hawaii has kept communities of bacteria enclosed in glass, and they have remained healthy and productive since 1967. There is no reason to think they may not be immortal despite being materially isolated from the global ecosystem. Similarly, Joseph Hansen of NASA has developed a series of experimental desktop biospheres consisting of several shrimp, algae, and other organisms in sealed orbs half filled with marine water. These last for years, and in some crystal balls the hardy animals have even reproduced. On a still larger scale, private and governmental space administrations in the Soviet Union, the United States, Japan, and other countries are developing the art of creating materially closed perpetually recycling ecosystems. Crucial not only to space travel and colonization, these miniaturized ecosystems could also protect endangered species, maintaining air, water, and food supplies, and allow, in the long term, the possibility of social, cultural, and biological quasi-independence on the ever more crowded and homogenized Earth.

If successful, controlled ecosystems will carry a powerful educational message about the need for cooperation of people with one another as well as with the other species that support the global habitat. And, if perpetually recycling ecosystems can be erected and maintained, a whole new scientific discipline may arise from the possibility, for the first time ever, of comparing "parent" and "offspring" biospheres. Former astronaut and physicist Joseph Allen points out that the quantum mechanical revolution that so marks modern physics derives from the comparison by Niels Bohr of helium and hydrogen nuclei: having more than a single biosphere to observe may likewise revolutionize biology.

Communication established between two semiautonomous biospheres may resemble in emotional impact the relationship of a mother or father to a daughter or son. Yet the "children" will teach: the safe modelling of potential ecological disasters within a new biosphere may provide dramatic warnings and even perhaps usable information on how to ward off the environmental catastrophes—from acid rain to pesticide contamination of foods—that potentially await us. New biospheres thus may serve as living whole-Earth laboratories or "control worlds," inaugurating differential reproduction on the largest scale yet.

The importance of the development within the biosphere of such enclosed ecosystems cannot be overestimated. Whether or not individual, national, or private venture capital models succeed or fail is irrelevant. What we see, rather, is the tendency of the Earth (or Gaia, or the biosphere) to re-create itself in miniature. Because we, from an evolutionary perspective, are natural and not supernatural creatures, the Earth is, through the high-tech expedient of modern world civilization, re-creating versions of the global ecosystem on a smaller scale. To some the view of an Earth biospherically splintered into semiautonomous ecosystems would be a technocratic blunder equivalent to the formation of a planetary Disneyland. But even if the Earth is saved as a single biosphere, such materially closed ecosystem technology will be necessary for extended human voyages into space or the settlement of off-world sites for emigration or long-term exploration. Thus, we do seem to be caught in precisely that historical moment when the Earth is begetting its first, tentative batch of offspring. That humankind is currently the only tenable midwife for Gaian reproductive expansion is a gauge of our possible evolutionary longevity and importance—provided that the violently phallic technology that promises to carry life starward does not destroy its makers first.

The "Gaia hypothesis" is at once revolutionary science and an ancient worldview, with the power to spur not only scientific research but religious debate. If we take it to its logical extremes, it says not only that the Earth is alive but that it is on the verge of producing offspring. From a strict neo-Darwinian perspective, this may be a mystery, for how can a giant organism suddenly appear *ex nihilo* and then just start reproducing?

Yet, from a broader philosophical perspective, the reproduction of the biosphere makes perfect sense. We are animals whose reproduction is an elaboration of the reproductive efforts of cells: the organismic and reproductive antics of the Earth have not appeared in an evolutionary vacuum. Gaia's weak, immature attempts at "seed" formation and reproduction result from the sheer numbers of organisms reproducing at the Earth's surface. What before occurred in the living microcosm of cells is now transpiring in the larger world of animal communities. The Malthusian tendency to increase exponentially in a limited space beyond the resource base apparently may account for more than just the evolution of new species: it leads also to the appearance of individuality at ever greater levels and scales of analysis.

This essay broaches what might be termed a Nietzschean ecology. That is, it attempts to hint at an art of biology whose unveiling may be as important as biology itself, at least as regards biological understanding as it applies to the "individual" in his, her, or its restless search for meaning. (Academicians, guard your territory!) The appearance of closed "offspring" biospheres from the original open biosphere repeats or continues the process by which "individual" plants, fungi, and animals appeared from communities of microbes. As the folk saying goes, Plus ça change, plus c'est la même chose: the more things change, the more they stay the same. As Nietzsche scrawled in one of his notebooks: "Everything becomes and recurs—forever!"

As we have seen, even a false idea may color our views of the world, and where there is a chance of changing the world, there is the chance of bettering it. Gaia is such an idea, yet one with the added punch that it may be proved true. (Oscar Wilde observed that "Even true things may be proved.") It was interesting to watch the debate develop in March 1988 as the Geophysical Union met in San Diego to "test" for the first time among polite scientific society the general validity of Lovelock's hypothesis. In fact, as everyone saw in the epistemology session (and any sort of philosophical discussions is rare at scientific meetings these days), it was fairly easy to show that Gaia is not, strictly speaking, testable. Whether one took him to be a very naïve epistemologist or an extremely sophisticated sophist, James W. Kirchner was correct when he compared the postulate that the Earth is alive to Hamlet's theory that "all the world is a stage." There is no way of proving or disproving such general notions. Kirchner pointed out that Gaia is not a valid hypothesis because it does not say something we can verify or falsify, something such as (Kirchner's example), "There are footlights at the edge of the world."

In fact, Gaia is not a hypothesis. It is, like evolution, a metaphysical research program. The idea that the Earth is alive is extremely fruitful, able to suggest many scientific models and lines of inquiry. Yet ultimately it is unprovable, a matter, at bottom, of faith. It is, after all, a worldview. What positivists miss in their attack on Gaia is that they are also up to their necks in metaphor and metaphysics. There is no avoiding metaphor and metaphysics. When worldviews collide, weak ones are obliterated in the encounter. In my view, what happened at this conference was an encounter of worldviews. But it was no head-on collision. Rather, the old panbiotic or animistic worldview (at the center of the "Gaia hypothesis") sneaked its way into mainstream discussion. In a direct confrontation, the Gaian worldview would have been eaten alive by the prevailing worldview (atomistic science and its Platonic "laws" as absolute reality). But by disguising itself as a testable hypothesis, Gaia was smuggled into a prestigious scientific discussion. We would never expect the discussants at a serious scientific conference to bring up as the main question their own view of reality. But this is, in effect, what happened. Like the Trojan horse, the Gaian worldview sneaked past the well-armed guards of metaphysical realism ("science") by disguising itself as a hypothesis. And now the worldview Gaia, having lodged itself inside the worldview metaphysical realism, is impossible to extract without damage to both. Our entire conception of life and its environment is being called into question. What is life? Technology? The environment?

Perhaps another Greek myth, because it has not strayed onto the dangerous battlefield of truth, better sums up the present philosophical situation: Once Narcissus stood and eyed the still waves that reflected his own image. He had never seen himself before. He became infatuated. And now we gaze in the looking glass of satellite imaging technology. Again we see the water. Again . . . but what is "ourselves"? And who—or what—is *this body*?

Notes

1. The quotation is from *We Are Three, New Rumi Translations,* by Coleman Barks. (Athens, Georgia: Maypop Books, 1987). Jalal ad-Din ar-Rumi Rumi (1207 to 1273) was a Sufi love mystic who wildly spun around as he delivered his mu-

sical verses, which were transcribed by assistants. He was the first "whirling Dervish," and it is claimed that his poetry read aloud in the Persian original is so musical it sends listeners into a trance by its aural quality alone.

2. The technique of leading people in certain directions and then "pulling the rug out from under them" resembles the method of the sleight-of-hand artist. Both the deconstructionist and the magician present signs that are typically organized or mentally ordered into a narrative of events. A difference is that, whereas the exponent of legerdemain presents approximately the minimal number of sensory stimuli to arrive prematurely and mistakenly at a certain impression of reality, and this impression is then revealed to be "wrong" (that is, clearly only an image), after the performance of the "trick," the deconstructionist uses language as the medium for the presentation of mirages that are more or less continuous; the deconstructionist does not entertain like the magician with a series of discrete and contained surprises, but reveals rather that the attribution of "finished" images and mirages from unfinished signs and stimuli proceeds unceasingly. The difficulty with deconstruction is that it shows offstage, whereas traditional magic shows onstage. But this difficulty has to do with the "broadening" of the stage, the spilling over of theater into the realms of everyday life: It cannot be gotten rid of by dismissing as unreadable all deconstructive prose. Clearly, the conclusions arrived at through the use of language, and especially of "language with ordinary words," may be as bogus as the conclusions arrived at through the motions of a sleight-of-hand artist—and especially one manipulating not apparatus onstage (where the theatrical element is expected), but small ordinary objects such as cards and coins in the home space so normally above suspicion.

3. We say all this keeping in mind that our language—and our science—bears within it its own deeply embedded and usually unexamined set of metaphysical assumptions. Derrida has unequivocally shown this. Just as Nietzsche did not need thermodynamics to be affected by the idea of eternal recurrence, one need not justify the culturally marginal notion of a living Earth by reference to or with the sanction of a cultural mainstream, a tradition of knowledge not at home with such ideas. Nonetheless, the possibility of scientific sanction indicates the reality of the approach of this notion into the mainstream.

4. Part of the problem with the whole concept of evolution—and all narrative "explanations"—may be the unexamined reliance upon the unprovable assumption of linear time, a logocentric assumption. The verb tenses of languages perpetuate the assumption of temporality. The relation of language to the bias of linear time is here dubbed "chronic." In fact, the relationship of lifeforms may be better seen as four- or multidimensional, in which case the evolutionary unfolding in linear time is better seen as only a "slice" through true spacetime.

15

A Good Four-Letter Word

Dorion Sagan
and Lynn Margulis

The Gaia view ranges from the dubious poetic conception of a reproducing Earth to the later, more modest, and therefore more scientifically acceptable, formulations of Margulis and Lovelock. We have no wish to homogenize religious and scientific views of what Gaia might mean. Nevertheless, it has become clear that no simple scientific or even well-intended metascientific statement will encompass Gaia in its richness or define it in its fullness. A rigorous scientific definition of Gaia is a first step. But Gaia, a way of knowing, is not simply one worldview among others (an interdisciplinary scientific approach combining astronomy, atmospheric chemistry, biology, biochemistry, remote-sensing technology, and thermodynamics). Gaia has been appropriated and is now finding social support outside the realm of science proper. A cursory sociological study would reveal that it has been attacked not only for being unscientific and "untestable," but as antihuman polemics, green politics, industrial apologetics, and even as non-Christian ecological "satanism." Such a diversity of enemies indicates that the power of Gaia goes beyond science; it vindicates Lovelock's

FIGURE 15.1. Gaia: the hollow spherical fringe, 20 kilometers high, at the Earth's surface. South polar projection drawing by Christie Lyons.

intuition that his idea was so important he needed to give it a "good four-letter word." Lovelock, before calling it Gaia, referred to the concept as "life as seen through the atmosphere," or "a cybernetic planetary system with homeostatic tendencies." The name "Gaia" was gladly offered by Lovelock's English country neighbor, the novelist William Golding. In Greek mythology, Gaia, a personification of the Earth, was mother of the Titans. The alternative spelling, Gaea, had already taken root in "scientific" English words such as geometry, geology, geography, and Pangeae.

Gaia science is no more exhausted by the negative approaches of its critics than it is by the gee-whiz adulation of its adherents. Outside science, Gaia has become the darling of the green or ecology movement—

a kind of full-bodied intellectual cognate to the widely disseminated but strictly visual image of the "whole Earth" (sic) from space. Essayist-physician Lewis Thomas has successfully identified the blue-and-white satellite photographs of Earth with the Gaian view by claiming that the space perspective gives instant recognition that the Earth, clearly a living being, is more than just another planet. Contrasting with the dead-as-old-bone moon, the Earth is the only "exuberant thing in this part of the cosmos" with the "organized, self-contained look of a live creature, full of information, marvellously skilled in handling the sun." The meditation upon the Earth as a living being, part of a philosophical monism that regards the entire cosmos as, in a sense, alive, reawakens premodern but not entirely prescientific sentiments. The depth and breadth of Gaia theory offers an excellent opportunity for historians and philosophers of science to chronicle a scientific revolution in the making; it is a striking reminder that science in the initial spasms of its birth cannot always be rigorously distinguished from prescientific, magical, or pseudoscientific systems of thought.

Science, according to standard anthropological thought, evolves out of religion just as religion grows out of magic. The 19th century contemporary and critic of Darwin, Samuel Butler, warned that the scientist is augur, medicine man, the priest in his most modern guise and, while useful, requires us to watch him very closely. From this perspective, science is only religion that is not recognized as such: it is a system of beliefs still being actively reworked, not yet settled into the ground of primordial assumptions, a system of beliefs not yet learned so perfectly and repeated so often it has been forgotten about and entered the realm of unconsciousness.

Gaia as a potential grand unified theory of biology is ripe for study from a wide range of disciplines, not just intrascientific or standard Anglo-American philosophy of science. Such disciplines might include deconstructive or literary critical approaches, phenomenology, and even psychoanalytic and feminist approaches (what does it mean to inhabit a de-deified but still immanent "female" Earth?).

Metascientific Gaia

Several billion trading, settling, warring, citifying, reproducing, largely technological human beings inhabit the surface of planet Earth. It appears that we must, to survive in present numbers, adopt some version of the Gaia hypothesis: only science has the status as a belief system necessary to induce human behavioral changes on a global scale. Gaia science operates out of the metaphor that the planet is not just a home (Greek *oikos*, the root of ecology) but a body. A body differs from an inert place in that it is sentient and reactive; indeed, whereas the difference between referring to Earth as a "living planet" and a planet that is "alive" may seem minor, debating it has caused dissension and distress among biologists and geologists.

Admission of a live Earth leaps toward the scientifically forbidden territory of animism—of personification, anthropomorphism, and narcissistic magical beliefs that have long been overcome by the progress of "objective" science. Gaia has narrowed the space, or expanded the continuum between life and nonlife, the organic and inorganic, the animate and inanimate. In Gaia theory, for example, the atmosphere becomes part of the biosphere, a sort of global circulatory system; the microbial rich soils are no longer inert substrata but rather living tissues at the planetary surface. Bolder still, the living biosphere provisionally encompasses not only the atmosphere and its clouds, but plate tectonics, the regulation of ocean salinity, and animal-like planetary thermostasis over three billion years. This new-found attention to our surroundings entails a change of values, giving our technical civilization a chance to recognize, alter, and even revert human impact upon the environment.

Scientific Gaia

Seeking remote detection of life on other planets, Lovelock developed the Gaia hypothesis because he recognized that his method offered startling insights into the nature of life here on Earth. Lovelock realized that one does not have to visit Mars to tell that it is lifeless. Chemistry and physics alone adequately model the Martian environment. Yet the same physics and chemistry fail to describe the Earth's atmosphere. Although invisible to us, our atmosphere is so out of chemical equilibrium that it would be easy for a hypothetical Lovelock doppelgänger to intuit life on Earth from Mars. Gases that should quickly, explosively react with each other, such as oxygen and nitrogen, or methane and hydrogen, maintain stable concentrations. If life evolves on the surface of a planet, Lovelock argues, because it must be in a continuous state of material exchange with the gases, liquids, and solids of its environment, life must become planetary in scale. Atmosphere and oceans are the conduits for the system embedded in them. Environment and organism then form not just a house but a body.

The "textbook" view of life is that it comprises millions of independent beings that inhabit inanimate surroundings. In Gaia theory, by contrast, the air and the ground are not independent inorganic chemicals; rather, the sediments and atmosphere are part of an entire living system. From the Gaian perspective, human air pollution on a global scale perturbs not just the atmosphere but affects all the biota. Feedback between the biological and geological realms is so intense that considering one in isolation from the other is an exercise in frustration. Meteorologists still state that they do not even consider the chemistry or the biology of the Earth. Atmospheric chemists claim meteorology lies beyond their territory. Neither science refers much to biology. Such territoriality is inimical to understanding the planetary body.

Whereas Gaia has been labelled untestable and "unscientific," the hypothesis nonetheless has spurred many lines of research into global biogeochemical processes. The important climatic role of dimethyl sulphide, a compound emitted by certain species of algae that may be involved in planetary temperature regulation, would doubtless not have been discovered without the impetus of a Gaian perspective. Perhaps Gaia theory as a whole is not open to validation or disproof but is rather, as philosopher Karl Popper said of Darwinian evolution, "a metaphysical research programme".

Phenomenological Gaia

Attempts to grapple with the implications of the Gaia hypothesis cannot bypass a phenomenological approach. What does it mean to inhabit a living organism? How are experiences, formerly attributed to mechanical causality or movement in an inert environment, described once we recognize the omnipresent responsiveness of the nonhuman environment?

Mythological Gaia

Like psychoanalysis, Gaia theory renews the dialogue of science (*logos*) with myth (*mythos*). Much of the sociocultural force Gaia attains is owed to the mythologies that still speak meaningfully to us. In our age, characterized by nihilism, monotheism, and "God is dead," Gaia is optimistic and positive. Its success must be attributed, at least in part, to its label. Subtler connections supplement this obvious link of Gaia theory to Greek intellectual bedrock. The "feminization" of a patriarchal god into an Earth mother, from a sky-based deity to an atmospherically veiled yet measurable entity: these are in need of rigorous mythological analysis. Trans-ecologically, the Gaian recognition of the Earth as no mere place or home but as a body startlingly recalls Narcissus, who, peering into the water, fell in love at first sight with his own never-seen-before reflection. A similar phenomenon occurred with humanity's first hypnotic views of the Earth outlined by black space. Narcissus drowned; will Gaia's focus on the Earth body lead to a similar fate?

16

THE BIOTA AND GAIA

One Hundred and Fifty Years of Support for Environmental Sciences

LYNN MARGULIS
AND GREGORY HINKLE

Though often held up as "pure" and "independent," our science is indelibly embedded in the language, religion, and social organization of our past. As a society we define our scientific goals through spending priorities revealed, for instance, in the funding of grants. When we compare the state of funding for environmental science in the mid-nineteenth century (and the underlying aims of the granting institutions at that time), we find remarkable, albeit disturbing, similarities in the funding of science in the United States late in the twentieth century.

The Hypothesis

The Gaia hypothesis is a recombinant derived from the lively imagination of James E. Lovelock and the National Aeronautics and Space Administration's (NASA)'s search for life on Mars (Lovelock 1988). Indeed, without planetary biology, the new science comparing Earth with its nearest neighbors, Mars and Venus, Lovelock would likely never have invented Gaia. As an answer to our more vocal critics, a more subversive and perhaps more appropriate title for this chapter would be "Gaia, Greed, and Glory" with the subtitle "Grants and Gaia." The chapter is divided into two sections. First, we applaud Jim Lovelock and his recognition that the "Earth is alive." We reaffirm Gaia as a creative, scientifically productive hypothesis. However, we express the Gaia hypothesis in an alternate way and assume the role of unofficial spokespeople for the silent majority of life on Earth, the microbes. Second, we discuss the past 150 years or so of financial support for "Gaian studies."

Rather than state "Earth is alive," a phrase that confuses many and offends others, we prefer to say that Gaia is a hypothesis about the planet Earth, its surface sediments, and its atmosphere. We describe the Gaia hypothesis as follows: the Earth's surface is anomalous with respect to its flanking planets, Mars and Venus. The surface conditions of Mars and Venus can be adequately comprehended by physics and chemistry. With respect to certain attributes, the Earth is, from the vantage of physics and chemistry alone, inexplicable. The Earth's physical and chemical anomalies, given new concrete knowledge about Mars and Venus, have become obvious. They include the presence of highly reactive gases (including oxygen, hydrogen, and methane) coexisting for long times in the atmosphere, the stability of the Earth's temperature (that is, the long-term presence of liquid water) in the face of increasing solar luminosity, and the relative alkalinity of the oceans. The pH of the Earth is anomalously high. When compared with its barren neighbors, Earth's surface chemistry is aberrant with respect to its reactive gases, its temperature, and its alkalinity. These discordant chemical and physical conditions have been maintained over geologic periods of time. Lovelock's concept, with which we entirely agree, is that the biota (that is, the sum of all the live organisms at any given time), interacting with the surface materials of the planet, maintains these particular anomalies of temperature, chemical composition, and alkalinity. Therefore, to understand the Earth's surface we must understand the biota and its properties; we can no longer rely only on physical sciences for a description of the planet.

We have today thirty million distinguishable types of organisms (this may underestimate the number of living species by a factor of 100 or more). Each organism interacts with its local environment. Each organism requires the activities of other organisms, not only for obtaining water, minerals, nutrients, and food, but also for removing its solid, liquid, and gaseous wastes. No individual, no matter how large or small, lives in a vacuum, nor can any feed off its own wastes. Indeed, every organism interacts with one or another gas; that is, each takes up one or more kinds of gas from the atmosphere and each emits a different quality and quantity of gas to the atmosphere. All metabolizing organisms exchange gases at all times. Many types, especially bacteria, protists, and plants, interact directly with surface rocks and minerals: many soil or mud-dwelling animals such as earthworms or brachiopods produce, remove, or dramatically change the properties of sediment. The Gaia hypothesis forces us to consider the cumulative, that is, global, effects of these local phenomena. How can a diverse biota, for such a long period of time, maintain within certain limits the temperature, the reactive gas composition, and the acidity and alkalinity of the Earth's surface? Though much remains to be done. Gaian mechanisms of regulation are now being recognized and studied (Charlson et al. 1987; Barlow and Volk 1990). The greatest hindrance to the study of Gaia is the fragmentation of science into a proliferating number of disciplines, departments, buildings, journals, and societies. Of course, physics, especially geophysics, chemistry, atmospheric sciences, astronomy, engineering, software and instrument development, and still other scientific fields, are absolutely essential to the study of the planets. While we have long recognized that these disciplines are required for the study of Mars and Venus, the Gaia hypothesis forces us to conclude that to study the Earth, the results and insights from all of the subfields of biology, especially microbiology, are required. The conclusion is inescapable: geophysicists and atmospheric scientists must study biology and biologists must know something of geophysics and atmospheric science. For too long, we have had atmospheric chemists wondering, Where does all that methane come from?, and biologists ignorant of where all that methane goes.

Earth-based and space-borne studies of cloud-covered Venus have described a very dry planet surrounded by a CO_2-rich atmosphere (>95 percent) with dense mists of sulfuric acid. Ignoring the claims of American space scientists that the surface of Venus was too harsh an environment for the cameras, detectors, and other equipment of a soft-landing remote sensor, the Soviets have successfully landed at least sixteen *Venera* spacecraft on the surface of Venus. Their *Venera* results have confirmed what was inferred from telescopic, ground-based astronomical studies: Venus, with its dry, CO_2 atmosphere, containing sulfate-particle–induced clouds, has an oxidized surface. When we train our analyzers on Mars we are again impressed by the extraordinary dryness of the red planet and the >95 percent relative concentration of CO_2 in the Martian atmosphere. (Despite humanity's industrious efforts, the Earth's concentration of atmospheric CO_2 is still <0.04 percent). And whereas the Earth's surface has water-deposited sedimentary rocks over about 70 percent of its surface, Mars' surface has very little evidence for water-deposited sediments, except for some exceedingly ancient dry "river" beds. Indeed, the loose regolith over most of the surface of Mars is presumably the product of meteoritic impact craters and volcanic debris. From the *Viking* landers and orbiters in 1975 and 1976, the results from ground-based telescope studies have been confirmed: Mars, like Venus, is a dry, CO_2-rich, thoroughly oxidized world. Both planets lack organic matter, they are dead—or they never harbored life at all.

If we look at the Earth with the same sort of space-borne, remote-sensing technologies, we are first aware of the great abundance of water on the surface. We can carefully choose places like the island of Hawaii, however, where a regolith of volcanic debris dominates and the evidence for life is not at first glance obvious (Figure 16.1). Were only a simple camera lowered from space onto such bleak Earth environs, there would be no palpable evidence for the presence of life. Knowing this, Lovelock, in his

FIGURE 16.1. Hawaiian scene: the Earth without life? (Photo courtesy of Carmen Aguilar-Diaz.)

first glimpse of Gaia, recognized that through its exchange of gases, the biota—regardless of its visible structure—leaves an imprint on the composition of the atmosphere. The cumulative effect of the biota's gas exchanges are planetary-scale chemical anomalies. As Lovelock tried to explain to his colleagues, these anomalies are the signature of life (Hitchcock and Lovelock 1967; Lovelock 1972). Whereas Venus and Mars presumably were formed from stellar media of very similar composition as the Earth, both are now dry, CO_2 planets with only trace quantities of water and oxygen in their atmospheres. The atmospheres of both Venus and Mars contain nitrogen as N_2 in relative gas concentrations of <3 percent (Figure 16.2). Venus and Mars, as Lovelock is fond of saying, have atmospheres composed of "spent" gases. The gases (CO, CO_2, N_2) are "spent" in the sense that they no longer react with each other because they already have reacted. The atmospheres of Mars and Venus can be modeled without any obnoxious intrusion by biology. When the same analysis is extended to the Earth, many anomalies surface: Earth's atmosphere contains 20 percent oxygen (O_2), an explosively reactive compound (see Figure 16.2). The 80 percent nitrogen gas (N_2) in the Earth's atmosphere forms a reactive mixture with the 20 percent oxygen; when sparked by lightning,

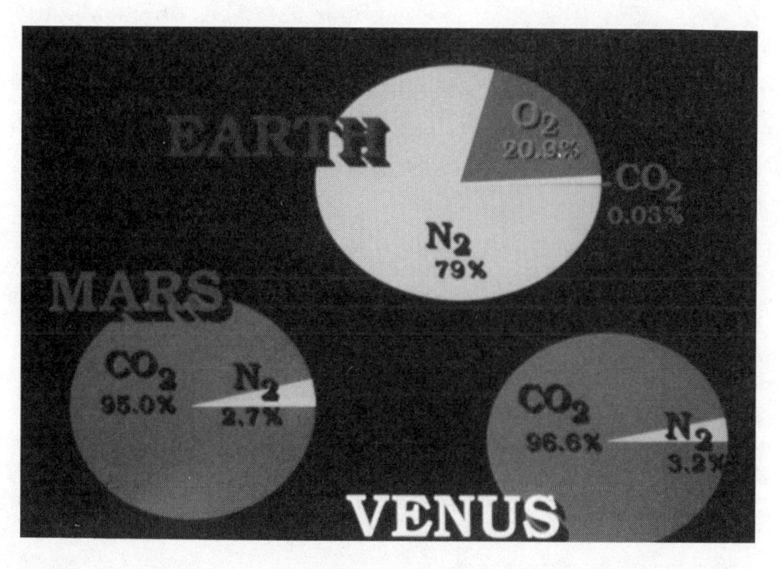

FIGURE 16.2. Earth, Venus, and Mars: relative concentrations of atmospheric gases. A 35mm Express drawing in collaboration with Jeremy Sagan.

oxygen and nitrogen form the stable ion nitrate. Because there are many lightning storms at any one time, abundant opportunity for the atmospheric formation of nitrate (NO_2) exists on Earth. Yet the molecular species N_2 and O_2 prevail (Table 16.1). Furthermore, these and many other chemical anomalies have existed in Earth's atmosphere for at least a billion years. Not only is the Earth's atmosphere an inherently reactive mixture, but the CO_2 concentration is astonishingly low (<0.04 percent). As CO_2 has been pumped out of the Earth's atmosphere and into calcium carbonate, the concentration of gaseous CO_2 has decreased, probably by several orders of magnitude since the formation of the planet. Calcium and magnesium carbonates form fossiliferous limestone and dolomite, the majority of which were generated from the remains of shelly organisms. Unlike Mars and Venus, Earth has an atmosphere inexplicable without an awareness of the biological production and sequestration of such mobile elements as hydrogen, carbon, oxygen, and nitrogen.

Earth's atmosphere can be explained; only a certain set of elements are distributed in a chemically bizarre fashion. The set includes carbon, hydrogen, sulfur, phosphorus, nitrogen, and others, including many metals. The explanation is related to the fact that these elements strongly interact with life. That the Earth's atmosphere is not explicable by chemistry and physics alone is one of Lovelock's great insights. Given 20% oxygen in our atmosphere, there are what first appear to be inane quantities of other gases: too much nitrogen, far too much methane (by more than thirty-six orders of magnitude), far too much nitrous oxide, far too much ammonia, far too much methyl iodide, and far too much hydrogen (Table 16.2). We could continue listing these; at last count over forty biogenic gases have been measured in the Earth's atmosphere (Levine 1989). Because all of

Table 16.1. Planetary Atmospheres

	Venus	Earth	Mars
Carbon dioxide, CO_2 (%)	98	0.03	95
Nitrogen, N_2 (%)	1.7 (ve)	79	2.7 (vi)
Oxygen, O_2 (%)	Tr (ve)	21	0.13 (vi)
Methane, CH_4 (%)	none	0.0000015	none
Water, H_2O (m*)	0.003	3000	0.00001
Pressure (atm)	90	1	0.0064
Temperature (°K, °C)	750, 477	290, 17	220, –47

*Depth of water in meters over the planet if all water vapor precipitated out of the atmosphere. vi = detected by Viking, ve = detected by Venera.

Table 16.2. The Atmosphere Problem[a]

Gas	Abundance	Expected Equilibrium Concentration	Discrepancy	Residence Time (y)	Output (10^9 tons/y)
Nitrogen	0.8	10^{-10}	10^9	3×10^6	1000
Methane	1.5×10^{-6}	$<10^{-35}$	10^{29}	7	2000
Nitrous oxide	3×10^{-7}	10^{-20}	10^{13}	10	600
Ammonia	1×10^{-8}	$<10^{-35}$	10^{27}	.01	1500
Methyl iodide	1×10^{-12}	$<10^{-35}$	10^{23}	.001	30
Hydrogen	5×10^{-7}	$<10^{-35}$	10^{23}	2	20

[a]Assumes 20 percent oxygen. y = years.

these compounds are burned by molecular oxygen in the atmosphere, there must be a constant and mammoth output or the concentrations of all these gases would soon be undetectably minute. What produces these enormous quantities of nitrogen, methane, hydrogen, nitrous oxide, and ammonia? The correct answer to all such questions is bacteria. Some of these processes are exclusively bacterial; others are derived through bacterial interactions with other organisms. Nitrate is reduced to nitrogen and nitrous oxide by bacteria; methanogenesis is limited to certain anaerobic bacteria; ammonia is formed by bacterial breakdown of urea and uric acid; seaweeds, or possibly their surface bacteria, emit methyl iodide; hydrogen is a product of bacterial fermentation. The answer to the question of what produces the anomalous chemistry of our atmosphere is almost always bacteria. Like the honey or wax of a beehive, the Earth's atmosphere, though certainly not alive, is largely a byproduct of life.

The principal anomalies of the present Earth relative to our neighbors are too much atmospheric oxygen and too little CO_2. On Venus, the CO_2 is virtually all in the atmosphere, giving the planet an atmosphere ninety times as dense as the Earth's. The tenuous atmosphere of Mars (0.6 mbar) is mostly CO_2 as well, but whether carbonate rocks are present on the surface of the red planet is unknown (see Table 16.1). Because Earth, Venus, and Mars all started some 4.5 billion years ago with roughly the same chemical makeup, the lack of CO_2 in the Earth's atmosphere today suggests that a depletion of atmospheric CO_2 has been occurring throughout geologic time. (In fact, the long-term decrease in CO_2 will soon, geologically speaking, broach an environmental crisis for most plants [the C3 plants] as carbon becomes their limiting nutrient [Lovelock and Whitfield 1982]). CO_2 has not disappeared from the Earth; removed from the air, most of the CO_2 can be accounted for by the biogenic precipitation into the sediments, many of which have been diagenetically altered to form marble, dolomite, and other carbon-containing rocks. These rocks were and are produced by the activities of microorganisms removing atmospheric CO_2 while producing molecular oxygen in the process of oxygenic photosynthesis. Given dilute concentrations of the proper salts, air, and of course light and water, oxygenic photosynthesizers sequester CO_2 and produce oxygen. This phenomenon is explicable by biology and completely unpredicted by the blind laws of chemistry.

Darwin recognized that all populations, given unlimited resources, had the capacity to grow exponentially. He called the many "checks" that keep all populations from ever reaching their reproductive potential "natural selection." However, Darwin failed to recognize the enormous impact the growth of populations have on their environment—that the environmental effects of growing and metabolizing populations of organisms are themselves potent agents of natural selection. In accentuating the direct competition between individuals for resources as the primary selection mechanism, Darwin (and especially his followers) created the impression that the environment was simply a static arena for "nature, red in tooth and claw" (Tennyson 1850). Darwin thus emphasized the separation of organisms from their environment. From bacteria to redwood trees, phytoplankton to beavers, the growth and metabolism of all organisms modify their environment. The Russian scientist Vladimir I. Vernadsky (1863–1945), who initiated the field of biogeochemistry, recognized that the divorce of the environment and the biota was artificial. Like most of his work, Vernadsky's comprehension of the interplay between the environment and the biota is almost unknown in the West. With the important exceptions of the great American ecologist G. Evelyn Hutchinson and the founder of biomineralization science, H.A. Lowenstam, few western scientists recognized the inceptive contributions made by Vernadsky to modern ecological thought. By insisting on the competition between organisms as the main source of selection and by ignoring the chemical reciprocity between the biota and its environment, neo-Darwinists have amplified these errors of omission.

Of course, Darwin's grand vision was not wrong, only incomplete. Ignorant of ecology, biogeochemistry, and geology, neo-Darwinism relinquished the ability to ask meaningful questions about the effects the evolution of life has had on the planet Earth. Neo-Darwinism's current funk over altruism reflects a failure to comprehend that every organism is dependent on a huge diversity of life as a source of respiratory gas, water, and food and as a sink for waste products. Do the planktonic algae in the oceans (those that release the sulfide gases that later seed the atmosphere with sulfate particles, which act as condensation nuclei for cloud generation) act altruistically? Are cyanobacteria "public-spirited" in ridding themselves of their wastes—metabolically produced oxygen that happens to be necessary for the continued existence of all oxygen-respiring organisms? Are those bacteria that do not produce sulfide gases or oxygen "cheaters" and thus at a reproductive advantage? Richard Dawkins' (1982) claim that the Gaia hypothesis cannot be true because there is no evidence for competition between Earth, Venus, and Mars is reflective of neo-Darwinism's preoccupation with the romantic, Victorian conception of evolution as a prolonged and bloody battle. As a hypothesis, Gaia integrates the evolution of the biota with the well-documented transformations in the surface and atmospheric chemistry of the planet through geologic time. That the Gaia concept cannot be framed by the stilted terminology of neo-Darwinistic population biology is not surprising because Gaia is a hypothesis based in sciences that neo-Darwinism proudly ignores.

The study of Gaia intrinsically involves disciplines as disparate as atmospheric chemistry and microbial physiology. Lacking the social means to focus on the Earth as a living planet, we have just recently recognized the need for a mission to planet Earth. Without such interdisciplinary activities, most Gaian phenomena will remain unstudied. Indeed, the current division into disciplines often impedes science: as Lovelock has noted, academic apartheid, interferes with an orderly study of the Earth as a planet. This kind of impedance is less familiar to the community of planetary scientists who have worked together to investigate Mars or Venus as entire objects of study. But such well-planned planetary-scale scientific collaboration has never been the case in the fragmented study of the history of Earth.

Although many details remain to be worked out, the Gaia phenomenon is a collective property of the growth, activities, and death of the myriad of populations that comprise the biota. Gaia involves exponential growth rates of living populations, and feedback related to the tendency of all life to respond to changes in the environment and to die when conditions exceed limits of survival. Sensory systems (to light, water, gravity, gas concentrations, pH, oxidation-reduction states, etc.) are diverse and well characterized in many organisms. The amplification of responses to changes in the environment in this feedback system lies in the exponential reproductive potential of all populations. The diversity of metabolic responses of organisms to environmental cues is related to the resiliency and change through time. The properties of living systems depend crucially on their history. They establish their own regulatory system, which is maintained within their boundaries. These features make autopoietic (living) systems different from cybernetic ones.

Gaia and Grants

In the 1820s, the Bridgewater Treatises were commissioned by Reverend Francis Henry Egerton VIII, Earl of Bridgewater. A noble clergyman, according to Gillespie (1969), Egerton assiduously neglected his parish. A champion of science, the Earl of Bridgewater charged the executors of his will, the Archbishop of Canterbury, the Bishop of London, and the President of the Royal Society, with the duty of selecting eight scientific authors capable of demonstrating (here we quote Bridgewater's will) "the Power, the Wisdom and the Goodness of God, as manifest in the Creation, illustrating such work by all reasonable arguments, as for instance, the variety and formation of God's creatures in the animal, vegetable, and mineral kingdom; the effect of digestion, and thereby of conversion; the construction of the hand of man and an infinite variety of other arguments; as also by discoveries ancient and modern, in arts, sciences, and the whole extent of literature" (Gillespie 1969).[1] Bridgewater's money and will supported many scientific activities. The Bridgewater Treatises were intended to offer a working epitome for the main branches of the natural sciences. The Treatises were expected to demonstrate the higher meaning of the order of nature and to ennoble empirical discovery into morality. These Treatises, products of the best British minds of the day, were to bring out the evidence of unity and design. The arguments were to show that a single-minded universe could not have risen by chance; it was statistically impossible for such an infinity of occurrences to work together for good without divine direction. Necessity established, it remained only to demonstrate benevolence—that is, to paraphrase the well-known Oxonian William Buckland (a founder of university geology courses of study): God's benevolence was shown via the proximity of Britain's iron ore to her coal and limestone. The Providence who ordained that the vegetable cycle

*All quotes referring to these Bridgewater Treatises are taken from Charles Coulston Gillespie's wonderful book *Genesis and Geology*.

coincided with the solar year was that same Providence who furnished man with a hand to work and a codfish with an eye that could see underwater. Because he gave money for these activities, Lord Bridgewater was greatly successful in obtaining the finest scientists in the British Isles to work on "the wisdom and the goodness of God, as manifest in the creation." Here we briefly discuss the contributions of five of the eight scientists whose work Bridgewater funded.

Professor and Reverend William Buckland was assigned geology and mineralogy. His appointment was to describe the clever position of the Earth in the solar system and the Deity's adequacy in his production of his durable creations. According to Buckland, "In all these we find such undeniable proofs of a nicely balanced adaptation of means to ends." We believe Buckland's interpretation of the word "adaptation" has not changed since 1830. Buckland continues, "Of wise foresight and benevolent intention and infinite power, that he must be blind indeed, who refuses to recognize in them the proofs of the most exalted attributes of the Creator." Buckland, in acceding to the wishes of his "granting agency," was trying to demonstrate a system of perpetual destruction followed by continual renovation that at all times tended to increase—in his terms—the aggregate of "animal enjoyment" over the entire surface of the "terraqueous globe."

Peter Mark Roget was dealt animal and vegetable physiology. He did not like to use the word "God" because he found it indecorous. Instead, he said, "In order to avoid the too frequent, and consequently irreverent, introduction of the Great Name of the SUPREME BEING into the familiar discourse on the operations of his power, I have . . . followed this common usage of implying the term "nature" as a synonym, expressive of the same power." The Reverend William Whewell was assigned astronomy and general physics. Whewell wrote that it was "impossible to exclude from our conception of this wonderful system, the idea of a harmonizing, a preserving, a contriving, an intending Mind of a Wisdom, a Power, and Goodness far exceeding the limits of our thoughts." So Whewell not only succeeds in finding what he was looking for, but he couches his conclusions in the same words as his request from the granting agency.

William Prough, according to Gillespie (1969), was an important chemist. He was handed chemistry, meteorology, and the function of digestion. He said because we know comparatively little about chemistry and its laws, it is more apt to represent the Deity as a free agent. Chemistry demonstrates that all preceding creations were only anticipatory to the creation and governance of man or, he asks, ". . . What would have been the use of this elaborate design without man as its ulterior object?"

William Kirby was assigned "the animal kingdom." He felt he was particularly fortunate in the ease with which the manifestation of the power and wisdom of goodness of God can be shown because—he thought—the animal kingdom offered conclusive demonstration of both the Fall of Man and the subsequent exertions of Creative Power. No one could suppose—he noted—that Adam and Eve in their pristine state of glory were prey to such disgusting later creations as lice, fleas, and intestinal worms, which now befall the sinner's lot! (Gillespie 1969).

These were the criterion for the selection of science projects in the 1820s. We now cite from a published booklet of the United States National Science Foundation's (NSF's) guidelines for criterion for the selection of research projects in 1988.*

1. **Intrinsic merit.** This is the most important criterion according to the National Science Foundation. This criterion is used to assess the likelihood that the research will lead to new discoveries or fundamental advances within its field of science.

2. **Utility or relevance.** This refers to the likelihood that the research can contribute to the achievement of an extrinsic goal, one in addition to that of the research field itself, and can thereby serve as the basis for new or improved technology.

3. **Effect of the research.** The research should contribute to a better understanding of improvement of the quality, distribution or effectiveness of the nation's [the United States] scientific and engineering research, education, and manpower base.

4. **Integration.** The National Science Foundation looks forward to "using and integrating the resources of all institutions in the support of science and engineering in their contributions to society and to this nation."

So, although we have dropped loyalty to God and Christianity, we still have loyalty to the field and loyalty to the nation as the two highest criteria. Just these three short statements show the intrinsic contradictions in pleasing the granting agencies and working on the Gaia hypothesis.

Indeed, the overall guidelines are less objectionable than those in given fields—and, of course, any investigator must apply through some given

*Criteria for the selection of research projects by the NSF taken from the National Science Board policy statement NSB-79-100, p. 9.

field of study. Here, for example, is a short version of guidelines to the NSF subfield "population biology and physiological ecology," which Lovelock has told us is intrinsic to Gaian studies of the atmosphere because of the role of exponential population growth in Gaian control systems. The purpose of the field of population biology and physiological ecology is to find the genetic basis for "adaptive traits." Adaptive traits adapt organisms as means to an end, according to William Buckland. The term *adaptation* is still prominent in the 1988 brochure. The atmosphere, in the organization of the NSF, has nothing to do with population biology and physiological ecology. These sciences, very far away from biology, are classified as belonging to the study of the physics, chemistry, and dynamics of the Earth's upper and lower atmosphere. If the Earth's lower atmosphere is deeply involved with microbiology and population biology but the charge to atmosphericists is to study the physics, chemistry, and dynamics of the Earth's lower atmosphere, and to ignore all of biology, how can one submit proposals to study Gaian phenomena and be funded? It is not possible. The guidelines go on to endorse research providing further insights into the physical and chemical characteristics and processes that produce such geological features as hydrocarbon and ore deposits. Throughout the meetings of the Chapman conference of the American Geophysical Union, reference was repeatedly made to the concept that coal, hydrocarbon gas, and many types of ore deposits are related to evolutionary biology, especially microbiology. The ignorance of these interrelationships is institutionalized by those who make scientific policy at the NSF and other governmental agencies responsible for funding science in the United States.

Our points are as follows: (1) Gaian science is ignored because there is no way to apply for financial support that will involve integrated study; and (2) we have legacies of religious, social, and historical points of view that have not disappeared just because it is 1991. These legacies are so intrinsic to our thinking and the way we attend to our scientific business that unless we are aware of our social embeddedness as scientists, we simply cannot proceed with the science required to verify or reject the Gaia hypothesis.

Vladimir Vernadsky always thought in global terms, although not, of course, in modern language; certainly he did not compare directly the surface features of Earth, Venus, and Mars. Vernadsky wrote, not in Russian, but in the *Transactions of the Connecticut Academy of Sciences*, a key paper on biogeochemistry (1944). Just before he died, in 1945, he published a paper in a magazine well known to most scientists, Sigma Xi's *The American Scientist*. G. Evelyn Hutchinson, professor at Yale University and a colleague of George Vernadsky, Vladimir's son, introduced Vernadsky's paper to this journal. These two contributions, one to the *Transactions of the Connecticut Academy of Sciences* and the other to *The American Scientist,* together present in English the general intellectual outlook of one of the most remarkable scientific leaders of the present century (Grinevald 1988). Thus Vernadsky's vision, which preceded, of course, Lovelock's development of the Gaia concept, was not ignored merely because his work was unavailable in English. Although most of his books were published in Russian, Vernadsky's *The Biosphere* has been available in French since 1929.* Furthermore, he wrote, "In everyday life, one used to speak of man as an individual living and moving freely about our planet, freely building up his history until recently the historians and the students of the humanities and to a certain extent even the biologists consciously failed to reckon with the natural laws of the biosphere, the only terrestrial envelope within which life can exist. Basically man cannot be separated from it; it is only now that this solubility begins to appear clearly and in precise terms before us. Man is geologically connected with the biosphere, its material, and energetic structure. Actually no living organisms exist on Earth in a state of freedom. All organisms are connected indissolubly and uninterruptedly, first of all, through nutrition and respiration, and secondly, with the circumambience material and its energetic medium" (Vernadsky 1945).

We anticipate with enthusiasm the recognition of Lovelock's worldview of the environment modulated by life and Vernadsky's worldview of life as a geological force by the NSF, NASA, National Center for Atmospheric Research (NCAR), Department of Energy, Department of Defense, Office of Naval Research, and private foundations on which we scientists depend—as Buckland depended on the Earl of Bridgewater—for our funding. We wait, that is, for the era of the Gaian biosphere and its appropriately funded science to arrive (Sagan 1990a and Chapter 14).

*The first English translation of the full text is due to be published in 1997.

17

GAIA AND THE COLONIZATION OF MARS*

LYNN MARGULIS
AND OONA WEST

*Dedicated to the memory of Heinz A. Lowenstam (1913–1993)

The Gaia hypothesis of James E. Lovelock holds that the surface temperature, chemistry of the reactive gases, redox state, and pH of Earth's atmosphere and surface sediments are homeorrhetically maintained by the metabolism, behavior, growth, and reproduction of living organisms. (Homeostasis is physiological regulation around a fixed set point, like control of adult mammalian body temperature around 37°C, whereas homeorrhesis, a parallel concept, refers to regulation around a changing set point, like temperature regulation in a developing mammalian embryo.) The term Gaia, the name of a daunting Greek goddess, is, in Lovelock's view, simply "a good four-letter word referring to the Earth." She is also Ge or Gaea (for example, the Geos satellite, geology, geography, or in Pangeae).

Gaian environmental regulation is achieved largely by the origin, exponential growth, and extinction of organisms, all related by ancestry and physically connected by proximity to the fluid phases (water and air) at Earth's surface. Organisms in communities form changing ecosystems that have persisted since the Archean. The interactions of organisms, driven by solar energy, produce and remove gases such that chemistry of non-noble gases, temperature, and alkalinity are actively maintained within limits tolerable to life.

Within this conceptual framework, biological as well as physical sciences become appropriate to the analysis of Earth's atmosphere and geologic history. Especially pertinent is the role of the microbiota (bacteria, protoctista, fungi) in Earth surface gaseous exchange that involves the recycling of those chemical elements (for example, H, C, O, N, P, S) absolutely required by life.

The Gaia Idea

Product of the lively imagination of a British atmospheric chemist and the international space program, the Gaia idea has come of age. The atmospheric composition of Earth signals unmistakably that the third planet is living: flanked by the dry, carbon dioxide–rich worlds of Mars and Venus, one invokes either physiological science or magic to explain Earth's wildly improbable, combustive, thoroughly drenched troposphere (see Table 16.1, p. 212). The Gaia hypothesis, in acknowledging this atmospheric disequilibrium (Margulis and Lovelock 1974) has opted for physiology over metaphysics.

More than twenty-five years worth of scientific contribution is listed in the Reference section. Many scientists are unaware of the extent of the serious literature and the potential contribution of the Gaia idea for integrating evolutionary, meteorological, sedimentological, and climatological data. Unfortunately, nonscientific Gaia literature (which tends to be anti-intellectual and hysterically toned New Age commentary) has received so much press attention and contentious comment that much of the primary science remains unknown.

Despite the fact that an "Earth system science" approach is vigorously encouraged for the solid-earth sciences, mention of the G-word (Gaia) still causes apoplexy in some scientific circles. This is remarkable, considering the broad parallelism of these approaches to understanding Earth processes. The U.S. National Academy of Sciences (NAS) (National Research Council, 1993) report on future directions of research in the solid-earth sciences advocates "a new approach to studying Earth processes, in which the Earth is viewed as an integrated, dynamic system, rather than a collection of isolated components" (statement by Frank Press in his introductory letter). This report calls for an understanding through integrated study of physical and biological processes and sees as desirable a process-oriented global approach to understanding Earth. Despite avoidance of the term, a Gaian approach is advocated by the NAS.

The Gaia hypothesis, rejected by some as the fantasy of New Age crystal swingers, has been largely misunderstood by the scientific community. For example, George C. Williams (1992) perpetuates confusion by unconscionably maligning Gaia: "It [the idea that the universe is especially designed to be a suitable abode for life in general and for human life in particular] had to be abandoned in its earlier forms with the triumph of Copernican astronomy . . . but some scholars still find it possible to argue that the Earth, at least, can be regarded as especially suited for human life. . . . [The] main modern manifestation [of this idea] is in the gaia concept of Lovelock and Margulis (1974a)."

The Gaia hypothesis demonstrates how life sciences are essential to understanding Earth, while revealing the inadequacy of evolutionary theory developed in the absence of climatological and geological knowledge. The Gaian viewpoint is not popular because so many scientists, wishing to continue business as usual, are loath to venture outside of their respective disciplines. At least a generation or so may be required before an understanding of the Gaia hypothesis leads to appropriate research.

Vikings of '76

When the Viking mission to Mars returned its data, some members of the scientific community thought that "planetary biology" or "exobiology" were doomed because the absence of Martian life rendered them sciences with no object of study. Lovelock and his colleagues thought just the opposite: now that data from Mars were available, speculations comparing the planets could be replaced with knowledge. It became certain that the bleak Martian landscape is devoid of life (Figure 17.1), whereas life is not only a planet-wide phenomenon but in today's solar system living beings are limited to Earth's biosphere.

Gaia has been called "Goddess of the Earth," or the "Earth as a single living being." These are misleading phrases. Because much scientific work mentioning Gaia suffers from problems of misunderstood terminology, we offer this physiologically oriented statement of the Gaia hypothesis.

FIGURE 17.1. View of the Martian regolith from the *Viking* lander (in foreground). The surface is thought to be red from ferric iron. Courtesy of NASA.

We reject the analogy that Gaia is a single organism, primarily because no single being feeds on its own waste nor, by itself, recycles its own food. Much more appropriate is the claim that Gaia is an interacting system, the components of which are organisms. Nowhere is this more evident than in examples of biotic influence on important geological processes (Table 17.1; Westbroek 1991).

The two landers and orbiters of the 1975–1976 Viking missions to Mars yielded data that complemented earlier Earth-based observations of that planet. Organic compounds were absent: the concentration of total organics, if present, must be fewer than one part per billion. The gas-chromatographic detection of oxygen was not due to life but to the release of O_2 from moistened peroxides, and the incorporation of radioactive CO_2 was due to cosmic radiation, including UV photochemistry, and not to photosynthesis. Once the reactants were spent, no new change was detected by these experiments. The conclusion is inescapable: no evidence exists for present life on Mars. The same is true of Venus.

As far as we know, the Gaia phenomenon is limited to Earth. Can it be extended by colonization of Mars? Comparison of Earth with Mars helps highlight both the nature of Gaia and implications of the idea for the study of Earth.

Extraterrestrial Germs

To prevent both lunar and Martian spacecraft from carrying microbes, "clean-room" techniques were applied. Even sterilization of the outside and much of the inside of the Viking spacecraft was undertaken. Ethylene oxide gas flooded the accessible components to ensure microbial cleanliness; this increased the total cost of the Viking mission by about 10 percent. During the U.S. Apollo missions to the moon in the 1960s and 1970s, fears of possible "back-contamination" were rampant: extraterrestrial "germs" might "contaminate" Earth. This issue is sure to arise again if there is any future return of materials from Mars. Since meteorites from Mars have landed on Earth, such fears seem silly, more a manifestation of pulp science fiction than a well-reasoned treatment of scientific probabilities.

Although investigators such as Rothschild (1990) have suggested that Martian life may still be found in oases, perhaps as permafrost bacteria or

Table 17.1. Biologically Mediated Geologic Phenomena

Example	Importance	Lithospheric Reservoirs and Examples
1. Phosphorus cycle	Essential for all life: component of DNA and RNA nucleic acids, and ATP, and other nucleotides; phospholipid membranes and the calcium phosphate of bones. Because phosphate is a major growth-limiting nutrient, the P cycle is completely biologically mediated (Madigan et al. 1996; Filipelli and Delaney 1992).	Earth's crust and deep-sea sediments; guano islands Atmospheric phosphine (PH_3) is negligible.
2. Calcium-carbonate deposition	Essential for formation of hard parts in shelled marine animals and many testate protoctists, e.g., foraminifera. Helps maintain pH balance in the oceans. As zlimestone, it is an important sink for CO_2.	Stromatolites Coral reefs Deep-sea carbonate ooze (foraminifera and coccoliths)
3. Organic matter deposition	Leads to development of anoxic conditions and CH_4 production, so that carbon is released to the atmosphere, thus preventing complete loss from the biosphere, leading to maintenance of elevated O_2 levels (Watson et al. 1978). Fossil fuels.	Oil shale and other organic-rich shales Coal, peat, oil, tar sands

Table 17.1. Biologically Mediated Geologic Phenomena (*cont'd.*)

Example	Importance	Lithospheric Reservoirs and Examples
4. Methanogenesis	Atmospheric composition of Earth (e.g., presence of methane, ozone) is inexplicable in the absence of life (Watson et al. 1978; Table 1).	Trapped natural gas, swamp and marsh gas Arthropod intestines Ruminant mammals
5. Regolith consolidation	Unconsolidated sediments are bound by biotic communities, e.g., mucilage coating of bacterial mats (Margulis and Stolz 1983).	Mud Unlithified sediment
6. Erosion acceleration	Weathering rates increased by biologically mediated erosion, bacterial endoliths, fungal hyphae, plant roots, and lichens.	Lithosphere-atmosphere-hydrosphere interfaces
7. Microbially mediated mineral formation (biomineralization)	Genesis of important mineral deposits. Interpretation of modern and ancient environments.	Banded iron formation Witswatersrand gold deposits Bog iron Rock varnish Manganese nodules

even as "endoevaporites" in isolated salt crystals, the chances of finding isolated life there are vanishingly small.

The Gaia hypothesis provided a framework for evaluation of Martian results. Life maintains its immediate environment and appears on Earth only as a planet-wide phenomenon. Life may have been sparse when it first appeared or may be sparse when it is dying out, as Lovelock emphasizes, but between these two endpoints life must be luxuriant. Why? Because of life's intrinsic tendency to grow, expand, and populate at exponential rates and its ability to travel. Therefore, a question of the 1990s is, Can life expand to Mars? This question, Can Mars be colonized?, is identical to that of, Can Gaia reproduce?

All organisms are connected through the atmosphere, and life as we know it on Earth is a global phenomenon, utterly dependent on sunshine. Hardy terrestrial forms, such as halophiles or sulfur-loving acidophilic archaebacteria, ammonia-oxidizing chemolithotrophs or carbonate-precipitating stromatolite-forming cyanobacteria, are extremes connected to, and tolerated by, a ubiquitous planetary biota. There are no virtuoso individualists. Martian life, if present, would by analogy to Earth most likely be found in communities.

Although it is theoretically possible that subvisible life will be found in the nether reaches of Martian deserts, it remains far more likely that the Martian wasteland is as dead as it appears. If so, one scientific challenge is to enact in reverse the very process that was once so feared: to deliberately contaminate or, as is now said, to "seed" Mars with life from Earth.

Ecopoiesis

The quest for life on Mars began (by telescope) long before the Viking missions, and it will not likely end with the deployment of rovers on the planet early in the next century. After acceptable confirmation that Mars is uninhabited, the next task might be to "seed" the red neighbor with propagules from Earth. (Many will justifiably argue that the resolution of more pressing Earth-based problems should be a far greater priority: curbing the human tendency to convert the surface of Earth to urban ecosystem or fostering and documenting the diversity of life.)

The first and perhaps most crucial task in making Mars habitable is to increase its surface temperature. Proposals for heating Mars have ranged from engineering dreams of melting the ice caps with giant orbiting mirrors or covering the surface with black lichens, to schemes of rocketing greenhouse chlorofluorocarbons (CFCs) into the atmosphere. Recent proposals tend to be more detailed and slightly more feasible, yet share with their forerunners a profound, simultaneous strength and weakness: although such schemes are ambitious enough to excite the imagination, making captivating layouts in popular science magazines, they are too grandiose and vague to be practical (Kluger 1992).

For example, even if several millions of tons of new, UV-resistant CFCs could be produced annually in situ from the surface of Mars, leading to a release of carbon dioxide and to planetary temperatures of 22°C, then what? Even if oceans appeared from ice trapped in the lower latitudes because a way had been found to return to the atmosphere the CO_2 now trapped in surface carbonates, what now? The density (and therefore livability) of a Martian atmosphere is probably intrinsically limited by the weakness of Mars' magnetic field. In the absence of magnetic deflection of solar wind, a Martian atmosphere would quickly be ablated. Even if genetically engineered plants and microbes were created to produce oxygen and other gases at hitherto miraculous rates, it still could take, as Christopher McKay (of the NASA Ames Research Center) estimates, about a thousand years to build an atmosphere to stable levels of oxygen in carrier gases breathable by eukaryotic microbes, let alone humans.

Although the new science of geophysiology and the success of biotechnology with microorganisms may have incited us to fantasies of planetary design, colonizing Mars so that humans might walk in the open along its canyons remains a distant fantasy. One should distinguish here between ecopoiesis (Haynes 1990, 1992; the inundation of a formerly uninhabited surface with viable living systems) and terraformation (McKay 1987; the recreation of Earth on another planetary surface). For the foreseeable future, ecopoiesis but not wholesale terraformation seems a possibility for Mars; the former is, however, a prerequisite for the latter (McKay et al. 1991). Ecopoiesis would not make Mars into an extraterrestrial paradise, so much as it would transform it into a global cesspool—colorful, perhaps, but rich in mephitic vapors. The early history of Earth, after all, and the present state of the gas giants in the outer solar system are characterized by a chemistry that more resembles sewer gas than food. Though alien and inhospitable to mammals, these reduced sulfurous carbon-rich volatile compounds were crucial to the origin and early evolution of life.

The only dependable way to make a planetary surface livable may be to repeat the evolutionary colonization process that occurred on Earth, which began with hydrogen, methane, ammonia, formaldehyde, sulfides, nitriles, and simple sugars. Shortly after life appeared, noxious gas exchanges among anoxygenic phototrophic bacteria and their dependents ensued. Sped up on Mars, the outcome of a rushed and deliberate Martian colonization process is likely to be highly unpredictable—possibly even tragic.

Will we humans, godlike, wave our wand? Do we really think, in our naivete, that strewing our scientific instrumentation over the red surface of Mars via robots in a geological wink of an eye will produce a New Blue Earth? Far more probably, Mars will be colonized slowly and gradually, and not by humanity but through humanity, facilitated by robots. For the foreseeable future it seems likely that the only human presence on Mars will be via the developing technology of telepresence. The landing of the two remote-sensing, remote-controlled, human-connected *Viking* landers in 1976 proves that the process of colonization has already begun. Unlike Neil Armstrong's epochal "one step for man, one giant leap for mankind," the ecopoiesis of Mars's surface has no instantly recognizable moment. The launch of human-built life detectors to Mars, the "telepresent" sensory cameras that radio their signals back to eager humans at mission control, space-crew first landings, early orbiting Mars stations, and the eventual habitation of the red surface by emigrants of a variety of species—all are part of a gradual process of ecopoiesis. All would be likely to occur haphazardly, with very little conscious planetary bioengineering.

The distinction between altering one's body to "adapt" to any inhospitable environment and altering the environment itself is largely specious from a Gaian viewpoint. As organisms evolve, both their bodies and the environment change irreversibly. Such change occurs through technology, which is not a uniquely human phenomenon. Animate and inanimate nonhuman technologies abound, for example wasp nests, humidified and airconditioned termite mounds, or the immense lithified limestone reefs fringing tropical islands.

Gaia's Propagules

Life packages its precious contents: production of heat-proof bacterial endospores, dinomastigote cysts, formation by trees of seeds and hardened fruits, rubbery eggs of snakes, or the tough egg cases of rays. Among the most remarkable of such propagules are the "tuns" of tardigrades or the salt-tolerant, dustlike eggs of brine shrimp (Figure 17.2).

To enable any Earthlings to dwell on the surface of Mars, bubble-like enclosures probably will be required that house a complexity of species in self-supporting recycling systems, in principle, like the stated goals of the exorbitant Biosphere II project in Arizona's Sonoran desert. This incipient Earth propagule (which "germinated" and released its contents in September 1993) contained eight "biospherians." The 17-acre facility allegedly was "materially closed" in the autumn of September 1991 to all but its enormous intake of external electrical power. It is clear that at

FIGURE 17.2. Propagules: clockwise from top left, bacterial endospores, dinomastigote resting cysts (in paleontological literature as hystrichospheres), walnuts, and possible future biosphere. Drawing by Sheila Marion-Artz.

present we are far from establishing any biospheres on Mars. The energy needed for the mere sustenance of any biospheres, let alone their use as bases for any bio-industrial modification of the planet, will require on-site nuclear power. However, as soon as adequately closed artificial biospheres are established, for example, to serve as base camps for CFC factories, global, terrestrial, biospheric Earth life will have de facto, if inconspicuously, colonized the surface of Mars.

Such an artificial biosphere, a radiation and desiccation-resistant form, is highly reminiscent of large-scale nonhuman evolutionary innovations far more continuous with the past than it seems at first glance. By packaging and miniaturizing the essentials for survival, life ventures out upon and ultimately makes a home for itself in formerly hostile terrain.

The ecopoiesis of Mars would likely be accomplished by interaction of many types of Earth organisms: bacteria, protoctists (mainly as algae), plants, and fungi will certainly play their roles. Indirectly, all life forms would be involved in planetary colonization, although at first multispecies bases will need to be constructed in an effort planned by exceedingly few, highly select, and passionately dedicated humans. Such bases are necessary to protect their inhabitants from an initially hostile external Martian world. Food plants must be grown and all wastes internally recycled.

That such enclosures of metal, glass, and plastic might be built by scientists, engineers, and other working people is hardly an argument for their absolute uniqueness: all previous technological advances in the evolution of life (for example, silica fretwork of diatoms, calcium phosphate bone and teeth in vertebrates, lignification leading to great height in plants, and the chitinous exoskeletons of insects and crustaceans) involved more than a single type of life and were prerequisite to the adaptive radiation of their inventors into new and formerly hazardous realms.

Humans have no exclusive hold on technology. Magnetite teeth in molluscs and wax synthesis by hymenopterans are technologies that preceded those of *Homo sapiens* by millions of years. Calcium phosphate teeth, barium sulfate gravitational sensors, and temperature- and humidity-controlled termite mounds were as much a prerequisite for cosmopolitan Cenozoic distribution of, say, rodents, charalean algae, and fungi-gardening termites as telephones and electric power are to human urban expansion. Silurian-Devonian emigration of life to the land, with its attendant problems of lack of support by water, depleted nutritional substrates, and its exposure to continuous solar UV radiation, demanded a dramatic repackaging of life's

FIGURE 17.3. Five-kingdom hand representing the major forms of life, all connected through nearly four billion years of "Darwinian time" at Earth's surface ("Vernadskian space"). In order of appearance (Ga = billion years ago) in the fossil record: Monera (bacteria or prokaryotae, 3.9 Ga), protoctista (algae, slime molds, ciliates and other microscopic eukaryotes and their larger descendants, 2 Ga), animalia (egg-sperm embryo forming diploids, 0.75 Ga), fungi (zygo-, asco-, basidiomycota, fungi imperfecti, and lichens that grow from fungal spores, 0.45 Ga), plantae (bryophytic or tracheophytic haplodiploids that develop from maternally retained embryos, 0.45 Ga). This illustration designed by Dorion Sagan is from the cover of *Five Kingdoms: An Illustrated Guide to the Phyla of Life,* (3rd ed.) by Margulis and Schwartz, 1997. (Available as a teaching unit from Ward's Natural History Establishment, Rochester, New York.)

resources—an incorporation into bodies of what at one time could be found only "outside"—in the mineral environment (Sagan 1992).

Such repackaging of living beings and their accoutrements might begin within recycling enclaves, "artificial biospheres." Above and beyond anything done later, the first of these bases on Martian terrain would already be colonization of Mars. Cosmic historians, in retrospect, might use establishment of such Martian base camps to date the reproduction of planetary life. Such "artificial biospheres" might be recognizable not merely as a human technology but as an expansion and metamorphosis of Earth's original biosphere by members of all of the five kingdoms of life (Figure 17.3). Gaia would have reproduced, challenging the objection of Doolittle (1981) that Gaia cannot be a life form because it is incapable of reproduction. Seen from afar, the settling of Mars would be akin to budding, a space-borne planting of a "sporulated" form of biospheric life—Gaia transporting propagules of itself to the surface of a new world.

Conclusions

A Gaian scientific world view is especially relevant in light of extensive human-wrought modification of the global environment and the talk about further missions to Mars. Although the fundamentals of Lovelock's Gaia hypothesis have not changed in 25 years, researchers still don't yet understand them. The Gaian approach critically enables research on Earth systems precluded by the patchiness of the "academic apartheid" from which Lovelock, as a young man, fled.

The Gaian concept of physiological surface regulation is unpalatable, especially to those who hold dogmatic ideas on Earth processes. Lovelock remarked (in the BBC program "Goddess of the Earth") that the Gaia hypothesis hasn't been controversial; it has just been ignored. But the scientific details, contained in the literature listed here (pp. 335–343), are becoming better known. We are hopeful that the full importance of the Gaia idea will continue to be more extensively understood by scientists and students, especially by geologists upon whom rests the future of Gaia-oriented scientific research.

18

FUTURES

DORION SAGAN
AND LYNN MARGULIS

Life and Machine Reproduction

Life may not progress, but it expands. Like an obsessive adult whose personality was formed in a forgotten childhood, humanity can be understood only as we make sense of our past. Earth life will have to evolve to live on other planets, or even, perhaps around other suns. And if as humans we survive, we will certainly change, becoming part of the future "supercosm"—the hypothetical continued expansion of life from Earth into the solar system and beyond. The huge increase in area and resources will unleash life's potential: the supercosm will be as different from Tokyo as Washington DC is from a bacterium.

Human beings are peculiar parts of the biosphere, the place where life dwells. The biosphere, the sum of life on Earth (the biota) and its surroundings, is part of us, and we have arisen from within it. As technologically dependent organisms, we have as much independence from the biota as a cancer virus has from the dividing cell in which it abides. Those twin delusions of human grandeur—our natural superiority and scientific objectivity—are conundra of projecting the techniques of human survival into realms where they do not belong. The trial-and-error method of science, the forming and testing of hypotheses, and the rapid transmission of science through culture are so similar to natural selection of hereditary variants, on the one hand, and to survival and growth via bacterial genetic transfer, on the other, that science can be considered as unconsciously imitative and well within the scope of older biotic process.

Indulging the human mind's penchant for categorical choices and in keeping with our assignment to assess the future, we indulge in forecasts for humanity. Either there will be a catastrophic nuclear war that destroys our technologies or our technologies will control themselves so that, with machines we begin reproducing in outer space. If the former occurs, people will vanish from the biosphere, global ecology will shift, and the biosphere will evolve in curious directions. It will not be a victory for humanity. However, if we survive our threat of nuclear war and become a multiplanet civilization, reproducing in outer space, this too will not necessarily be a victory for humanity. It will be a further expansion of the biosphere, a victory for the biota, for the nexus of all life, including machines.

The great ape *Homo* in his present state is a singularly technological creature. In truth, a human being may be thought of as an obligate technobe, a weak body entirely dependent on rapid harvesting of agricultural grasses, on milking, slaughtering, and packaging domesticated artiodactyls; on extraction of organic compounds, remnants of vast communities of photosynthesizers as fossil fuel oil from deep wells; on electromagnetic communication satellites, automobiles, and airplanes; in short, on machines.

Unfortunately for those who believe humanity is the apotheosis of life on Earth, the idea of reproducing machines is not a matter of scientific fantasy but a matter of fact in the present organization of the biosphere. Only the organic macromolecules DNA and RNA are capable of reproduction in the replicative sense: in one act of synthesis, they make complementary copies of themselves. All else—cells, boys, elephants, trees, McDonalds, and branches of the Chase Manhattan Bank—do not directly reproduce. Much molecular replication, cell growth, development, and construction is involved before two cells, two boys, two elephants, two trees, two McDonalds, and two bank branches appear in the biosphere where a single one was before.

Unfortunately, nature is not dichotomous in a way that matches our verbalizations. Nature does not conform to our definitions. Although there is an ineffable continuum between the living and the nonliving, we are beginning to understand the functions and organizations that are common to living entities. Living systems, from their smallest limits as wall-less bacterial cells to the entire surface of planet Earth, self-maintain. As living beings they are bounded systems—they retain their recognizable features, even while undergoing a dynamic interchange of parts.

We "modern humans" may never be the agents of the microcosm's expansion into space. Visual image processing in the form of eyes evolved many times; for example, it developed in protists, marine worms, mollusks (such as snails and squids), insects, and the ancestors to fish and mammals. Wings, likewise, evolved independently in insects, reptiles, birds, and bats: similar aerodynamic designs arose to meet the similar contingencies of the air. This tendency of organisms to evolve in similar directions despite the fact that they have different recent ancestors is called convergence. Convergence suggests that many kinds of beings will expand into space, just as many kinds have moved onto dry land and into the atmosphere. But like the first lungfishes, which came out of water but never evolved into the ancestors of land animals, these early flirtations with space may never be consumated by continued life there. The presence of nervous systems and community behavior in many sorts of animals suggests that if we people and our "urban" associates fail, other life-forms will evolve to cart the primordial microcosm into space. If human beings become extinct—or if, like the horseshoe crab or lungfish, we just happily remain in our present habitats—the biota may, for a time, remain confined to Earth. But remember it took humans (*Homo*) only a few million years to evolve (from *Australopithecus*). Even if all anthropoids—all humans, monkeys, and apes—became extinct, the microcosm would still abound in those assets (for example, nervous systems, manipulative appendages) that were leveraged into intelligence and technology in the first place. Given time to evolve in the absence of people, the descendants of raccoons—clever, nocturnal mammals with good manual coordination—could start their own space program. Sooner or later the biosphere is likely to expand beyond the cradle of this third planet.

It is an illuminating peculiarity of evolution that explosive geological events in the past have never led to the total destruction of the biosphere.

Indeed, like an artist whose misery catalyzes beautiful works of art, extensive catastrophe seems to have immediately preceded major evolutionary innovation.

Life on Earth answers threats, injuries, and losses with innovations, growth, and reproduction. The disastrous loss of hydrogen gas (H_2) from the gravitational field of the Earth led to one of the greatest evolutionary successes of all time: the use, not of H_2 but of water (H_2O) in photosynthesis. But this substitution of a necessary ingredient also led to a devastating pollution crisis. The accumulation of oxygen gas in the atmosphere, (a gas originally toxic to the vast majority of organisms) permanently changed the planet. The oxygen crisis that began only two thousand million years ago prompted the evolution of respiring bacteria. These microbes which used oxygen to derive biochemical energy more efficiently than ever before, eventually took over most of the world. Some of the oxygen-breathing bacteria became symbiotic, merging with different (oxygen-eschewing) bacteria to form eukaryotic cells, which, becoming sexual, evolved into fungi, plants, and animals.

The most severe mass extinctions the world has ever known, at the Permo-Triassic boundary 245 million years ago, were rapidly followed by the rise of mammals, with their sharp eyes and large receptive brains. The Cretaceous catastrophe, including the disappearance of the dinosaurs 66 million years ago, cleared the way for the development of the first primates, whose intricate eye–hand coordination led to technology. World War II ushered in radar, nuclear weapons, and the electronic age. And the holocaust of Hiroshima and Nagasaki over fifty years ago decimated Japanese industry and culture, unwittingly clearing the way for a new beginning in the form of the rising red sun of the Japanese information empire.

With each crisis the biosphere seems to take one step backward and two steps forward—the two steps forward being an evolutionary solution that surmounts the boundaries of the original problem. Not only meeting but transcending challenges confirms that the biosphere is resilient. The biosphere habitually recovers from tragedies with renewed vigor. Nuclear conflagration in our hemisphere here in the north would kill hundreds of millions of human beings. But it would not be the end of all life on Earth, far from it. As heartless as it sounds, a human Armageddon might prepare the biosphere for less self-centered forms of living matter. As different from us as we are from dinosaurs, such future beings may have evolved through matter, life, and consciousness to a new superordinate stage of organization, and in doing so consider human beings as impressive as we do iguanas.

Such a vision offers only metaphysical consolation. Barring direct fatal impact by an atomic weapon, only 10 micrograms (that is, 10 millionths of a gram) of radioactive fallout—the debris that explodes into the stratosphere, blows in the wind, and later settles down—is needed to kill a person. Current estimates put Russian and U.S. nuclear bomb arsenals at 10,000 megaton bombs apiece. As the late inventor Buckminster Fuller showed by dropping tiddledywinks on a giant map spread across the ballroom floor of the New York City Sheraton Hotel, 5000 bombs released at random on the globe would paralyze all the major cities. And given present arsenals, a full-scale nuclear war is expected to deplete from 30 to 60 percent of the ozone of the stratosphere. The dust and smoke of city fires would rise up and surround the Earth, first burning it but later leading to a severe drop in worldwide mean temperature.

Radiation could also accelerate worldwide plagues of AIDS-like and other diseases with compromising effects on the human immune system. Yet the health and stability of the microcosm might even be strengthened. The increase in radiation-induced mutations wouldn't change microbial evolution because a huge reserve of radiation-resistant mutants to supply the evolutionary process has always been present. *Micrococcus radiodurans* (now called *Deinococcus*), for example, has been found living in the water used to cool nuclear reactors. Nor would the destruction of the ozone layer, permitting entry of torrents of ultraviolet radiation, ruin the microbial underlayer. Indeed, it would probably augment it, because radiation stimulates the bacterial transfer of genes.

The accelerated nature of evolution in general and cultural evolution in particular makes it impossible to predict future evolutionary change, especially at long-range. If we simply extrapolate current trends, we arrive not at the future but at a caricature of the present. For example, when the telephone was invented, in an anecdote told by science fiction writer Arthur C. Clarke, it was predicted that, in the not-too-distant future, every city and town might even have the use of one. When helicopters were invented, on the other hand, there were commentators who saw the day when every suburban household would park its private twirly vehicle near its automobile in a heliport-garage. Respectable scientists, writing in technical journals with full citations to the professional literature and mathematical equations, predicted that the surface of the Moon was covered with commercially exploitable levels of oil. Some stated that desert lichens, seasonally turning green grew to nearly cover an entire hemisphere of the planet Mars with each moist summer. Other scientists predicted thick dust layers would so impede a lunar landing that explorations of the Moon would be impossible and ought not be attempted. We certainly do not pretend to have private knowledge of the future, but we do prefer to contemplate possibilities based on an awareness of our long-term past.

Beyond short-term technological fads are the long-term trends of life—extinction, expansion, symbiosis—which seem universal. We, the species *Homo sapiens,* will reach extinction, with or without a nuclear war. We may, like ichthyosaurs and seed ferns, leave the annals of Earth history without an heir, or we may, like choanomastigote protists, australopithecines, and *Homo erectus* mammals (the respective ancestors of sponges and of us) evolve into distinct new species.

No matter what our progeny evolves or devolves into, however, if it remains on Earth eventually it will be scorched alive. By an astronomical reckoning, the sun has a total life span of only about ten billion years. After all the sun's primary hydrogen burns up as fuel, nuclear reactions that convert lighter to heavier atoms are expected to take over. As the radiating sun expands into a red giant, our dying star will shine as it has never shone before. The luminous body is expected to generate such immense heat that oceans will boil and evaporate.

As the Earth is scorched, its oceans boiled to steam by the final outbursts of a waning sun, only living forms that have wandered beyond this home planet or have protected themselves in some way, will be salvaged. As is the wont of life, the habitats of life's predecessors will be brought into the homes of life's future. The insertion of past dwelling places into new ones is an intense sort of conservatism, a deep-rooted refusal to change that is observed in tropical bee hives, naked mole rat tunnels and Russian emigré populations at Ville Franche-sur-mer (Mediterranean France) or San Francisco's valleys. Such a monomania for preservation may be just what is needed to rescue future organisms, and with them, life itself, from the fate of an exploding sun.

We already see hints that the boundaries of life, human and cockroach, pet and grain, are expanding. Populations, industries, universities, and suburbs have rapidly grown, but none has grown indefinitely without causing severe resource depletion and environmental transformation. Natural selection, which simply refers to different rates of survival among growing reproducing entities, whether of monkeys or of McDonalds, can prune or frighten. Population growth is limited and populations are beyond good and evil. All living forms grow in response to the availability of space, food, and water. When too numerous, all organisms either perish or transcend themselves.

Organisms that transcend themselves always find new ways to procure "Lebensraum" (room to live), carbon, energy, and water. All this expanding beingness produces new wastes and new needs for space, food and water. The increasingly abundant production of new wastes stresses those that made it. Life itself expands without much remorse and creates its own new problems; life forces new solutions. One can imagine an example. Pollution might be created by the venting of new chemicals in the outer solar system as part of a program of resource acquisition by future corporations. Such toxic wastes might even reach Earth. On Earth, new microbes able to tolerate or make use of such wastes, might be forced to evolve. This, in turn, would establish a living partnership that stretched millions of miles, from Earth to the moons of Saturn.

To grasp the potential of life in the future, we must look at life in the past. The dramatic evolution of humans cannot be separated from the coevolution of our microbial ancestors, the bacteria that constructed our cells and those of our food species of plants and animals. In coevolution, over thousands of years partners change genetically. Inherited partnerships evolve together as new proteins and developmental patterns emerge. When ultimately the partners totally depend on each other, they become larger new entities. No longer is it valid to consider them independent or separate individuals. The agricultural grain corn, *Zea mays* provides one striking example of coevolution. Corn has evolved in a few hundred human lifetimes, during the past six thousand years. Corn on the cob no longer withers naturally as do the seed-releasing flowers, the teosinte grasses from which modern corn evolved. Corn now must have its thick husk removed by human hands in each and every generation. Cows must be milked, chickens fed. Now the reproduction of our food sources—corn, cattle, and chickens—is tied to our own. Corn cannot complete its life cycle without people; these organisms form a part of us. Once an inconspicuous, self-sufficient grass on the Mexican plateau, the plant teosinte has been selected by hungry peoples and has been grown for larger and larger kernels. It has become a major staple for humanity. Like the electric wires in the elevators of Manhattan and Los Angeles, the luxury of yesterday has become the necessity of today.

The prodigious increase in the human population depended on plants, and probably will continue to depend on them and their bacteria-derived chloroplasts if we are to move into space. It took a thousand hectares during the last interglacial period to support a single Old Stone Age hunter. Over 10,000 times less agricultural land is required to support a modern Japanese rice farmer than his hunting and gathering predecessor. Thus for every hunter that once roamed the island of Honshu, over 10,000 inhabitants in a Tokyo suburb may thrive. Like the cells of the microcosm before us, human beings must coevolve with plants, animals, and microbes. Eventually, we will probably aggregate into cohesive, technology-supported communities that are far more tightly organized than simple or extended families, or even nation-states or the governments and subjects of superpowers.

Because new symbioses tend to form during evolution and any organism is always a member of a community of different species, no single life-form or member of one species alone could ever colonize space. Humans seem well suited to help disperse the Earth-based biota, and they may occupy a prominent place in the supercosm—just as mitochondria, oxygen-using bacteria, now permanently inside cells of plants and animals, helped the mosses and ferns, amphibians and anthropoids settle dryer land. But for us humans to play the prominent role in the expansion of life into space, we must learn from the successful communities of the microcosm. We must move more rapidly from antagonism to co-existence. We need to treat the members of species whose health is of interest to us as fairly as a small farmer does his egg-laying chickens and milk cows. Unlike poaching rare animals for their pelts, garishly displaying horned heads over a mantelpiece, shooting birds for sport, or bulldozing rain forests, such fair treatment means cohabiting the plains and forests with our planetmates. Contrary to his hunting ancestors, the small farmer of today does not destroy a chicken or cow for a single feast, but nourishes populations of his animals, consuming their milk and eggs.

This sort of change from killing nearby organisms for food to helping them live while eating their dispensable parts is a mark of species maturity. It is why agriculture, in which grains and vegetables are eaten but their seeds are always stored, is a more effective strategy than the simple plant gathering. The trip from greedy gluttony, from instant satisfaction, to long-term mutualism has been made many times in the microcosm. Indeed, it does not even take foresight or intelligence to make it: the brutal destroyers destroy themselves—those who interact more successfully inherit the living world.

Even with an understanding of our origins, our view of our future blurs the further we look. But, as the visionary poet William Blake wrote, "What is now proved true, was once only imagined." There are many imaginable ways by which people might evolve into a species distinct from *Homo sapiens*. The simplest would not only be the accumulation of random mutations but by sexual recombination of preexisting genes. Although all human beings belong to the same species, population extremes may be noted. A Pygmy woman, for instance, may not be able to give a Watusi man a baby because her pelvis is too small. This example illustrates the natural variety present in any species, which may, over time, give rise to divergent species unable to interbreed because of outward changes resulting from inner ones: altered symbionts, behavior, rearrangements of chromosomes, changes in mitochondrial genes, duplications of nucleotide sequences in the DNA, or others.

But cells can now be fused in forced fertilization and the simple accumulation of vast numbers of changes in DNA base pairs can now be engineered. The genetic "writings" of future biotechnologists ultimately may be new organisms. The use of sets of bacterial genes—or at least the funding for such use—has already become commonplace. Through biotechnology those pieces of DNA called plasmids are inserted into bacteria and thus quickly replicated. Genes coding for proteins, even human proteins, may be replicated via association with plasmids.

The fascinating question of direct intervention in human evolution is approachable from several separate fronts: traditional natural selection (deforestation, animal and plant breeding) as well as newer techniques: biotechnology, computers, and robotics. Given that evolution accelerates, it must be only a matter of time before these approaches converge. Geologically speaking, we refer to exceedingly brief time periods, even within our children's lifetimes.

Computer science has been one of the most rapidly growing fields in the history of technology. From vacuum tubes to transitors and semiconductors, the information-handling elements of computers have miniaturized tens of thousands of times in only several decades. Their switching speed, the time required to switch on to off in a binary code, has decreased from twenty to a billion times per second.

As computerized records, books, and other devices become commonplace because the raw, siliceous, and miniaturized components of computers are so inexpensive, society will transform. The trend for money to become increasingly electronic will continue. Education will become easier as teaching gadgets enter the market. Beyond the "paperless office," there will occur what the computer expert Christopher Evans called "the death of the printed word." Traditional printed books will become as extravagant—and as expensive—to people of the future as first editions or hand-printed manuscripts seem to us. Books will appear to be immensely laborious undertakings. Each bulky mass of ink-spotted paper will take on the antiquated aspect of the Mainz Bible of Johannes Gutenberg. Because the complex nature of future societies is bound to be dependent on and monitored by computer intelligence, social movements, financial transactions, and exploratory discoveries will be recorded in machine memories. Because retrieval of computer-stored events will be far more faithful than movie "re-creations" or historical novels, it will be possible to relive history. Through technology, life's ancient ability to preserve the past in the present, its mnemonic fidelity, will vastly improve. This memory phenomenon, aided by cinema, written history, electromagnetic records, and other computer technology, is still accelerating.

Because silicon chips with thousands of bits of memory can pass through the eye of a needle today, microprocessors—tiny computers—are now lightweight enough to insert into machines, making them robots. Robots have great potential for the future. In 1976 the robotic part of the *Viking* spacecraft performed a task no human being could have done. Landing on the ultraviolet light-bombarded, frozen, and suffocating surface of the red planet, it stretched its mechanical arm, drew in a sample, and analyzed the dry and oxidized Martian regolith. Other robots are more mundane. Metal robots with many arms fasten tires to cars with a productivity rates that far exceed that of their human counterparts. The assembly line itself is becoming assembled. Robots in Japan, make parts for other robots.

As computers and machines come together in the new field of robotics, so robotics and bacteria may ultimately unite in the so-called biochip. Based not on silicon but on complex organic compounds, the "biochip" becomes an organic computer. Manufactured molecules, like photosynthesizing plants, would of course exchange energy and heat with their surroundings. Energy would be converted not into cell material, but into information. The possibilities inherent in such a development are awesome. "Living" computers could trade millions of hydrogen atoms per second and perhaps be integrated into conscious organisms. At this distance in the future the imagination is overwhelmed. The outcome of information exchange between computer, robotic, and biological technologies is not foreseeable. The most outlandish predictions, in retrospect, will seem naive.

What are possible fates of *Homo sapiens* in the next centuries? Let's explore two of many. As we have seen, the nucleated cells of all animals, fungi, and plants contain genes packaged as chromosomes. Species are known to evolve by several means, including chromosomal rearrangements, the accumulations of mutations in DNA, and symbiosis. Chromosomes undergoing heritable changes can cause jumps in evolution larger than those caused by nucleotide base-pair mutations. Symbiotic leaps can, in a few generations, establish new species. Such modes of variation should operate on populations of people. Abrupt chromosomal changes, such as those involved in karyotypic fissioning, have led to many new species of mammals. Karyotypic fissioning is the name of a process in which chromosomes break apart at their centers. Many species of Cenozoic mammals, compared with their ancestors, show half-chromosomes, broken at their centers. Dr. Neil Todd has shown how karyotypic fissioning has led to the evolution of dogs from wolves, pigs from boars, and even the humanlike apes from their apish ancestors. Combined with incest, karyotypic fissioning, in principle, may lead to new species of humans. The conquerors of the supercosm, if they are our descendants, or at least the descendants of some of us, are likely to have even more fissioned chromosomes than we do now and to have new traits, such as the ability to move easily, grow, and reproduce under decreased gravity.

Future humans may even be green, a product of symbiosis. An example of such a symbiotically produced species of human is *Homo photosyntheticus,* the imaginary cure to the heroin problem suggested by the algae expert, Ryan Drum. *Homo photosyntheticus,* he claims, are descendants of green heroin addicts with shaven heads into which a thin layer of algae has been injected. Strung out under the lights, such green hominids do not have to be addicts, but Drum suggests that because they would be fed by their internal resources, they would be far less of a social burden.

Evolution has already witnessed nutritional alliances between hungry organisms and sunlit, self-sufficient bacteria or algae. *Mastigias,* a Pacific Ocean medusoid, a peaceful coelenterate of the Man-of-War type, helps its photosynthetic partners by swimming toward the areas of most intense light. They, in return, keep it well fed. This could happen to our *Homo photosyntheticus,* a sort of ultimate vegetarian who no longer eats but lives on internally produced food from his scalp algae. Our *Homo photosyntheticus* descendants might, with time, tend to lose their mouths, becoming translucent, slothish, and sedentary.

Symbiotic algae of *Homo photosyntheticus* might eventually find their way to the human germ cells. They would first invade testes and from there enter sperm cells as they are made. (This is hardly outrageous: insect bacterial symbionts are known to do exactly this. Some enter sperm, and some are transmitted to the next generation via eggs.) Accompanying the sperm during mating, and maybe even entering women's eggs, the algae—like a benevolent venereal disease—could ensure their survival in the warm, moist tissues of humans.

In the final stages of this eerie scenario, we envision groups of *Homo photosyntheticus* lounging in dense masses upon the orbiting beaches of the future, idly fingering green seaweeds and broken mollusk shells. Electronically connected to their bank accounts, they would have no incentive ever to hurry.

We have suggested two possible paths of the evolution of humans. They are fanciful, perhaps, but the lessons of the past tell us that even if our details are absurd dramatic changes are inevitable. We can think of other peculiar possibilities. One is cybersymbiosis, the evolution of parts of human beings in future life-forms. People in this scenario are as crucial to the development of the supercosm as bacterial interaction was to the macrocosm. If we do transcend the fate of mammalian extinction and survive in an altered form, we may persevere not as individuals but as remnants. We can imagine ourselves as future forms of prosthetically pared people—with perhaps only our delicately dissected nervous systems attached to electronically driven plastic limbs and levers—lending decision-making power to the maintenance functions of reproducing spacecraft.

19

A Pox Called Man

Lynn Margulis

My title is from Nietzsche, who said 100 years ago that 'The Earth is a beautiful place, but it has a pox called man.'

Firstly, I would like to consider philosophy and philosophers. Scientists resent philosophy. I think this is because they are afraid that philosophers will reveal what scientists really do. I agree with Kierkegaard's assertion that the less support an idea has, the more fervently it must be believed. A totally preposterous idea requires absolute unflinching faith.

I suggest that our culture is teeming with preposterous ideas, believed with unflinching faith by scientists and everyone else, and that some of these actually corrupt our potential concern for the Earth.

Modern science has given very important insights about life, but our culture prevents us from accepting and utilizing those insights. I use four main examples: the Earth from space; the Chewong peoples of the forested regions of Malaysia; the organisms of the microcosm, most of which are ignored by biologists; and lessons from Gaia, the Vernadsky/Lovelock view of life beyond biology.

Nietzsche understood by philosophy "a terrible explosive in the presence of which everything is in danger." Scientists are terrified both by philosophy and philosophers. They tend to denounce philosophy as "soft" or deny its relevance, when in fact it has much to say about what scientists do.

The Earth from Space

The image of the Earth from space (Figure 19.1) transformed all the cosmonauts and astronauts. They have tried to explain their philosophical shift to the public, but they feel that no one listens. Frank White published a book containing interviews with all living cosmonauts and astronauts, which takes us with them out into space, looking in. For example, Eugene Cernan, the last person to walk on the Moon, says

FIGURE 19.1. The Earth rising over the lunar landscape. Courtesy of NASA.

> When you are in Earth orbit, looking down, you see lakes, rivers, and peninsulas such as Florida or Baja California. You quickly fly over the changes in topography like snow-covered mountains or deserts or tropical belts—all very visible. You pass through sunrise and sunset every ninety minutes.
>
> When you are in Earth orbit you get a new perspective. One minute you are over the United States, the next minute, you are over another area of the world. You can see from pole to pole and ocean to ocean without even turning your head. You literally see North and South America go around the corner as the Earth turns on an axis you can't see and then miraculously Australia, then Asia, then all of America comes up to replace them. You ask yourself the question, "Where really am I in space and time?"
>
> You don't see the barriers of color and religion and politics that divide this world. You wonder, if you could get everyone in the world up there, wouldn't they have a different feeling—a new perspective?

The astronauts and cosmonauts are all trying to convey that same message. Scientists studying other planets—Mars, Jupiter, Venus—study them as wholes, but we who study the Earth do not. Why don't we? We don't because we are children of a Judeo-Christian, Muslim, neo-Darwinist, or some other kind of religion. These religions are absurdities in that not only are they muddled, but they are dangerous for our relationship with the Earth and our nonhuman planetmates. The cultural background in which we have been brought up precludes our learning about the Earth as a whole planet. When scientific results clash with cultural and religious unstated "truths," science demurs.

Remote sensing capabilities, for example, tell us a good deal about the scale of the macrocosm. Figure 19.2 is a remarkable Landsat image of an Amazonian river (running from right to left) near Rondonia, Brazil. What are these lines? A variant of crop circles? A dehumanizing housing estate? At ground level we can see that these lines are actually roads surrounded on each side by destroyed strips of forest (Figure 19.3). The enormous rate of forest clearing makes it a global phenomenon.

We feel no pain as we saw off our noses to spite our faces because the tree carnage is done in the name of progress, saleable forest products, and more "Lebensraum" for desperate Brazilians.

FIGURE 19.2. Satellite image of stripes in the Amazon rain forest, Rondonia, Brazil. (Courtesy of C. J. Tucker, NASA.)

FIGURE 19.3. One stripe of the Rondonia Amazon rain forest at ground level. (Courtesy of C. J. Tucker, NASA.)

The Elders

David Suzuki and Peter Knudtson's wonderful book, *Wisdom of the Elders* (1992), shows that the pattern in many traditional cultures is more conducive to learning about biology than is ours. An example is the *med mesign* concept of the Chewong peoples of Malaysia. *Med mesign* means "different eye." Each type of creature sees the world through its own eyes. There is a tiger way, a water-snail way, a monitor lizard way, and a people way. *Med mesign* refers to the way each sees the world through his or her own perception. The Chewong identify with fellow life-forms, illuminating their moral obligations in their attitudes and daily activities.

In one Chewong tale, a family is relentlessly pursued by a ravenous tiger. Bongso is the Chewong spiritual leader and hero. He is gifted with the ability to see in every other creature's world without losing the perspective of his own. He eventually succeeds in saving the terrified family by impaling the tiger with a trap made of sharp spears deep in the forest. As the villagers look on, he blows sacred smoke on the slain beast's head and asks it, "Why did you want to eat us?"

And the tiger looks up, and with his last breath replies, "All I saw was meat. All I saw was meat. All I saw was meat." The tiger saw the fleeing family with *med mesign.* Indeed, each creature sees the truth looked at through its own eyes. The Chewong see meat as the wild game that they stalk in the forest. The man-eating tiger looked at mothers and babies, and all it saw was meat. All our culture sees is cash. All we see is cash. We only see cash. In the meantime, the forests are burned down, the rivers and oceans are polluted, children are neglected, and people starve.

The Organisms of the Microcosm

What insights can we garner from biology? From science? What new insights come from studying the subvisible organisms of the microcosm?

The only ultimately productive beings are the cyanobacteria. These green geniuses convert sunlight into organic matter and release gases to the atmosphere. Many of them happen to be trapped inside plants. Productivity is a bacterial, especially cyanobacterial, virtuosity now, and it always has been. Ultimately, a nation's gross national product can only be biological, not industrial.

From school biology we know about biological variation, character changes, DNA changes, and symbiogenesis. We know about the inheritance of variation and biotic potential—more individuals are produced than can possibly survive in the populations of all creatures at all times. We know that the most efficient way of getting rid of organisms, cockroaches, for example, is not to kill them one by one but to completely alter their habitat: to promote them, give them more habitat. We know about the effects of crowding. We know that the garbage never goes out, it just goes round and round. We know that matter is not lost, rather it circulates. We know that people's cells do not harbor former free-living photosynthetic bacteria that still actively photosynthesize whereas those of plants do.

We know that there are natural limits to all population growth. This cannot be taught because our culture tells us that humans dominate the Earth. And the culture only sees cash. We know that crowding causes destruction. We know it causes fighting and other extremes of behavior. Whenever mammals are crowded, aggressive behavior results: even herbivores cannibalize their fellows if severely crowded and starved. We know these things. Why can we not do something about it? Because our cultural presumptions contradict this knowledge.

Other lessons come from the microcosm. We know that the living world is not just inhabited by animals and plants. Plants are virtually identical to animals from the view of microbiology. This divides life into the bacterial world and everything else. We know that life began three and a half billion years ago, whereas animals appeared fewer than 700 million years ago. Most evolution has not involved animals at all, and yet nearly all our studies of evolution are of animals (Figure 19.4).

The protoctists, about 250,000 species, are mute and powerless. Yet they invented nearly everything of interest to evolutionists. Development of sexes, cell fusion, and intracellular motility are protoctist phenomena. The Protoctista constitute a fifth kingdom alongside plants, animals, fungi, and bacteria (Figure 19.5). Protoctists consist of the nucleated cells (eukaryotes) lying outside the fungi, plants, and animals. Symbiogenesis, my favorite subject, is involved in the speciation of all protoctists and many other eukaryotic organisms. Our cultural world is divided into "plants, animals, and germs," all presaging continued powerlessness for protoctists (Margulis, McKhann and Olendzenski, 1993).

In Western Australia, in Shark Bay, there are cyanobacterial ecosystems still in existence in areas too hypersaline for most other organisms. On this Australian coast, fascinating structures have been created by com-

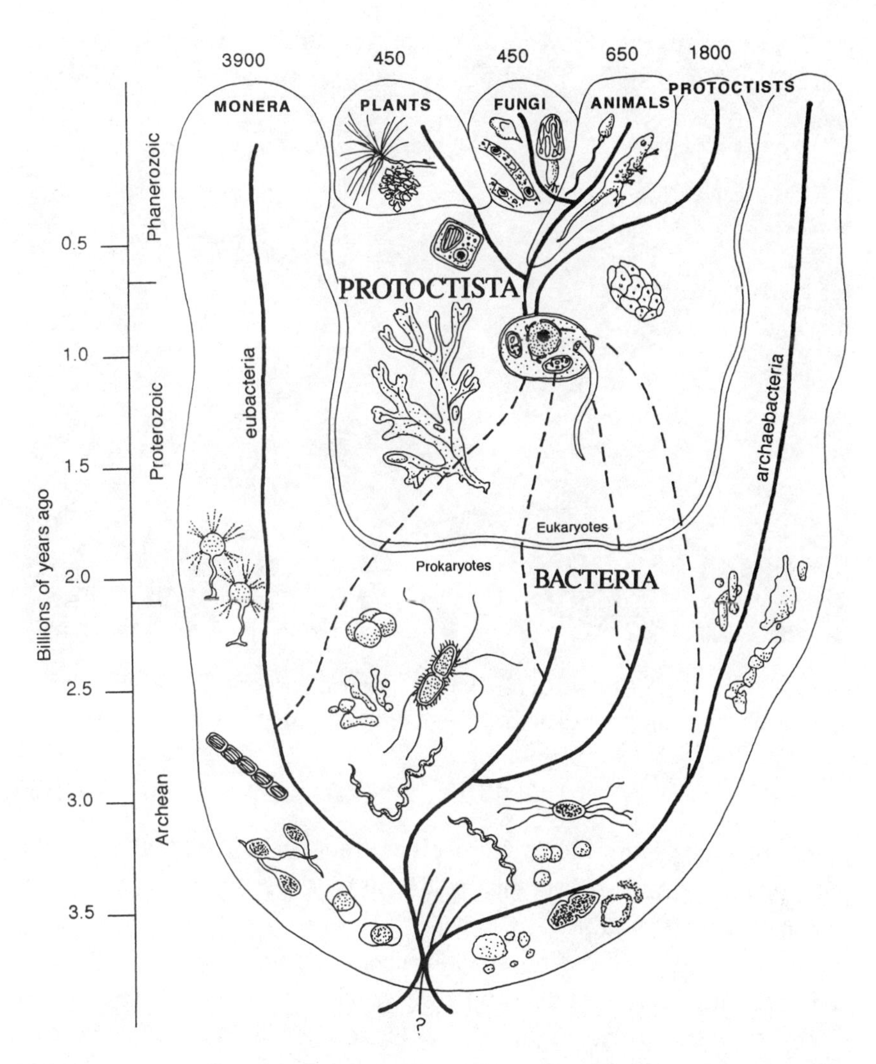

FIGURE 19.4. Evolution of the five kinds of life. The origins of life is indicated by the question mark. Drawing by K. Delisle.

munities of cyanobacteria (Figure 19.5). Their counterparts have been studied in fossils that are at least 2000 million years old. Looking underwater, the oxygen bubbles, the waste of cyanobacteria, can be observed to be released, breaking towards the surface. In deserts, microorganisms, unlike humans, can survive for days, months, or even years. As soon as water comes, many types resume photosynthesizing. They convert sunlight into organic matter, leak oxygen, and form communities. These microscopic communities are relatively stable, unlike those formed by humans. Studies

FIGURE 19.5. Stromatolites of Shark Bay, western Australia. These limestone rocks made by communities of microbes can be considered "bacterial skyscrapers."

of marine mud communities suggest that cyanobacteria and other accompanying microorganisms stabilize sediment to form community structures that allow the growth of many larger organisms.

It is now the 1990s and we still have not completed the eighteenth century Linnaean task of describing the species of life on Earth. This deficiency is especially evident in the three kingdoms that include microorganisms. My laboratory has spent 18 years studying 2 mm of what looks like dirty sand to most scientists. The microorganisms inhabiting the sand illuminate features of other beings. They grow, produce gaseous waste, and alter their environment. Predation occurs even in bacteria. Symbiogenesis leads to new forms. Great sensitivity to environmental changes abounds.

Bacteria and protoctists are not primitive, nor necessarily unicellular or simple. Bacteria can carry out every biological process known in the biosphere, except talk. We feel we are independent of microorganisms and that they should be eradicated, but this view is just part of our inflated human arrogance. Ralph Waldo Emerson, the nineteenth century poet, summed up the view most people still retain of the evolutionary process.

> Striving to be man, the worm
> Mounts through all the spires of form

Nearly everyone believes what our culture teaches: evolution has clearly come to its final summit, namely man.

Lessons from Gaia

Vernadsky is well known in Russia, but virtually unknown outside. He was a cartographer, crystallographer, and a fine scientist. He gives us an insight into ideas behind biology. His book *The Biosphere* was first published in 1926, but has yet to be fully translated into English.* Vernadsky viewed life as a complex organic mineral, animated water. He avoided the word *life* and used *living matter* instead. Gravity pulls things down, but living matter gradually pulls things across the Earth, he said. The biosphere is as much a manifestation of the Sun as it is of earthly properties.

> Ancient religious institutions which regarded terrestrial creatures, especially human beings, as 'children of the sun,' were much nearer the truth than those which looked upon them as a mere ephemeral creation, a blind and accidental product of matter and earth forces.

Vernadsky wrote this in 1944, in English, but he still has been magnificently ignored. Why? Because his insights are at odds with our cultural unstated assumptions.

The Gaia hypothesis of J.E. Lovelock, originally developed independently of Vernadsky, holds that the surface temperature, chemistry of the reactive gases, redox state, and pH of the Earth's atmosphere are homeorrhetically maintained by the metabolism, behavior, growth, and reproduction of living organisms. Homeostasis is a physiological regulation around a fixed point, like the control of adult mammalian body temperature around 37°C, whereas homeorrhesis, a parallel concept, refers to a regulation around a changing set point, like temperature regulation in a developing mammalian embryo.

Gaian environmental regulation is achieved largely by the origin, exponential growth, and extinction of organisms. All life is related by

*See footnote p. 220.

ancestry and physically connected by proximity to the fluid phases (water and air) at the Earth's surface. Organisms in communities form changing ecosystems that have persisted since the Archean period (3900 to 2500 million years ago). The interactions of organisms, driven by solar energy, produce and remove gases such that the chemistry of non-noble gases, temperature, and alkalinity are actively maintained within limits tolerable to life.

Within this conceptual framework, biological as well as physical sciences become appropriate to the analysis of the Earth's atmosphere and geologic history. Especially pertinent is the role of the microbiota—bacteria, protoctista, and fungi—in Earth's surface gaseous exchange that involves the recycling of those chemical elements (for example, H, C, O, N, P, S) absolutely required by life.

The product of the lively imagination of a British atmospheric chemist and the international space program, the Gaia idea has come of age. The atmospheric composition of the Earth signals unmistakably that the third planet is living: flanked by the dry carbon dioxide–rich worlds of Mars and Venus. One invokes either physiological science or magic to explain Earth's wildly improbable, combustive, thoroughly drenched troposphere (when compared with Mars and Venus). The Gaia hypothesis, in acknowledging this atmospheric disequilibrium, has opted for physiology over miracles.

Many scientists are unaware of the 25 years' worth of serious scientific Gaia literature and the potential contribution of the Gaia idea for integrating evolutionary, meteorological, sedimentological, and climatological data. Unfortunately, some other Gaia literature, and the hysterically toned New Age commentary that accompanies it, has received so much press attention and contentious comment that much of the primary science remains unknown.

Despite the fact that an "Earth system science" approach is vigorously encouraged for the solid earth sciences (see Chapter 17, p. 221), mention of the G-word (Gaia) still causes apoplexy in some scientific circles. This is remarkable, considering the broad parallelism of these approaches to understanding Earth processes. Despite avoidance of the term, a Gaian approach is advocated by the US National Academy of Sciences

> A new approach to studying Earth processes [is needed], in which the Earth is viewed as an integrated, dynamic system, rather than a collection of isolated components.

The Gaia hypothesis, rejected by some as the fantasy of New Age crystal swingers, has largely been misunderstood by the scientific community, yet it demonstrates how life sciences are essential to understanding the Earth. Part of its failure to be accepted comes from its revelation of the inadequacy of evolutionary theory developed in the absence of climatological and geological knowledge. The Gaian viewpoint is not popular because so many scientists, wishing to continue business as usual, are loath to venture outside their respective disciplines. At least a generation may be needed before an understanding of the Gaia hypothesis leads to adequate research.

Gaia has been called "Goddess of the Earth" or the "Earth as a single living being." These are misleading phrases.

I reject the analogy that Gaia is a single organism, primarily because no single being feeds on its own waste nor, by itself, recycles its own food. Much more appropriate is the claim that Gaia is a huge ecosystem, an interacting system, major components of which are organisms. Nowhere is this more evident than in examples of biotic influence on important geological processes, as described in Peter Westbroek's charming book *Life as a Geological Force*, 1991.

Gaia is noisy. If we listened carefully we could hear as our 30 million different species of planetmates sing to us. Can you make out the words of the song? "Got along without you before I met you, gonna get along without you now."

Many species, especially those in the four nonanimal kingdoms, do not need humans to take care of them and would not blink if we drove ourselves to extinction tomorrow. The assertion made by some politicians and propagandists that by preserving biodiversity we can somehow preserve the whole planet's life is just a further example of unabated human arrogance. Species conservation, as Niles Eldredge suggests, is primarily a matter of aesthetics and always has been:

> It is essential for our survival to conserve the global ecosystem, which translates into conserving as much as possible of the natural ecosystems of the world. It isn't really a question of species survival at all (except in our case). It is quite true that only those of us who love nature will be hurt if the spotted owl of the old forests of the Pacific Northwest is really driven to extinction through destruction of its habitat. Targeting individual species for survival is, in part, an act derived more from aesthetics than economics. But logging interests are quite right when they accuse conservationists of

> not wanting to save the owl so much as they want to save the forest itself. The forest—those magnificent stands of Douglas firs and other tree species—stands for the habitat, the ecosystem, itself.

Biodiversity of microbes is essential to nutrient cycling and therefore to plant life, but the Earth could survive perfectly well a return to the pre-Phanerozoic microbial scum. Although mammalian aesthetics would be devastated, indeed survival would be undermined, were large forms of life to be permanently extinguished, Gaia would continue to prosper as she did "before she met us."

Culture

Every scientist does research in a cultural context. We cannot help it and it is very hard for investigators of today to see the extent to which our research is culturally dictated. But if we go back more than a hundred years, aided by Charles Gillespie's book *Genesis and Geology,* we see more clearly. In 1829 the Earl of Bridgewater pledged £8000 in his will to any great man who would study "the Power, the Wisdom and the Goodness of God, as manifested in the Creation." The eight works thus produced became known as the Bridgewater Treatises. Of the fortunate recipients, four were clergymen, four were physicians. Three of the eight had concerns about the Earth. I want to talk about one of these three.

Professor William Buckland, one of the first to lecture formally on geology in the nineteenth century, was entrusted with the sixth treatise. Buckland introduced his Oxford University course with these words: that "the indications of the power, wisdom and goodness of the Divinity will be demonstrated from the evidence of design in His works, and, particularly from the happy dispensation of coal, iron and limestone, by which the Omnipotent Architect or Divine Engineer has assured manufacturing primacy to his British creations." In his treatise he went on to describe how "a system of perpetual destruction, followed by continual renovation, has at all times tended to increase the aggregate of animal enjoyment, over the entire surface of the terraqueous globe." The Earl's money had been made available to those who would study "the Power, the Wisdom and the Goodness of God. . . ," exactly what Buckland was doing.

We laugh at Buckland's contribution, yet he was merely working within his cultural context. He was pleasing the equivalent of his granting agency. Now let us laugh again at the National Science Foundation (NSF), which funds much of the scientific research carried out in the United States. If an investigator does what they, or other similar institutions around the world want, his loyalty cannot be to the Earth and its 30 million or so inhabitant species because they do not fit the goals of the national institutions. The bulletin of the NSF outlines some of the guidelines under which the grants are awarded.

> The science has intrinsic merit leading to the advance in that field.
> The probability that the research leads to new or improved technology.
> Will it improve the quality of this nation's man-power base?
> Will it integrate resources to contribute to society and to the nation?

Jingoism and field chauvinism prevail. For an example of another kind of nonsense, look at the geology section in any recent NSF bulletin: "Will the research provide insight into the physical and chemical processes that produce such geological features as hydrocarbon deposits?" But hydrocarbon deposits are biogenic; in fact, as I describe later, they are protoctistical. So there is no formal way for biologists to work in this area because of the ignorance of the NSF. They have never heard of protoctists. Just as the Earl of Bridgewater did, the NSF officers determine the kind of work that people do. Scientists, like everybody else, have their outlook blinkered both by their cultural context and by those who pay for research allotments. Is there another way?

I think science itself is just one way of knowing. The way in which it informs can be multiplicitous and widely used, to further the aims of the Earl of Bridgewater, the NSF, or for other goals. Science is simply a nondictatorial way of directing interactions with the material and energetic world. Science is a way of enhancing sensory experience with other living organisms and the environment generally. Everything is observed by an observer, but that observer exists within a cultural context.

Our culture measures scientific activity in the workplace by the rate of cash flow per square foot. Investigators are rewarded when we bring in students or grants, build buildings, buy chemicals—all of which increase the rate of cash flow. Biologists, geologists, and natural historians want to nurture the rest of the biospheric inhabitants. They see that there is something out there other than people and that the others are essential to the maintenance of human culture, maybe even human existence. So we are confronted by an ultimate contradiction: we want to nurture the 30 million species with which we share the planet, but our culture insists that the world is made for humans. The criterion for "scientific" success is the rate at which we convert the rest of the biosphere to urban ecosystems. Many of the conclusions of biological science cannot be encompassed by a culture that puts humans at the center of all things and only values the conversion of the biosphere into human habitat, including new biology buildings.

Conclusions

Kierkegaard said that the less support an idea has, the more fervently it must be believed. Totally preposterous ideas require absolute unflinching faith, including those discussed here by Suzuki and Knudtson (1992).

FIGURE 19.6. The Earth at night. (Compiled by Woodrull T. Sullivan and the Hansen Planetarium, University of Washington, USA. Credit: © 1985 W. T. Sullivan, Ill.)

Our beliefs are so fervently held, so intimately embedded in our perspective, that we can't even explicitly acknowledge them. By now I hope you will agree that

we have a cultural system
that ignores the air and water
and our biological heritage.

We have a society
that believes garbage goes out,
not around;
mistakes linen paper and metal disks
for food, searching the world at their demand;
and rewards scholars as they increase the rate of cash flow.

We suffer a culture that wants to convert the whole Earth
into its own image of God; an angry urban landlord.
Of course our culture resists the lessons of life.
Of course our culture dismisses bacteria, protoctists, and
fungi as germs and disdains the stranger.
It knows no other way.

The bright dots in the photograph (Figure 19.6) are cities full of people. "The Earth is a beautiful place," said Nietzsche, "but it has a pox called man."

PART IV

EVOLUTION AND EVOLUTIONISTS

20

Big Trouble in Biology

Physiological Autopoiesis versus Mechanistic Neo-Darwinism

Lynn Margulis

The Current Dilemma

More and more, like the monasteries of the Middle Ages, today's universities and professional societies guard their knowledge. Collusively, the university biology curriculum, the textbook publishers, the National Science Foundation review committees, the Graduate Record Examiners, and the various microbiological, evolutionary, and zoological societies map out domains of the known and knowable; they distinguish required from forbidden knowledge, subtly punishing the trespassers with rejection and oblivion; they award the faithful liturgists by granting degrees and dispersing funds and fellowships. Universities and academies, well within the boundaries of given disciplines (biology in my case), determine who is permitted to know and just what it is that he or she may know. Biology, botany, zoology, biochemistry, and microbiology departments within U.S. universities determine access to knowledge about life, dispensing it at high prices in peculiar parcels called credit hours.

As Ludwik Fleck (1979) documented, professional knowledge conforms to political realities. Any attempt to breach the acceptable is summarily dealt with, occasionally by devastating criticism, but far more frequently by neglect and ignorance. Hence biologists receive Guggenheim Fellowships for calculations of the evolutionary basis of altruism or quantification of parental investment in male children, while the tropical forests are destroyed at the rate of hundreds of acres per day and very little funding exists for the study of live plants in their natural environments.

A single example of the current dilemma suffices here: Since the retirement of Professor R.E. Schultes at Harvard University (1986), professional education in the production of food, drug, and fiber compounds by plants from New World tropical forests ("economic botany") is virtually unobtainable in the United States, whereas lessons in neo-Darwinist religious dogma are exceedingly easy to find. Computer jocks (former physicists, mathematicians, electrical engineers, and so forth), with no experience in field biology, have a large influence on the funds for research and training in "evolutionary biology," so that fashionable computable neo-Darwinist nonsense perpetuates itself. I here try to explore some of the roots of this institutional malaise.

The big trouble in biology is directly related to big trouble in our social structure and its priorities. This is a big subject. I necessarily limit my comments to the consequences of one philosophical muddle, an aspect of the academic biologists' assumed truth. Science practitioners widely believe and teach—explicitly and by inference—that life is a mechanical system fully describable by physics and chemistry. Biology, in this reductionist view, is a subfield of chemistry and physics. The idea expressed by physicist Sheldon Glashow (1988) is commonly held even among biologists: "Just as chemistry is ultimately reducible to physics, so is biology ultimately reducible to chemistry." We compare this pervasive mechanistic belief of biologists, most of whom are smitten by physicomathematics envy, with a life-centered alternative worldview called autopoiesis, which rejects the concept of a mechanical universe knowable by an objective observer.

Most practicing biologists do not yet know about autopoiesis; as an organized group of scientists, they do not face the issue, What is life? No tradition in the organization of professional life scientists forces them to ponder life itself. What is life? simply is not a subject of inquiry, even at plenary sessions of ISSOL (International Society for the Study of the Origins of Life). Rather, biologists, convinced that the universe is mechanical, engage in the incessant search for "mechanisms": of life, the human body, and the environment. By mechanisms they mean sound, light, or chemical signals interacting with carbon-containing matter that determine how life works.

The mechanistic worldview has many problems, one of which is the failure of neo-Darwinist biologists to think physiologically in general and to recognize the principles of autopoiesis in particular. Biologists are failing to embrace alternatives to a mechanical universe run by their supposed superiors: physicists, chemists, and mathematicians. A few of the destructive consequences of this philosophy on the academy and its students are outlined here. Both experimental work and theoretical analyses within the life sciences are severely affected by this prevalent physics-centered philosophy. Biochemical research, evolutionary biology, and biological education are all suffering the consequences. Neo-Darwinism is simply one example of a mechanistic philosophy used for illustration in this essay.

Autopoiesis

First, then: What is this alternative to mechanistic neo-Darwinism? What is this new concept of life, physiological in outlook, called autopoiesis? Autopoiesis is a set of some six principles developed by Humberto Maturana and colleagues to define the living (Varela, Maturana, and Uribe 1974). Autopoiesis combines the Greek words auto (self) and poiesis (to make); indeed, the latter root also gives rise to poetry. It refers to the dynamic, self-producing, and self-maintaining activities of all living beings (Fleischaker 1988). The word *autopoiesis* tries to define life by indicating its most indispensable aspects. Properties of autopoietic systems (such as cells, organisms, and communities), along with some physical and chemical correlates of these properties, are listed in Table 20.1. The simplest, smallest known autopoietic entity is a single bacterial cell. The largest is probably Gaia—life and its environment-regulating behavior at the Earth's surface (Lovelock 1988). Cells and Gaia display a general property of autopoietic entities: as their surroundings change unpredictably, they maintain their structural integrity and internal organization, at the expense of solar energy, by remaking and interchanging their parts.

Metabolism is the name given to this incessant buildup and breakdown of subvisible components—that is, to the chemical activities of living systems. If physiology is the study of the functions of living organisms and their parts, then metabolism is the chemical manifestation of those functions. Metabolism can be defined as the sum of the enzyme-mediated

Table 20.1. Properties of Autopoietic Systems

Property	Aspects	Examples of Biochemical/ Metabolic Correlates
Identity[a]	Structural boundaries; identifiable components; internal organization	Membrane-boundedness; nucleic acids, proteins, fatty acids, and other universal biochemical components of living systems
Integrity[a]/unitary operation[b]	Single, dynamic functioning system	Sum of multienzyme-mediated networks and their connection to nucleic acid and protein synthesis
Self-boundedness	Boundary structure produced by system	Lipoprotein membranes; gram-negative, cellulosic, or other cell walls and their connections to primary metabolism
Self-maintenance/ circularity[a]	Boundary structure and components produced by the functioning of the system	Lipogenesis, carbohydrate synthesis, peptidogenesis, nucleic acid synthesis (polymerization), and their interrelations
External supply of component raw materials	External supply of H, C, N, O, S, P, and other elemental constituents	Enzymes that incorporate CO_2, N_2, and so on into cell material; ribulose biophosphocarboxylase (RuBPC'ase), succinyl carboxylase, nitrogenase, and so forth
External supply of energy[b]	Light or chemical energy supply: convertible into organic bond chemical energy	Chlorophylls: methanogen coenzyme F, bacterial rhodopsin, uptake and incorporation of sugars and other organic compounds into system

[a]Varela F.G., Maturana H.R., and Uribe R. 1974. Autopoiesis: The organization of living systems. Its characterization and a model. *Biosystems* 5:187–196.
[b]Fleischaker G.R. 1988. *Autopoiesis: System, Logic and Origins of Life.* Boston: Boston University, Ph.D. thesis.

network of chemical and energetic transformations of living systems. It is more easily understood as the incessant movements of matter that occur all the time in living systems and that cease when the system dies. Autopoietic systems metabolize, whereas nonautopoietic systems do not. Proteins, viruses, plasmids, and genes are all components of live material. When contained within the boundaries of animal, plant, or other cells, they may be required to sustain cells or organisms and their autopoietic behavior; yet proteins, viruses, plasmids, and genes, intrinsically incapable of metabolism, are never autopoietic in isolation. Metabolism includes gas and liquid exchange (breathing, eating, and excreting, for instance); it is the detectable manifestation of autopoiesis. Autopoiesis determines physiology and hence is the imperative of all live matter. Autopoietic entities, that is, all live beings, must metabolize. These material exchanges are the *sine qua non* of the autopoietic system, whatever its identity. Metabolizing bacteria, of many different types, directly interact with each other via nonautopoietic components (for example, plasmids, viruses). Together, all the bacteria on Earth form a worldwide living system—a huge autopoietic entity. Charles Darwin recognized the continuity of the entire system through time, whereas Sorin Sonea has emphasized its unity through space (Sonea, 1987, 1988; Sonea and Panisset 1983).

Autopoiesis, in principle, does not depend on any specific material substances. Life may not have to be made of water; proteins containing carbon, nitrogen, and hydrogen; nucleic acid; nor any other particular chemical compounds (see Table 20.1). However, on Earth, since all life today has a common ancestry, all is part of a water–protein–nucleic acid chemical system with continuity for more than three billion years. Thus, knowledge of the chemistry of autopoiesis of life on Earth provides us a framework to evaluate studies of living beings, especially research on the origin and evolution of life (Fleischaker 1988). The autopoietic point of view of dynamic integral systems, using specific carbon-chemical interactions as the basis of self-maintenance, sharply contrasts with the current mechanistic view of life—the parent of neo-Darwinism, which is so highly fashionable in today's academic circles.

Neo-Darwinism

Neo-Darwinism, or the "modern synthesis," is a scientific school, primarily in English-speaking countries, that has been in vogue among biologists from universities and colleges since the 1930s. This body of work claims to unite the early twentieth-century discoveries of heredity (transmission, or Mendelian, genetics) with concepts of Darwinian evolution. Mendelian genetics, sometimes disparagingly called beanbag genetics by its critics, is the study of the transmission of traits (eye color, height, enzyme activity) from one generation to another. Evolution, according to the neo-Darwinist oracle, results from the accumulation of random heritable changes (mutations) in individuals.

The monk Gregor Mendel, breeding sweetpeas in the monastery garden at Brno, Czechoslovakia, showed definitively that certain heritable traits are indeed transmitted from parents to offspring without dilution, corruption, or any other change. Darwinian evolution, on the other hand, asserts that inherited changes in characteristics of organisms are established in populations as the result of natural selection; it emphasizes the differential survival and reproduction of organisms with distinct hereditary endowment.

Using algebra based on the Mendelian formalism developed for animal populations, neo-Darwinists proffer formal mathematical explanations for the ways in which organisms evolve. Neo-Darwinism has produced a large body of professional literature that is the sacred text of most evolutionary biologists. Self-identifying neo-Darwinists control what little funding for evolutionary research exists in this Christian country. Since the seventies, leaning heavily on computer simulations, the neo-Darwinist religious movement has generated subfields called population genetics, behavioral ecology, sociobiology, and population biology. The priests and practitioners teach the Mendelian precept that discrete genes act independently and that the interactions of genes determine the characteristics of the organisms that are selected. Fanciful abstractions have been invented by the neo-Darwinists, many of whom are scientists who, beginning as engineers, physicists, and mathematicians, found biology "easy." Several of them (for instance, Richard Dawkins of Oxford, Robert Trivers of Santa Cruz, Robert May of Oxford, John Maynard-Smith of Sussex, W.D. Hamilton of Oxford, and George Williams of Long Island) have become famous darlings of life scientists today. I attribute their popularity in part to the soothing effects of their assertions of mathematical certainty.

Yet, as British molecular biologist Gabriel Dover (1988b), instructor of genetics at Cambridge University, says, "it is unlikely that true Mendelian genes exist which do not contain any internal repetition and whose mutant alleles rely solely on selection or drift for increased representation in the population." If, as Dover is claiming, the assumptions used by neo-

Darwinists are indefensible, we spectators hardly can expect the mathematics of the subfield biologists listed earlier to illuminate the histories of life. Those remaining biologists who actually live among and observe metabolizing animals, plants, and microbes have difficulty measuring the quantities or even understanding general concepts labelled and taken as directly observable by the aforementioned mechanistic practitioners (such as "sexual strategy" and "cladistics" [Patterson 1983], and "inclusive fitness," "evolutionary stable strategies," and "cost-benefit energetics" [Maynard-Smith 1978a,b; 1983]). These imponderable immeasurables, in my mind, have no reference in the real world. However, the use of such labels serves a crucial social purpose. It binds the users, a growing group of influential scientists and their students, into a cohesive "thought-collective" (Fleck 1979).

Neo-Darwinists, closet neo-Darwinists, and non–neo-Darwinists argue among themselves about "who selects" and "what is selected." These intellectual skirmishes become acrimonious (Dawkins 1976, 1982). Dover (1988), for example, attempts to extricate us from some of these evolutionary tangles when he writes: "The study of evolution should be removed from teleological computer simulations, thought experiments and wrong-headed juggling of probabilities, and put back into the laboratory and the field. . . . Whilst there is so much more to learn, the neo-darwinist synthesis should not be defended to death by blind watchmakers." (Dover is referring here to the neo-Darwinist arguments forcefully presented by Dawkins in his 1986 book.) Abner Shimony (1989), in calling natural selection a "null theory," exposes the gross inadequacy of the common oversimplifications. Although the contribution of Darwin himself is lauded and his memory cherished, the physics-centered philosophy of mechanism and its runt offspring neo-Darwinism (Maynard-Smith 1983) is causing the "big trouble" referred to in the title of this Chapter. Like most scientists, the neo-Darwinist practitioners see themselves in a simple search for truth, believing they leave philosophy to the philosophers. Of course, they espouse the philosophy in which they are immersed, no matter how strongly they protest, "neutrality," "objectivity," and "reason."

Neo-Darwinist Oversights

My view is that neo-Darwinist fundamentals, derivative from the mechanistic life science worldview, are taught as articles of true faith that require pledges of allegiance from graduate students and young faculty members. I include as examples of such fundamentals a nonautopoietic definition of life; a bodiless, linear concept of evolution; and an uncritical acceptance of the mesmerizing concept of adaptation. I paraphrase some of these examples from standard textbooks of genetics and evolution:

> Life, according to the neo-Darwinist gospel, is a collection of individuals that reproduce, mutate, and reproduce their mutations.
>
> Evolution, according to this same testament, is change over time in gene frequencies (by gradual accumulation of mutations) caused by natural selection in natural populations.

This standard neo-Darwinist doctrine asserts that mutations arise by chance. They are chemical changes that are heritable, that is, changes in the DNA sequence of any cell or of any organism comprised of such cells. Such chance mutations, perceived as physical determinants of life that govern the existence of the organism, are purported to be the source of all evolutionary novelty. (Critics of neo-Darwinism, although they have no well-developed alternatives, have long dismissed the probability that eyes, brains, and flight evolved by chance [Clark 1984; Reig 1987; Vorontsov 1980].*) Neo-Darwinists then explain the strong correlation between structures of organisms and their survival requirements with the soothing idea that organisms "adapt" to their environments.

These assertions seem to me to be misdirected, incorrect, or, at best, grossly inadequate. Indeed, the term "adaptation" is used by late twentieth-century biologists exactly as it was by the early nineteenth-century British geologist William Buckland to describe the clever position of the Earth in the solar system and the deity's adequacies in his production of durable creations:

> In all these [favorable circumstances] we find such undeniable proofs of a nicely balanced adaptation of means to ends, of wise foresight and benevolent intention and infinite power, that he must be blind indeed, who refuses to recog-

*The origin of radically new behaviors and structures is probably heritable discontinuities that are then modified by mutation (hereditary endosymbioses, karyotypic fissioning, and so forth). These ideas of Neil Todd are detailed in Margulis (1993), Margulis and Bermudes (1985), and Bermudes and Margulis (1987).

> nize in them proofs of the most exalted attributes of the Creator. (Gillespie 1969)

Although philosophers David Abram and Dorion Sagan are among the few to say so explicitly, such prevailing neo-Darwinist fundamentals, with their pre-evolutionary legacies, are frankly at odds with nonmechanistic, including Gaian, system-philosophies of biology (Abram 1985; Sagan, Chapter 14). Nonmechanists, such as Lovelock, Bermudes, and Dyer, incorporate dynamic, interactive physiological thinking, whether or not they are explicit about their autopoietic perspective (Lovelock 1988; Margulis and Bermudes 1985; Bermudes and Margulis 1987; Dyer 1989). The life-centered alternatives to mechanistic neo-Darwinism recognize that, of all the organisms on Earth today, only prokaryotes (bacteria) are individuals. All other live beings ("organisms"—such as animals, plants, and fungi) are metabolically complex communities of a multitude of tightly organized beings. That is, what we generally accept as an individual animal, such as a cow, is recognizable as a collection of various numbers and kinds of autopoietic entities that, functioning together, form an emergent entity—the cow. "Individuals" are all diversities of co-evolving associates. Said succinctly, all organisms larger than bacteria are intrinsically communities. In this nonmechanistic view, animal and plant physiology becomes a specialized branch of microbial community ecology (Margulis 1993). Individual animals and plants are not selected by natural selection because there are no literal "individual" animals or plants; "natural selection" just refers to the fact that biotic potential is not reached; the ability of populations of cells and organisms to maximally grow is always limited by the growth of different cells and organisms and their associated surroundings.

Although appropriately critical biologists such as Dover have reviled the defensive naivete of the "neo-Darwinist modern synthesis," they have not replaced it with a comfortable philosophical alternative (Dover 1988a). Hence, insofar as I know, the irreconcilable tensions between the autopoietic and neo-Darwinist views have not yet been articulated.

Fundamentalism and Fundamentals: The Fleckian Thought-Collective

Ludwik Fleck, beginning at the age of forty-seven (in 1943), directed a microbiology and immunology laboratory in Buchenwald until 1945.

Saved from the gas showers because he was useful to the Nazis, the Polish Jew Fleck (with his coworkers, primarily Polish physician, Marian Ciepielkowski; French serologist and professor, X. Waitz; Eugen Kogon, bacteriologist; and professor Alfred Balachowsky of the Pasteur Institute, and some German technicians) was put to work producing vaccine against *Rickettsia prowazeckii,* the causative agent of typhus. For two years, while thousands of prisoners were marched to gas chambers just beyond the laboratory doors, Fleck and his colleagues produced large quantities of totally ineffective "vaccine," which was routinely sent to German soldiers at the war zones. Fleck reserved the real vaccine, in exceedingly short supply, to protect himself, his family, and friends. Surrounded by lives in daily danger, Fleck paid close attention to how easily scientists and technicians mentally imbibe the prevalent "common myth." In the end, Fleck's roughly six hundred liters of harmless "vaccine" was never more than a placebo—with which about thirty thousand SS men at the front were injected.

Daily duplicity not only ensured Fleck's survival, but also substantiated his theory of scientific facts. The theory claims that all "scientific facts" are merely consensuses among socially interacting "card-carrying" scientists. Fleck's book develops the concept that "the fact" is a product of a complex social process beginning with individual observation or measurement and terminating with the integration of a stylized "true statement" into the knowledge of the society at large. A practicing microbiologist and scholar for the rest of his life, Fleck—active as a scientist, philosopher, and beloved human being—died in Israel in 1961, some twenty years after his Second World War experiences (Cohen and Schnelle 1986).

Probably the drama of his own experience confirmed for Fleck the validity of his thesis (Fleck 1979). A key innovator in the field of the sociology of science, Fleck invented useful methods to analyze scientific activity. He showed how certain words and phrases become banners for the immediate identification of scientific friend or foe. Typical modern-day Fleckian examples include Lamarckism, Lysenkoism, vitalism, mechanism, Darwinism, sociobiology, and even autopoiesis. Fleck documented the processes by which social activities (including attendance at scientific meetings, contributions to professional newsletters and journals, incorporation of common myths into textbooks, and other instruments of socialization) cement into cohesive groups otherwise unruly scientists and technicians. These groups—which Fleck called "thought-collectives"—are then recognizable. They can be evaluated by the process of identification and naming. Once identified and named, the thought-collective achieves the status of "professional tribe," as do today's neo-Darwinists, whose members are bound together by many ties, including those of common scientific language.

Employing Fleck's concepts, I list in Table 20.2 a small sample of words drawn from neo-Darwinism in general. Sample neo-Darwinist terms in current use by molecular evolutionists are listed in Table 20.3. These "technical terms," I claim, have little significance except to the people who identify themselves as members of the scientific disciplines named in the titles of the tables, that is, as members of the thought-collective. By contrast, the universal terms in Table 20.4 are concepts relatively independent of language and culture. The value of these quantities is easily measured by scientists now, as they were in the past. Because none of the neo-Darwinist "battle cries" (Fleck 1979) in Tables 20.2 to 20.4 are directly measurable, all quantification associated with them is indirect and necessarily involves various assumptions and unstated hypotheses. These terms, devoid of meaning outside the neo-Darwinist context, including the molecular evolutionary context, serve this never-mentioned quasi-religious purpose: they bind practicing biologists into Fleckian thought-collectives that protect sacred knowledge.

Table 20.2. Neo-Darwinism: Words Used as Battle Cries[a]

Adaptation
Altruism, altruistic behavior
Cheating, selfish behavior
Fitness, inclusive fitness
Genetic variation, diversity
Genotype, phenotype
Group selection
Individual
Kin selection
Levels of selection, units of selection, natural selection
Sexual selection, sexual reproduction
Species, race

[a]These kinds of criticisms of neo-Darwinist concepts and terminology have been made also in Lambert D.M., Miller C.D., and Hughes T.J. 1986. On the classic case of natural selection. *Rivista di Biologia—Biology Forum.* 79:11–49; and Hughes A.J. and Lambert D.M. 1984. Functionalism, structuralism and "ways of seeing." *Journal of Theoretical Biology,* III: 787–800.

Table 20.3. Molecular Evolution: Words Used as Battle Cries

Advanced, primitive organisms
Archaeobacteria, eubacteria, metabacteria
Conserved sequences
Eucytes, parkaryotes[a]
Higher, lower organisms
Molecular homology, convergence, divergence
Quickly evolving/slowly evolving molecules
Rooted trees

[a]Lake J.A. 1988. Origin of the eukaryotic nucleus determined by the rate-invariant analysis of RNA sequences. *Nature* 331:184–186.

Table 20.4. Universal Science: Terms and Their Units of Measure

Acceleration (centimeters per second per second)
Density (grams per unit volume)
Energy (ergs)
Heat (calories)
Length (meters)
Light intensity (einsteins)
Magnetism (electromagnetic units per gram)
Mass (grams)
Pressure (torr, atmospheres, bars, millimeters of mercury)
Temperature (degrees Kelvin, degrees Fahrenheit)
Velocity (meters per second)
Volume (length, width, height)
Time (seconds, years)

Why do members of the neo-Darwinist social group dominate the biological scientific activities in U.S. and other English-speaking academic institutions? Probably there are many reasons, but a Fleckian one is that the neo-Darwinist mechanistic, nonautopoietic worldview is entirely consistent with the major myths of our dominant civilization. Our rapacious civilization, identified by the fact that international currencies can be exchanged within it, has been characterized by William Irwin Thompson, that master social critic and analyzer of mythmaking (mythopoiesis), as follows:

> We have built up a materialistic civilization that is concerned almost exclusively with technology, power and wealth. . . .

> Each culture casts its own shadow, a shadow which is the perfect description of its own form and nature.
>
> The shadow which our technological civilization casts is that of Lilith "the maid of desolation" who dances in the ruins of cities. Now that we have made a single polluted city of the entire world, she is preparing to dance in the ruins of our planetary megalopolis. . . .
>
> To effect a reconciliation [with Lilith] man must not seek to rape the feminine and keep it down under him. If he seeks to continue his domination of nature through genetic engineering and the repression of the spiritual, he will ensure that the only release from his delusions can come from destruction. Lilith will then dance on the ruins of Western civilization. (Thompson 1981a,b)

The myths of our technical civilization are easily contrasted with those of some Native Americans. These great people from Beringia (the landmass present some ten thousand years ago, when glaciers bound huge quantities of water in what is now the Bering Strait) preceded all European and African migration onto these two huge American continents. Perhaps we can assume that Chief Seattle speaks for his ancestors and descendants when he says, "The Earth does not belong to Man, Man belongs to the Earth. All things are connected, like the blood which unites us all" (Campbell 1983).

In the world monetary civilization, geological and biological resources are perceived as infinite. Indeed, their very existence is assumed to be determined by human activities (such as market supply, labor, and so forth). Such myths of our technological civilization cannot accommodate an autopoietic-Gaian view of natural history, like that quoted here from Chief Seattle. The Native American perception, just as any nonmechanistic worldview, must be rejected by neo-Darwinists, in whom such views induce psychic dissonance. A world philosophy based on the recognition of the autopoietic and nonmechanical nature of life must upset the believers in the fundamental myths of our technological civilization. In the world of the Native American, humanity belongs to the Earth; in the world of the money machines, the Earth belongs to humanity. In the autopoietic framework, everything is observed by an embedded observer; in the mechanical world, the observer is objective and stands apart from the observed. In the autopoietic view, the only truly productive organisms are the green photoautotrophs (bacteria, algae, and plants capable of converting sunlight energy into the organic compounds of food) and a few of their bacterial chemoautotrophic relatives (some obscure forms of life, like those living at great depths in submarine vents, capable of converting geochemical energy into food); in the mechanical view, humanity is truly and infinitely capable of being productive. The autopoietic view, which accepts as given that green linen paper is not food and can never be food, also realizes that garbage never goes out, it only goes around; in the mechanical worldview, economics and politics are thought to be directly related to quantities of money and its distribution.

Central to the autopoietic view is the physiological idea that the material components of all life incessantly move: they cycle at the surface of the Earth in chemical transformation and physical transport that always depend directly on the energy from that brilliant star, our Sun. Humanity has very little to do with the fact that the matter of life is always transporting and transforming at the surface of the Earth. The Earth behaves physiologically and not mechanically. We people (*Homo sapiens,* only one of perhaps 30 million living species) accelerate but do not dominate the metabolism of the Earth system.

We people, for all our architectural maneuverings and hydroelectrical water reroutings, for all our cementing of grasslands and conversion of tropical forests into steak, can never be productive: we can only consume the organic products of the green autotrophs referred to earlier. Our use of energy for automobile and jet-plane locomotion and our consumption of food such as *Zea mays* (corn) and *Triticum* (wheat) is simply the playing out of our autopoietic nature as newly evolved, mammalian-weed apes (Margulis and Sagan 1997).

Physiologically oriented biology, studies of life that recognize that autopoietic entities are qualitatively different from other countable matter, tends to be ridiculed or ignored by current practitioners of neo-Darwinism. I suspect that neo-Darwinists, upon observing physiology and contemplating autopoiesis, suffer cognitive malaise. Their mathematized formulations systematically ignore physiology, metabolism, and biological diversity; they fail to describe the incessant, responsive, reciprocal effects of life embedded in environment. Suffering philosophical distress, physics-worshiping neo-Darwinists must reject autopoiesis and its attendant life-centered biology with the same zeal with which the Spanish true church, guarded by its Inquisitors, rejected the mescal- and peyote-eating religions of the Native Americans.

Until the present, only scientists outside the great wall of the English-speaking academy have espoused nonmechanistic, non–neo-Darwinistic philosophies. Such scientists develop Gaian philosophies (Lovelock 1988; see also Chapter 13) or are engaged in building secondary biospheres (Sagan 1987, 1990a). In the meantime, inside the monastery, in university life-science departments, victims are accumulating.

Who are the victims of these latter-day religious wars for the souls of the biological science practitioners? Primarily graduate students, young investigators, and teachers, in whom direct observations of life and experience in the field often foster an expansive autopoietic attitude. The study of physiology and immersion, especially in tropical nature, tends to lead students to a perception that the living planetary surface behaves as a whole (the biosphere, the place where life exists on the Earth). Yet the academy guards, using neo-Darwinism as an inquisitory tool, superimpose a gigantic super-structure of mechanism and hierarchy that protects the throbbing biosphere from being directly sensed by these new scientists—people most in need of sensing it. The dispensers of the funds for scientific research and education, and other opportunity makers, herd the best minds and bodies into sterile laboratories and white-walled university cloisters to be catechized with dogmatic nonsense to such an extent that many doctoral graduates in the biological sciences cannot distinguish a nucleic acid solution from a cell suspension, a sedimentary from an igneous rock, a kelp from a cyanobacterium, or rye from ergot. The English-speaking biology academy has lost sight of the biological priorities. Furthermore, young investigators or students, potential ecologists, botanists, and zoologists who stray from the neo-Darwinist fold are threatened with expulsion from this prevailing Fleckian thought-collective with its mechanistic thought-style. Were today's budding biologists to take seriously Thompson's mythopoiesis, Varela and Fleischaker's autopoiesis, and Lovelock's Gaian analysis, they would, en masse, have to walk out on the university (Thompson 1981b). In other words, if an individual with ambition to study nature rejects neo-Darwinist biology in today's ambience, he becomes a threat to his own means of livelihood—that is, to his own autopoietic integrity.

One lesson of the autopoietic concept of biology is that in general, for any organism, many potential threats to its autopoiesis exist. Examples include lack of food, restricted living space, and improper salt balance. A commonly employed name for any general threat to autopoietic integrity is "stress." All organisms—swimming bacteria, surf-battered algae, and hormone-exuding college students—can behave to reduce stress. All organisms respond in ways determined by their hereditary endowment and their environmental astuteness to lessen threats to the self-maintenance of their internal organization. Stress-purging, stress-avoiding, stress-reducing behavior is intrinsic to all autopoietic entities. Nonautopoietic entities do not respond, they are passive. Neither automobiles nor DNA molecules can resist stress.

From these comments it can be concluded that among academic biologists inside the convent walls, neo-Darwinist reductionism will prevail until the suddenness of a new planetary culture replaces the technological civilization to which Thompson refers. Only after the new civilization binds us consciously to our nonhuman planetmates, especially the truly productive green ones, can the physiology of autopoietic visionaries replace the mechanics of the neo-Darwinists inside the academic cloister. Alternatively, neo-Darwinism is expected to prevail until overpopulation (with its concomitant toxic water, polluted airways, and garbage) destroys technological civilization and its money-machine stockpiling thought-collective, of which neo-Darwinism is only a tiny part.

Neo-Darwinism and Gaia

Gaia is the idea that certain environmental surface properties of the Earth—for example, the temperature and chemical composition of the lower atmosphere—are directly controlled by the biota. (The biota is the sum of the organisms inhabiting the Earth: live animals, plants, and microorganisms. The biosphere, which extends some 8 kilometers above and 12 kilometers below the surface of the Earth, is the place where the biota resides.) The validity of the Gaia idea, of the self-regulating biosphere, has been forcefully argued by Lovelock (1988). Indeed, the Gaia hypothesis has been called a "grand unified theory" of biology (Sagan 1988); it also has been recognized as more a point of view than a scientific hypothesis (Sagan 1990b; Chapter 14).

In autopoietic language, Gaia is the largest unit we know of that displays the properties listed in Table 20.1. For those unfamiliar with the Gaia hypothesis, probably the best way of thinking about it is to contemplate the assertion that the atmosphere and surface sediments of the Earth are part of the living system. That is, life does not "adapt to" a passive physicochemical environment, as the neo-Darwinists assume. Rather, life actively "produces and modifies" its surroundings. The oxygen we breathe, the humid atmosphere inside of which we live, and the mildly alkaline ocean waters in which the kelp and whales bathe are not determined by a physical universe run by mechanical laws. In stark contrast with a mechanical, physics-centered world, the metabolizing biosphere is physiologically self-controlled. The breathable oxygen, humid air, and mildly alkaline oceans result from the growth of bacteria, plants, and algae that produce oxygen using solar energy; water transportation is driven by the activities of great forests, primarily of neotropical trees; and the neutralization of the acid tendencies of the planet is accomplished by the production of alkaline substances such as urea and ammonia by myriad sea creatures (for example, by urination and bad breath). These are simply three examples of Gaian Earth-surface regulatory activities. Many others exist (Lovelock 1988; Hinkle 1988).

The Gaian worldview is an autopoietic one; the surface of this third planet is alive with a connected megametabolism that leads to temperature and chemical modulation systems in which humanity plays a small and only very recent part. (After all, humanity as *Homo sapiens sapiens* evolved only some forty thousand years ago, long after the Gaian system, which is more than three thousand million years old, was completely in place.)

Neo-Darwinists, who ignore chemical differences between living beings, who never factor autopoiesis into their equations, and who consider organisms as independent entities evolving by accumulation of chance mutations, must hate and resist autopoiesis and the Gaian worldview.

If we can assume that consistency is a scientific virtue, then acceptance of a Gaian-autopoietic worldview requires that we reject the philosophical underpinnings of neo-Darwinism as it is currently practiced. Neo-Darwinism, in the Gaian perspective, must be intellectually dismissed as a minor, twentieth-century sect within the sprawling religious persuasion of Anglo-Saxon biology. As yet another example of a thought-style in the great family of biological-scientific weltanschauungen, past and present, neo-Darwinism (like phrenology and nineteenth-century German nature philosophy) must take its place (like British social Darwinism) as a quaint, but potentially dangerous, aberration.

The current dilemma, the big trouble of conflicting myths and thought-styles in professional biological science, is not likely to see resolution soon. Speaking for the practitioners of autopoietic-alternative worldviews, who recognize the embeddedness of all people in the great Gaian system, I must applaud the philosophy of Chief Seattle. With him we realize that "Man belongs to the Earth," and money, only green linen paper, is indigestible for all autopoietic entities like us who lack lignases (lignin-digesting enzymes). At the same time, we must face our social fate and scientific destinies. Regrettably, the destinies within academia of the proponents of physiology and autopoiesis probably more resemble those of Seattle and other Native Americans than those that await practicing neo-Darwinists.

After all, the glorious, greedy tribesmen of western Europe (the aggressors) and their African slaves from whom most people on this green new North American continent are descended and from whom we imbibed our myths of domination, are the true fathers of neo-Darwinism. These ancestors, sharing a racist and anthropocentric thought-style, easily confiscated the land and decimated the people to replace the nature-knowing Native Americans. Thus, any of us academic biologists who welcome a lively biology should be naive indeed if we conclude that the neo-Darwinist thought-collective will abrogate its powers and succumb to logic and reason without an intellectual battle to the death. The academic groves and wet field-stations, the university corridors and DNA-recombination laboratories, the governmental funds for missions to planet Earth, the ribosomal RNA-sequence data banks, the column chromatographs, the shuttle payload bays, and the contemplation of the Amazon River Basin will not be surrendered by the neo-Darwinists nor any other money-machine representatives until a punctuated discontinuity in thought-style penetrates their thought-collective from the outside. Circumstances beyond their control must lead the presently powerful to relinquish their strongholds. Forces beyond their present awareness must overtake these entrenched servants of greedy masters. Perhaps this is what Thompson, (1981a) means when he writes:

> When we have moved beyond the desolation of all our male vanities, from the stock market to the stockpile of rockets, we will be more open and receptive. Open and bleeding like that archaic wound, the vulva, we will be prepared to receive the conception of a new [planetary] civilization. . . .

21

THE RIDDLE OF SEX

DORION SAGAN
AND LYNN MARGULIS

Why are so many organisms sexual? What keeps them competitive with organisms that accomplish the same end through budding or fission? Most biologists explain the existence of traits in organisms on the basis of survival value to the individual or the species (Futuyma 1985). Yet the story of the evolution of reproductive patterns is not clearcut. Our ongoing efforts to understand this process illustrate a key area of research in evolutionary biology today (Margulis and Sagan 1986b).

At first and perhaps second glance, sex seems a superfluous and unnecessary evolutionary bother. To put it in the economic language in which biologists have described evolutionary science from its inception, the "cost" of sex—finding mates, producing special sex cells with half the usual number of chromosomes, and investing time in these activities—seems all out of proportion to any possible advantage.

Biologists have thought that sex remains because of the increased variety of zygotes that results from two parents. This variation, it was reasoned, allows sexual organisms to adapt faster to changing environments than do asexually reproducing organisms. Yet there is absolutely no evidence that this is true.

When the idea was tested by comparing animals that can reproduce either asexually or sexually, such as rotifers and asexually reproducing lizards, scientists found that as the environment varied, the asexual forms were as common as or even more common than their sexual counterparts (Bell 1982).

Biologists need a new perspective on this important problem. We believe that sexuality in animals is a product of a history in which sex became entangled with reproduction. Sexual animals have been successful for reasons not directly related to biparental sex. Thus, we think that it is not sensible to ask, What selection pressure maintains sex in an organism? Once animals and some other organisms became committed to a link between sexuality and reproduction, in many cases there was no turning back.

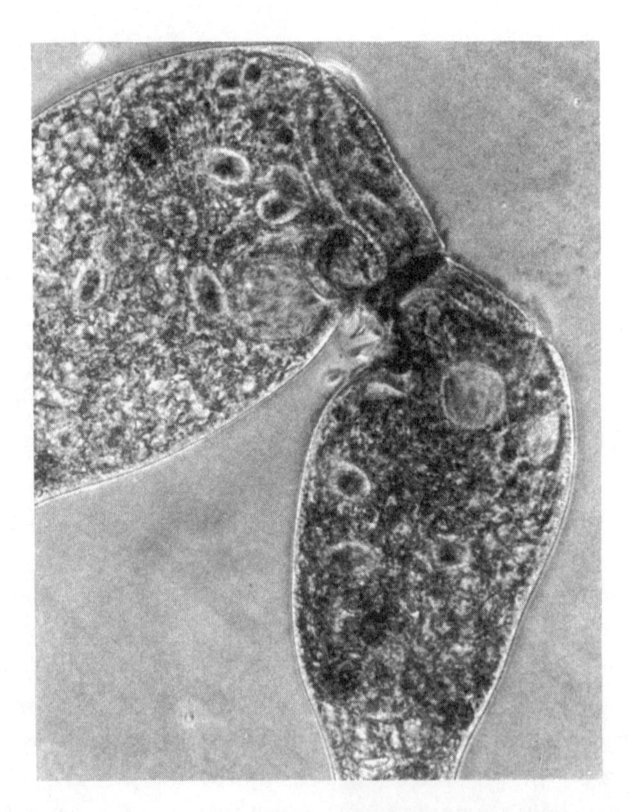

FIGURE 21.1. A photograph of sex in ciliates of complementary genders: the mates look just the same to us. The death of both individuals follows the process within four days.

How did these organisms become sexual in the first place? In discussing a topic so enriched by the imagination, we must define our terms. Sex in the biological sense has nothing to do with copulation; neither is it intrinsically related to reproduction or gender. Sex is a genetic mixing in organisms that operates at a variety of levels; it occurs in some organisms at more than one level simultaneously.

We, then, are defining sex as a union of genetic material to produce an individual from more than a single parent. The smallest known framework for sex, defined in this way, is the entry of nucleic acid into a cell. In bacterial sex, bacteria regularly exchange genes, for example, by passing genetic information in the form of viruses. By this definition, too, an influenza virus infecting humans is engaging in sex by inserting its genetic material into the cells of its host. In protoctists, plants, fungi, and animals, sex takes the form of a fusion of cells such that two nuclei from different parents join within a common cytoplasm. *Stentor,* two of complementary genders, in sexual embrace, are seen here in Fig 21.1.

Bacterial Sex

Bacterial sex may have begun over three billion years ago, when our atmosphere lacked free oxygen (Cloud 1983). Without oxygen, no ozone layer existed in the thin atmosphere to protect genetic material from ultraviolet radiation (Margulis, Walker, and Rambler, 1976). Data gathered by *Explorer 10* from Sun-like stars suggest that the output of light energy at that time may have been so great that it was a wonder that genetic information survived at all (Canuto, 1982).

Yet life did develop, under the pressure of constant bombardment from both benign visible and dangerous ultraviolet rays. Bacterial and viral sex must have soon followed, as a means of guarding and spreading needed genes throughout the threatened biosphere. Any organism that could not protect its scant genetic hoard in that age soon perished.

Early sex seems to have developed from a genetic repair system that could restore damaged DNA. The first repair may have happened by chance—a case of chemical desperation. Those cells that detected damaged DNA and excised it survived. As time went on such methods were refined. In photosynthetic organisms in particular, for which the radiation was both essential and lethal, repair became a way of life.

In standard DNA repair, an organism copies an intact strand to produce a healthy double-stranded molecule. This splitting and splicing is closely related to sex—the mechanism that allows the cell to accept DNA from a foreign source. Thus methods that first allowed survival in a radiated world evolved into sexual mechanisms.

Bacterial sex promoted both diversity and survival. New varieties arose as patterns for new proteins were shared and copied. Even today, toxins and ultraviolet light can revive eons-old solutions. Some bacteria respond dramatically to ultraviolet DNA damage. They immediately stop growing, release viruses or plasmids (if they have been harboring these small genetic entities; Sonea and Panisset 1980), and make error-ridden copies of their damaged cell DNA (the "SOS response") so that at least some descendants will survive.

Conversely, when bacteria lose the ability to deal with ultraviolet light, they often also lose their genetic recombination system. The "rec minus" mutant of *Escherichia coli* can no longer recombine; it is also hundreds of times more sensitive to death by ultraviolet radiation than its sexual relatives. The two processes, protection from ultraviolet radiation and genetic recombination, must be very closely related.

Thus the repair of ultraviolet light damage may have preadapted bacteria to sex. By rupturing genes, this energetic form of light put selection pressures on bacteria for the development of repair systems, some of which involved "adopting" DNA from neighboring cells. By the time the atmosphere developed a protective layer of ozone, splice-and-repair mechanisms had been integrated into the life of bacteria.

The genetic recombination that so fascinates genetic engineers today evolved first as a technique for DNA repair and then into the closely related sexual mechanism. Research on fertility factors, episomes, plasmids, infection, and conjugation all involve the recombining of genes; they are all forms of bacterial sex.

Meiotic Sex

The sexuality of familiar plants and animals—the sex that is hitched to reproduction—is not the genetic splicing of bacterial sex. Meiotic sex, found in eukaryotic organisms, is an entirely different procedure that evolved after bacterial (or prokaryotic) sex. Meiotic sex involves two reciprocal processes: the reduction by half of the number of chromosomes to make sperm, eggs, or spores, and the fertilization that reestablishes the original chromosomal number.

While bacteria have sex under certain conditions, they never need it to reproduce (Figure 21.2). Most animals and many plants, however, must undergo this complex process for the species to survive. How did meiotic sex evolve? Its origin seems tied not only to mitosis but to symbiosis (Fig 21.5) the history of which is a fascinating evolutionary puzzle in its own

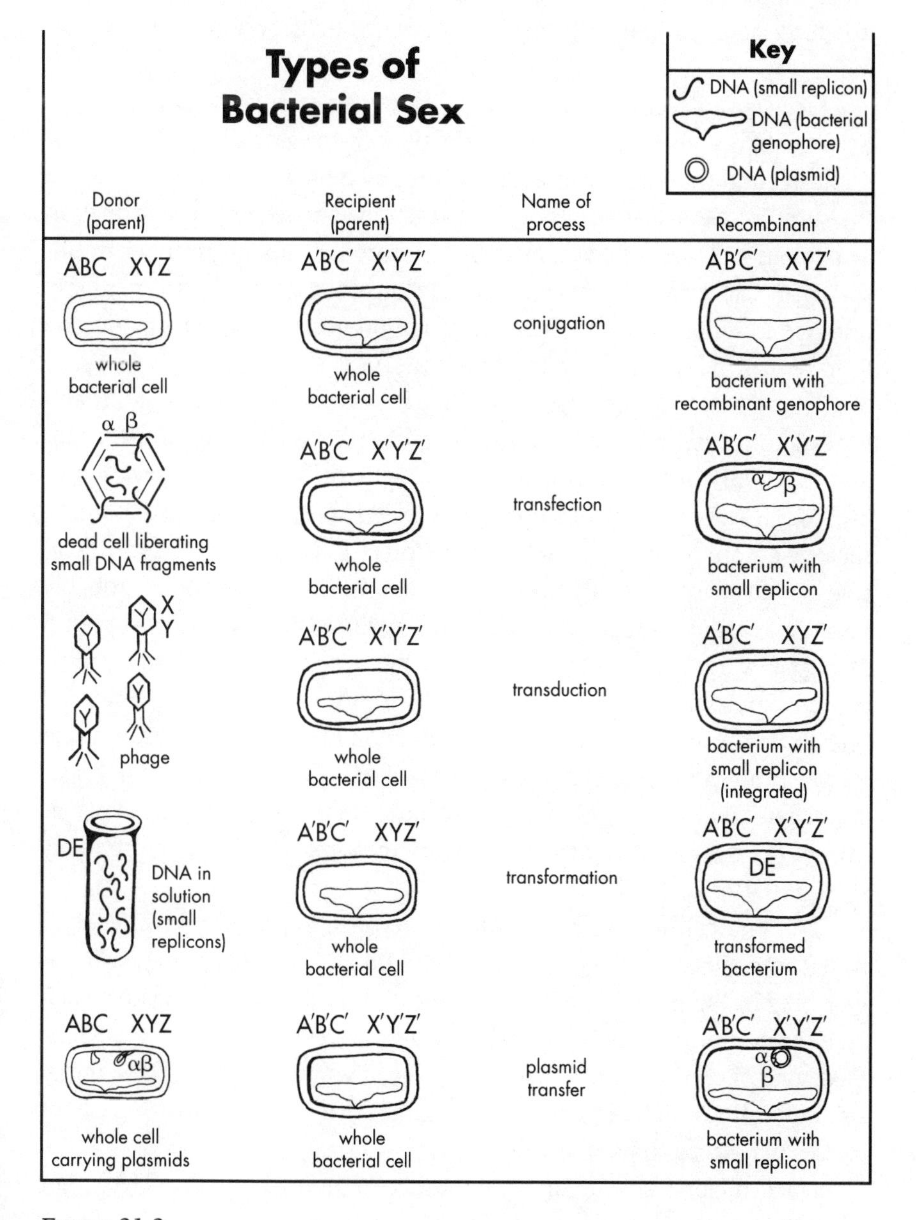

FIGURE 21.2.

right (Margulis 1981). (The symbiotic origin of mitotic cell division as precursor to meiosis is just too complex to detail here (see Chapters 3 & 4 pages 34 & 47). Let's just say that the ultimate effect of mitosis is the distribution of genetic information in DNA–protein packages, called chromosomes. This meticulous genetic delivery system handles hundreds of times more genetic information than bacterial cells. Its efficiency institutionalized it as the standard mechanism of cell division in plants and animals.

Meiosis as a form of cell division follows a pattern very similar to mitosis; it differs in that when a cell divides, chromosomal DNA does not replicate, and the kinetochores (structures attaching the chromosomes to the spindle) are delayed in their reproduction. The result is the formation of haploid cells, destined to meet and restore the diploid number in offspring. Because meiosis never occurs in organisms that do not regularly undergo mitosis, and meiosis is a variation on mitosis, meiosis is assumed to have evolved by modification of mitosis.

The two distinct phases of meiotic sex—chromosome reduction and fertilization (cell and nuclear fusion)—arose separately. They are still separate and only occur sporadically and irregularly in some eukaryotic microbes. However, in certain lineages, such as those ancestral to animals, meiotic chromosome reduction and the precise fusion known as fertilization became coupled (Figure 21.3). Chromosome reduction probably began as a delay in the timing of reproduction of both kinetochores and DNA. By waiting too long to divide in mitosis, the kinetochores pulled two chromosomes to each of the two offspring cells instead of the normal one chromosome. This resulted in a reduction in chromosome number in the offspring cell. If we regard these reproducing kinetochores as remnant spirochetes living in the chimera of a modern nucleated cell, this explanation of their duplication delay as the origin of meiosis seems more likely (Margulis 1993).

The first cell fusion, a precursor to fertilization, could have resulted from cannibalism, where one already-mitotic microbe ate another without digesting it (Figure 21.4). Microbes have no immune defense against such an internal grafting. This cannibalism would have led to diploidy, the doubled state of chromosomes that is "relieved" by meiosis. Regardless, meiosis and fertilization had to have become interlocked in a feedback cycle for today's patterns to have evolved.

Once meiotic sex became established it flourished (Fig 21.6). But why? Biologists must be careful not to jump to conclusions. The evidence

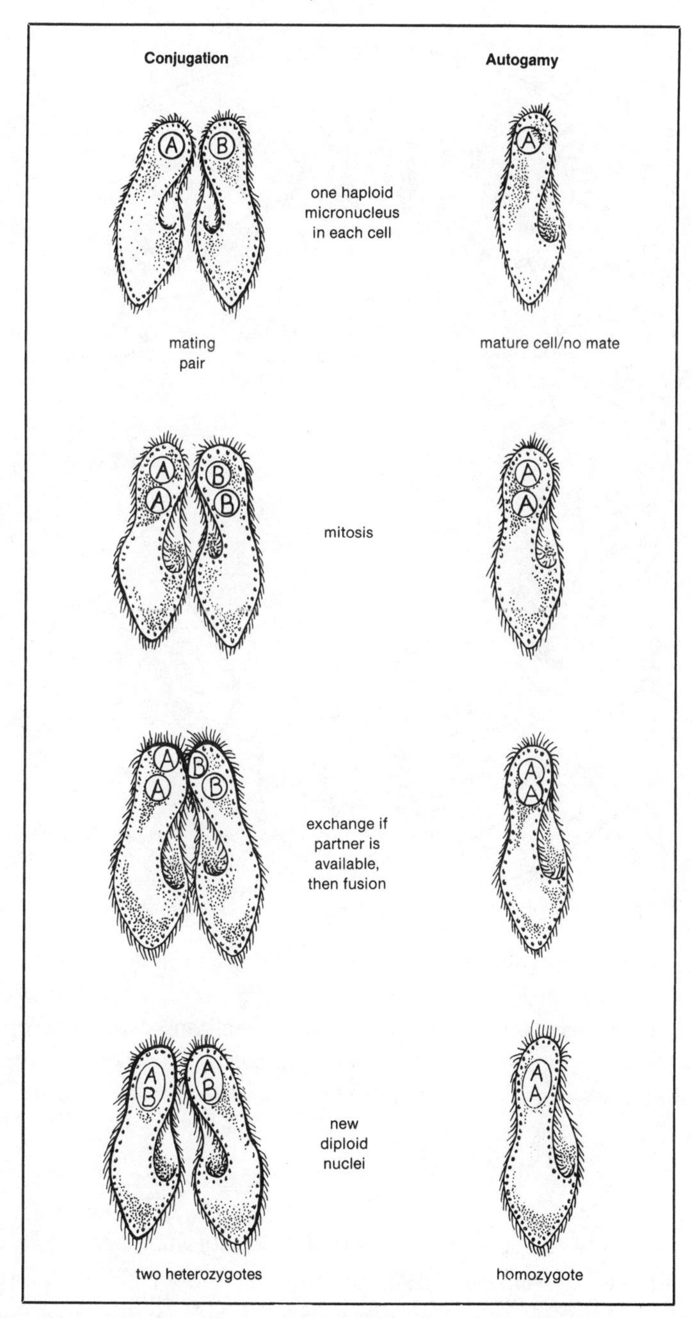

FIGURE 21.3. When paramecia are ready to conjugate, a complex process of meiosis and mitosis produces eight haploid micronuclei, all but one of which die (*top*). Mitosis produces two micronuclei. If a partner is available, one micronucleus will be exchanged and the pair will fuse. If no partner is available, the cell's own two nuclei will fuse. The cell achieves rejuvenation in either case.

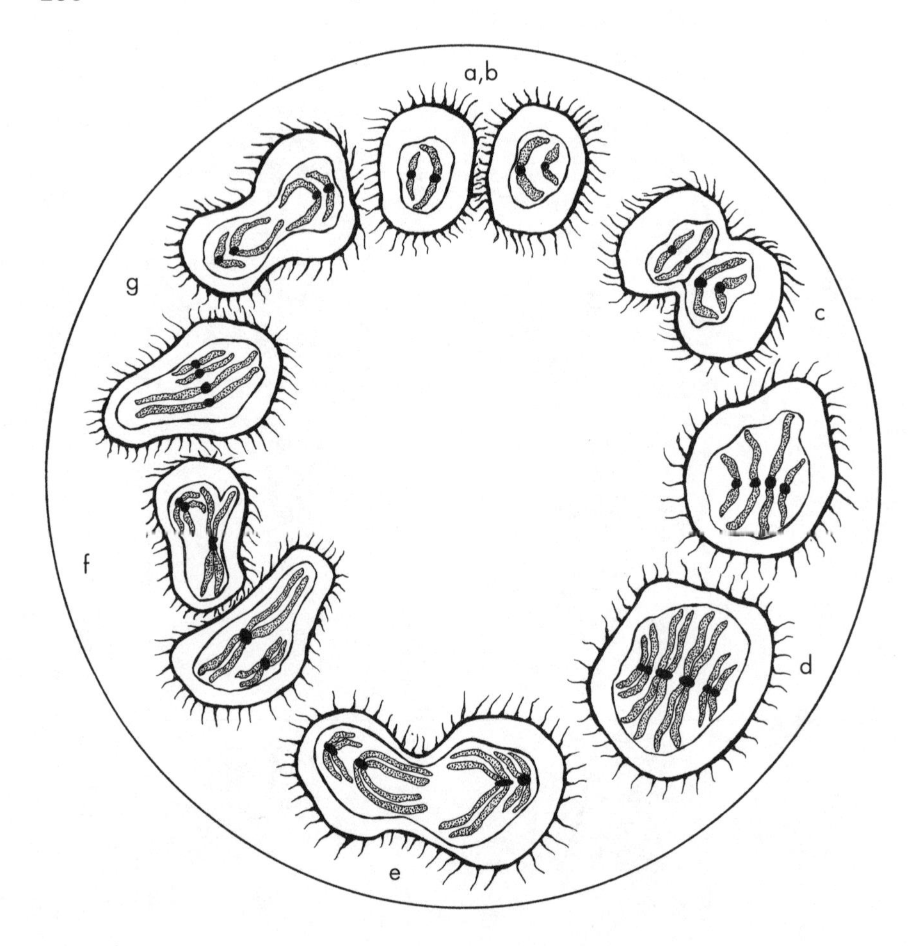

FIGURE 21.4. Meiotic sex may have originated from a form of cannibalism; consider the hypothetical cycle above. Clockwise, two starving protists (a,b) fuse to form a double organism (c). The chromosomes replicate (d) and the organism divides (e), but the "tardy" kinetochores lag behind (d–f). The kinetochores finally replicate (g) and the cell divides again (g) to form haploid organisms (a,b).

shows that meiotically sexual organisms are not automatically more varied or better adapted to changing environments than those lacking meiotic sex.

Many theories of sex are clearly fallacious. A recurrent but dubious interpretation describes sex as some sort of genetic rejuvenating mechanism. This theory is based on the observation that asexually produced protists, paramecia, survive for only months, while sexually conjugating strains survive indefinitely. Yet there is a counter example to the theory.

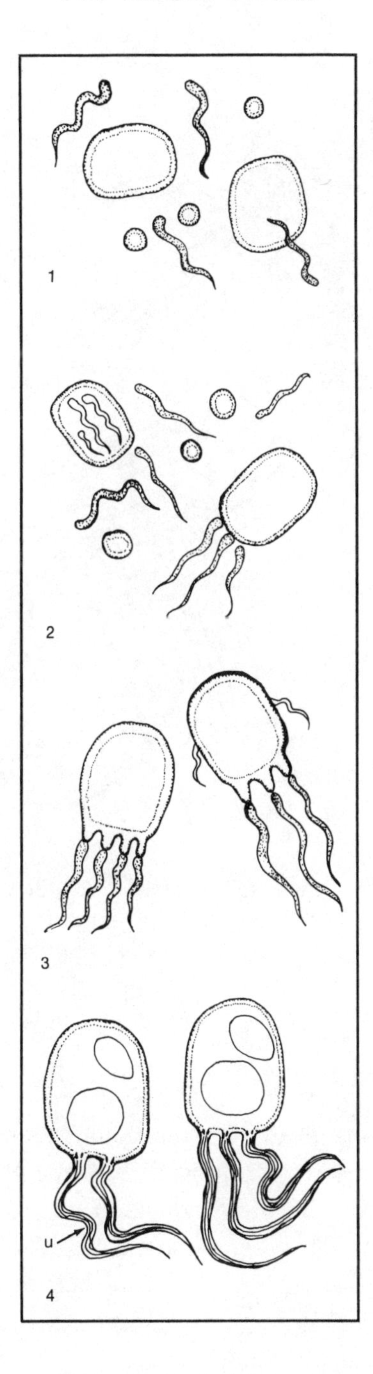

FIGURE 21.5. Motility in eukaryotes resulted from early symbioses between spirochete-like organisms and larger cells. The resulting undulipodia (cilia and "eukaryotic flagella," u) are far more complex than bacterial flagella. By analogy to modern spirochete associations, mitosis in eukaryotes originated through further evolution of such motility symbioses (Sagan and Margulis, 1993; See Chapters 3 and 4, pp. 35–58).

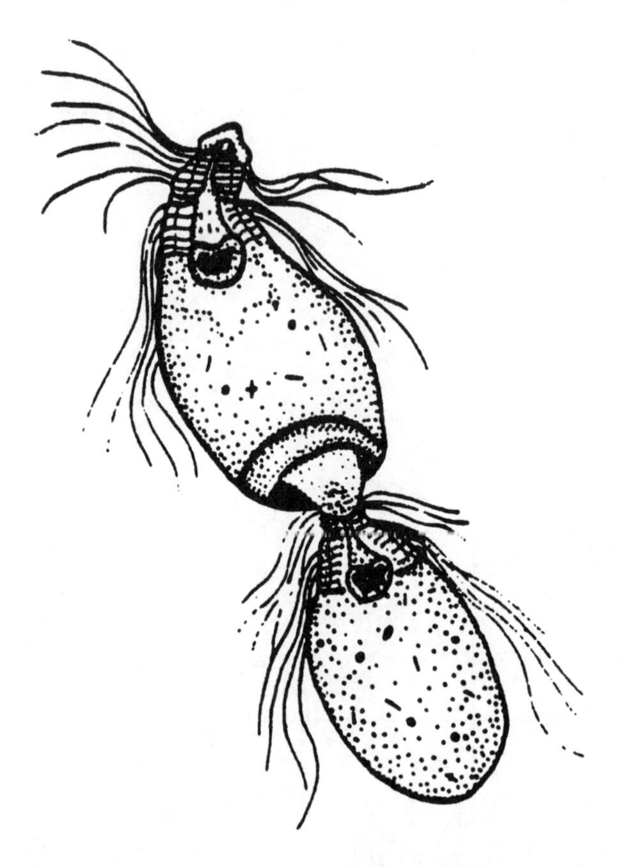

FIGURE 21.6. The words "male" and "female" have different meanings in the protist world. In mating *Trichonympha,* the organism that travels or enters its partner is arbitrarily defined as male (the lower right-hand organism). In others like *Stylonychia* (not shown) or *Stentor* (Figure 21.1), the mating ciliates are identical, and male-female designations are meaningless.

A paramecium that gets ready for sexual conjugation but finds no partner undergoes a process known as autogamy in which the nucleus of a single cell undergoes meiosis and the products of meiosis from the same cell fuse in the absence of any sexual partner. This totally inbred cell line survives just as long as do its conjugating relatives that undergo two-parent sex (Figure 21.3).

Paramecium aurelia, for example, has one large macronucleus and two small micronuclei. The enormous macronucleus, with thousands of gene copies, usually does all the work, making messenger RNA, while the diploid micronuclei do nothing. But during autogamy each diploid micronucleus divides twice meiotically, forming four haploid micronuclei. These four divide mitotically once, creating eight haploid micronuclei. Then, in nature's typically absurd style, all but one of the haploid micronuclei die. This last one divides mitotically to create two micronuclei with exactly the same genes. If a willing sexual partner is present, conjugation occurs and one of the two micronuclei is sent to the partner as another one is received from the partner. But—and this is a big but—if no partner is around, the two haploid nuclei fuse. No new genes have entered the paramecium. Indeed, this self-fertilization renders the organism entirely homozygous. Yet the paramecium is recharged, rejuvenated, able to reproduce again for generations.

Thus it is not the receiving of genes from two parents but the meiosis itself that often accompanies the gene exchange that is important in survival in *Paramecium aurelia.* We think that meiosis became tied to two-parent sex and that meiosis as a cell process, rather than two-parent sex, was a prerequisite for evolution of many aspects of animals.

Meiotic sex and tissue-level multicellularity both evolved well before 520 million years ago; meiosis seems intimately connected with complex cell and tissue differentiation. After all, animals and plants return every generation to a single nucleated cell. We believe that meiosis, especially the chromosomal DNA-alignment process in prophase, is sort of like a roll call, ensuring that sets of genes, including mitochondrial and plastid genes, are in order before the multicellular unfolding that is the development of the embryo. Meiosis has been maintained because it is connected to physiological necessities such as tissue development in the hosts. When the complex meiotic organisms survived, meiosis was taken along for the ride.

Putting these ideas of sexual origins together, our hypothesis is quite different from the accepted wisdom about the role of sex in evolution. Bacterial sex, a modified DNA repair mechanism, allows organisms to accept new genetic components as easily as one can catch a cold. Without it the more complex cells of animals and plants could not have evolved. Although the roll call processes of meiosis are crucial, the two-parent aspect of meiotic sex is an evolutionary legacy. Animals' complex tissues and organs, not their sexuality, are what was selected for. Some animals did, in certain cases, forgo two-parent sex, but they never gave up meiosis.

Ongoing work in this area opens up intriguing ideas, including the possibility of human reproduction that would circumvent the biparental sexual cycle by cloning an egg to make a person. If we are correct, biparental sex, but not meiosis, will be bypassed in such cloned people. Like many areas of biological research, conventional wisdom concerning sexual origins should periodically be reexamined.

22

Words as Battle Cries—Symbiogenesis and the New Field of Endocytobiology

Lynn Margulis

. . . and there is the additional consideration, that each of the elements whose fusion goes to make up the impregnated ovum, is held by some to be itself composed of a fused mass of germs.

Samuel Butler 1898

Our minds are incarcerated by our words. The biological term *symbiosis* has been used in a way that obscures not only its literal meaning but also the phenomenon's instrumental role in evolution. Biology textbooks define "symbiosis" anthropocentrically—as mutually helpful relationships or animal benefits, implying social contract or cost-benefit analysis by the partners. This definition is silly—symbiosis is a widespread biological phenomenon that preceded by eons the human world and the invention of money.

"Symbiosis" was defined first by German mycologist H.A. De Bary (1879) as "unlike organisms living together." The phrase "unlike organisms" soon came to mean members of different species.

Lichens, complexes of fungi associated with photosynthesizers (either cyanobacteria or green algae), have served as examples of symbionts. The lichen fungus *Cladonia* on a petri plate grows as fuzz; the lichen alga

Trebouxia in pure culture on the surface of agar as slime; but alga and fungus growing together as the British soldier is ground cover; superficially the lichen is a land plant (Figure 22.1).

Since the last century, scientists have recognized that symbiosis has the power to generate great biological novelty and discontinuity. I argue that symbiosis is far more innovative in the generation of biological novelty than is the accumulation of chance mutations, although the latter is more commonly credited as the basis of evolutionary change.

Students and teachers most often encounter the word "symbiosis" in its textbook definition. Authors of science texts typically describe symbiosis as follows:

- An internal partnership between two organisms in which the mutual advantages normally outweigh the disadvantages (Collocott 1972).
- An association that must always benefit at least one of the species, because otherwise it would soon dissolve (Minkoff 1983).

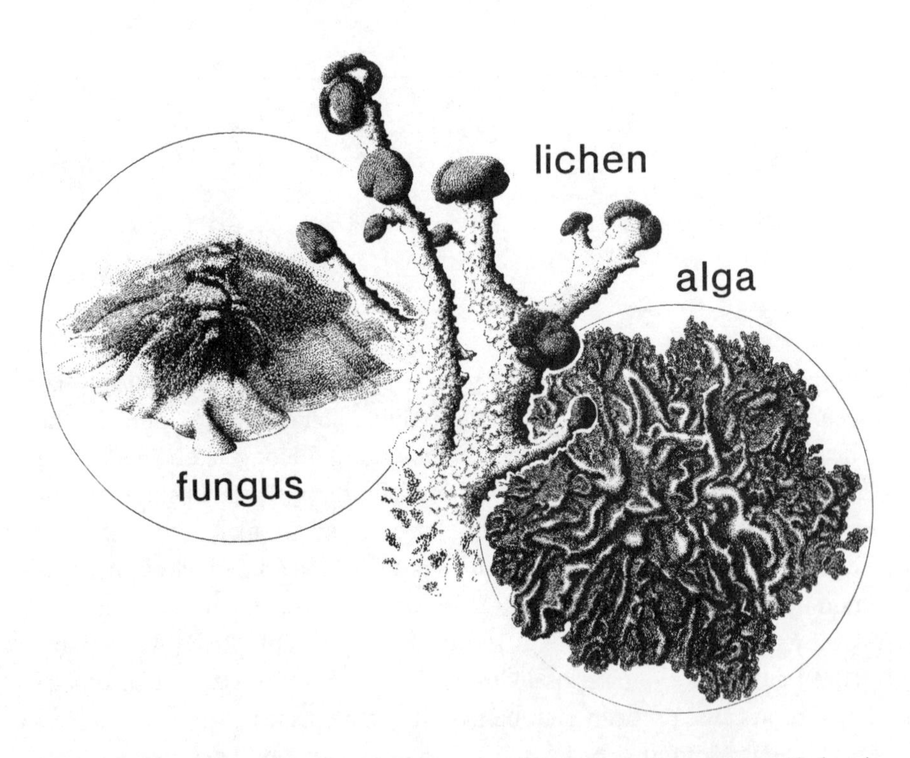

FIGURE 22.1. The two bionts (the green alga *Trebouxia* and the fungus *Cladonia*) merge to form the holobiont (*Cladonia cristatella*), the British soldier lichen. Drawing by Christie Lyons.

Even in current biological secondary literature, symbiosis is often taken to mean a "mutualistic biotrophic association" (Schiff and Lyman 1982) or a "mutually beneficial . . . relationship" (Avers 1989). However, the research scientists today studying symbioses embrace De Bary's original definition in modern guise: symbiosis refers to protracted physical associations among organisms of different species, without respect to outcome. These scientists reject the textbook and secondary analyses that gauge symbiosis on what might be considered customer satisfaction.

Symbiosis Does Not Equal Mutualism

If symbiosis is defined as a beneficial relationship between organisms of different species, it is difficult to distinguish it from mutualism. Recent biology texts use mutualism to refer to social relationships among organisms, of the same species or of different species, which need not be physically associated. Because symbiotic partners must be members of different species that are in physical contact with each other, according to textbook definitions, symbiotic relationships should be a subset of mutualistic relationships. Both symbiosis and mutualism are considered to be positive, or favorable, relationships, as opposed to negative relationships such as parasitism. But these relationships are contingent—especially sensitive to environment (Lewis 1973a).

And complications abound. In practice, temporal and spatial aspects of symbiosis often are not described in texts. Symbiosis researchers examine whether or not the partners experience prolonged, permanent, cyclical, facultative, or casual relationships (Starr 1975). Most writers focus instead on the impossible: proof that the association "benefits" the partners. Because the unassociated partners (for example, the alga or fungus) cannot be grown under the same conditions as the lichen, a strict proof of "benefit" cannot be made.

Attempts to clarify meanings have compounded the problem because measuring benefit (a unitless quantity), either in the field or laboratory, has not been not feasible. As the term symbiosis became nearly synonymous with biotrophic mutualism, new terms to indicate neutral relationships between physically associated organisms were invented (Lewis 1973b). Commensalism (from the Latin, meaning eating from the same table) describes two species of organisms physically associated with each other but deriving nutrients from a third (for example, clownfish and sea anemones feeding on bacterial symbionts). Phoresy is used to describe the carrying of one organism by another (for example, remoras by sharks).

A Note About History

Almost entirely unknown to English-language scientists, a Russian school of biology science in the early 1900s emphasized the role of symbiosis in evolution. Andrei Sergeivich Famintsyn (1835–1918) experimented with the isolation and growth of chloroplasts from plant cells. Konstantin Sergeivich Merezhkovsky (1855–1921) developed the "two-plasm" (cell-within-a-cell) theory, claiming that chloroplasts originated from cyanobacteria (blue-green algae). From this work, he invented the term "symbiogenesis," the "origin of evolutionary novelty via symbiosis." Finally, Boris Michailovich Kozo-Polyansky (1890–1957) suggested that cell motility originated by symbiosis. Thus each of these three Russian scholars, all of whom held esteemed positions in Russian academia, contributed fundamentally to our understanding of the concept of symbiogenesis (Khakhina 1979; Figure 22.2).

In the early 1900s in the United States, by contrast, there was little research on symbiosis. Anatomist I.E. Wallin (1883 to 1969) was prolific and enthusiastic in his early years when he stated his principle of symbionticism (1927), by which he stressed the importance of obligate microbial symbioses in the origin of species. But his ideas were rejected and ridiculed (Mehos 1992), and for the last forty years of his life, while working at the University of Colorado Medical School, Wallin avoided symbiosis research.

Paul Portier, a French contemporary of Wallin, also emphasized the importance of symbioses in evolution (Margulis 1981; Portier 1918). Although Portier was supported by the King of Monaco, he too was aggressively attacked. The French scientific community, led by the microbiologist August Lumiére (1919), helped demolish western enthusiasm for the role of symbiosis in evolution (Mehos 1992).

Mutual-Aid Biology

Human social concerns have inextricably permeated discussions regarding the participants in symbiosis. These concerns have contributed to the misconstruing of the term. Belgian biologist-politician P.J. Van Beneden (1873) first used the term "mutual aid" in describing "repayment" for

FIGURE 22.2. The Russian symbiogeneticists: Boris M. Kozo-Polyansky, Andre S. Famintsyn, Konstantin S. Merezhkovsky, and the symbionticist from Colorado Ivan E. Wallin. See Sapp, 1995 and Khakhina, 1992 for details.

services among "lower animals." Wholesale extrapolation from "the society of men" to "the community of animals" became especially evident in Peter Kropotkin's *Mutual Aid* (1902). A Russian prince exiled to London, Kropotkin sought answers to questions of human relations in nature:

> Mutual aid is met with even amidst the lowest animals, and we must be prepared to learn some day, from the students of microscopical pond-life, facts of unconscious mutual support, even from the life of micro-organisms. (Kropotkin 1902, p. 10)

Kropotkin's analyses of animals, "savages," "barbarians," medieval city-dwellers, and modern society all extend his theories that

> . . . mutual aid is as much a law of animal life as mutual struggle, but that, as a factor of evolution, it most probably has a far greater importance, inasmuch as it favours the development of such habits and characters as insure the maintenance and further development of the species, together with the greatest amount of welfare and enjoyment of life for the individual, with the least waste of energy. (Kropotkin 1902, p. 6)

To Kropotkin and many subsequent scholars, the idea of symbiosis and mutual aid—cooperative forces in evolution—was to be contrasted with the idea of competition—a negative force leading to the struggle for existence. Kropotkin's work accentuated both the confounding of mutual aid with symbiosis and the imposition of human social analysis on descriptions of organismal interaction.

Most western scientists have regarded symbiosis and mutualism as political slogans, therefore choosing not to focus experiments on these biological phenomena. For most of this century, then, symbiosis research was divorced from cellular, molecular, and evolutionary biology.

Evolutionists and most other biologists—both experimental and theoretical—still consider symbiosis analyses to be remote to evolutionary analyses (Keller and Lloyd, 1991). Symbiosis is ignored, or only defined, in the major textbooks of evolution (for example, Avers 1989; Ayala and Valentine 1979; Ehrlich and Holm 1963; Futuyma 1986; Kimura 1983; Minkoff 1983).

Only two English-language biology textbooks use symbiosis as their organizing principle. One, designed for undergraduates, is an excellent introduction to symbiosis (Ahmadjian and Paracer 1986); it describes dozens of associations by taxa. The second, an erudite and useful graduate text, is dedicated to the experimental analysis of symbiosis (Smith and Douglas 1987). But neither book evaluates symbiosis as a major mechanism of generating heritable variation in evolution.

Obscurity and Funding

Symbiosis remains an obscure, primarily botanical subfield of biology and, at least in the United States and United Kingdom, is still not funded per se. In contrast to mainstream zoological pursuits (for example, parasitism and infestation [which are associated with disease and thought to require urgent scientific investigation and high levels of funding], cladistics, or systematics), the healthy, positive, perhaps even feminine connotations of symbiosis and mutualism have suggested that research on these topics is relatively unimportant. Indeed, this term-contentiousness has impeded research. Most of my colleagues[1] would agree that mention of symbiosis in a grant application tends to deny funding.

This prejudice leads to limited support for symbiosis research. There have been studies, assisted by the Office of Naval Research, of bacteria harbored in the light organs of luminous fish. Agricultural research funds have fostered analyses of leguminous plant associations with nitrogen-fixing bacteria. And zoologists have been encouraged by oceanography and marine science programs to study algae of coral reefs. However, in these investigations symbiosis is seldom considered to be a means of generating inherited variation in evolution.

Evolutionary Novelty

Whereas all biologists agree that mutation (base-pair changes, deletions, duplications, and transpositions) is a major source of evolutionary novelty, few emphasize the importance of other mechanisms. These alternative mechanisms include karyotypical alterations (polyploidy, increase in number of chromosome sets; polyteny, increase in amount of DNA per

[1]Personal communication from J.W. Hastings, Harvard University, Cambridge, MA; B. Kendrick, University of Waterloo, Canada; L. Muscatine, University of California, Los Angeles; K.H. Nealson, Center for Great Lakes Study, University of Wisconsin, Milwaukee; J. Sapp, University of Melbourne, Australia; D.C. Smith, Edinburgh University, Edinburgh, Scotland; and R. Trench, University of California, Santa Barbara.

chromosome; and Robertsonian fusions, chromosomal translocations). Raikov, a Soviet cytologist, has stressed polyenergy (the increase in number of homologous genomes in a nucleus) as a mechanism of evolution in ciliates and other protoctists (Raikov 1982).

Karyotypic fissioning refers to the phenomenon, in mammals, in which an extra centromeric synthesis in a fertile member of the population leads to a doubling of the number of chromosomes because each single metacentric is converted to two telocentrics. Because no total change in the amount of DNA per karyotype occurs, fissioning tends to be benign with respect to viability and fertility. In spite of a great deal of evidence in its favor, the importance of karyotypic fissioning in mammalian evolution has been almost exclusively argued by Neil Todd, publisher of the *Carnivore Genetics Newsletter* and adjunct professor at Boston University (Margulis 1993; Todd 1970).

The acquisition of additional genomes as a mechanism of evolution of prokaryotes has been widely discussed, and it was evaluated as an extremely important force by Sonea and Panisset (1983). The special case of homologous genome acquisition known as meiotic (or eukaryotic) sex is, of course, described in most English-language textbooks on evolution. Yet the intimate relationship between sex and symbiosis and their analogous components (Figure 22.3) is overlooked.

The analogies between the processes of recognition, fusion, and emergence of new individuals in both sex and symbiosis are obscured by differences in the terminology for the two processes, as can be seen in Figure 22.3. Few scientists are aware of the acquisition and migration of foreign nuclei in the establishment of the dikaryon (in red algae rendered heterokaryotic) and the occurrence of nuclear parasitism as an evolutionary phenomenon. Research on acquisition of foreign nuclei has been done almost exclusively by phycologists Lynda Goff of the University of California at Santa Cruz and Annette Coleman of Brown University (Goff and Coleman 1987).

The best-understood examples of morphogenetic innovation and speciation come from studies of symbioses, including lichens (Honegger 1991). Well-documented cases of new species emerging include fish with luminous bacteria (McFall-Ngai 1991), weevils that lose bacteria (Nardon and Grenier 1991), and the amoebae that survive bacterial infection (Jeon 1991). The amoebae incorporate former food bacteria, which form new intracellular organelles. As a result, more complex new species of free-living amoebae emerge (Sagan and Margulis 1987).

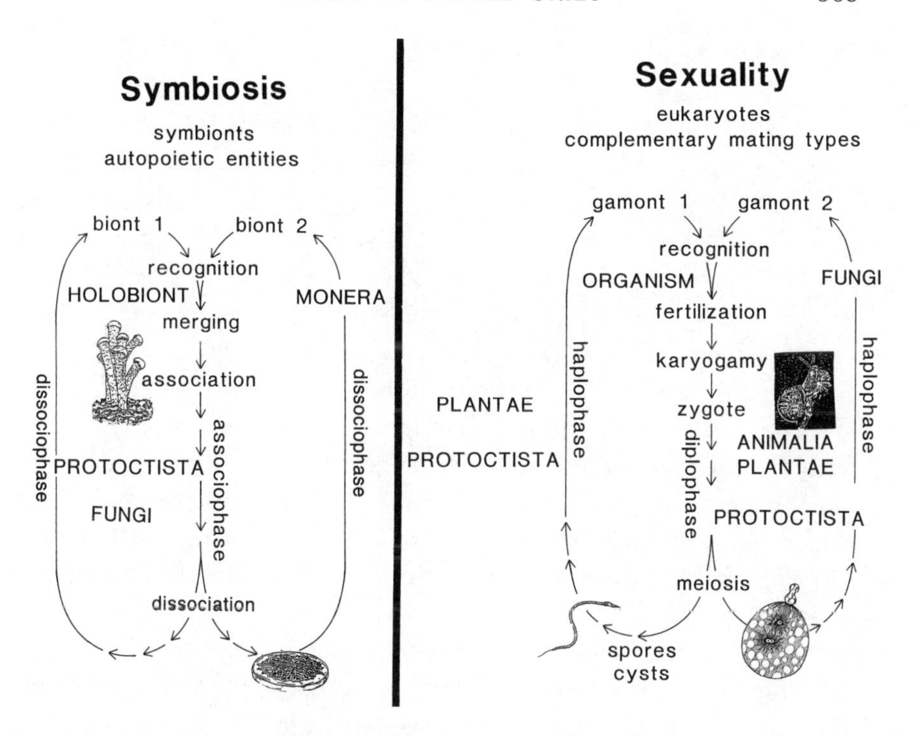

FIGURE 22.3. Cyclical symbiosis and meiotic sexuality are analogous. Both symbiotic and sexual partners must sense and recognize each other; their bodies (bionts and gamonts) or representative cells (biont cells and gametes) must merge (fusion, conjugation, or fertilization); integrating mechanisms (association and karyogamy) must establish and maintain the integrity of the new individual (holobiont and zygote), and, at a subsequent time in the life cycle, dissociation or reduction by meiosis ensues to form bionts or haploid gametes. Although meiotic sexuality is a more ritualized process than cyclical symbiosis, both are likely to be maintained by selective pressure on the unassociated bionts or the haploids under certain recurring environmental conditions.

Future Symbiosis Research

Before the founding in 1983 of the International Society for Endocytobiology (ISE) by the two German scientists, Werner Schwemmler (an insect-bacteria symbiopgeneticist) and H.E.A. Schenck (who studies the chemistry of *Cyanophora* and other algae), the fields of intracellular symbiosis and evolutionary studies had separate histories. Virtually all of recent western evolutionary biology had emerged as neo-Darwinism from population genetics.

Endocytobiology is defined by the ISE as the study of "intracellular space as oligogenetic ecosystem." The ISE regards all intracellular symbionts as objects of its study (Schenck and Schwemmler 1983; Schwemmler and Schenck 1980). This newly defined field is rooted in descriptions of bacterial symbionts and their correlation with studies of eukaryotic cell organelles (Lee and Fredrik 1987). The history of the field is recorded in the original scientific literature collected by Dyer and Obar (1985).

The ISE, by publishing three international colloquia (two held in Germany and one in New York; Lee and Fredrick 1987; Schenck and Schwemmler 1983; Schwemmler and Schenck 1980) and articles in their journal, *Endocytobiosis and Cell Research,* has begun to unite biologists from disparate traditions in common pursuit of cell origins. A new journal, *Symbiosis* (founded in 1985), published by Miriam Balaban and edited chiefly by lichenologist Margolith Galun in Rehovot, Israel, offers an outlet for scientists who experimentally investigate the molecular and cellular bases of symbioses.

Now a symbiosis of journals is under discussion: *Symbiosis* and *Endocytobiosis and Cell Research* are planning a protracted physical association. The new name is currently under discussion: it will probably be called *Endocytobiology and Symbiosis.** This journal would, for the first time, unite those scientists studying all kinds of symbioses with those studying intracellular organelles and cytoplasmic heredity.

The appearance of this new journal will offer neo-Darwinist evolutionists and experimental endocytobiologists a regular forum for professional interaction. Most of the scientists who consider themselves endocytobiologists do not attend meetings in general evolutionary biology. Thus, as is usual in the sociology of science, potential conflict, as well as integration, is limited by lack of communication (Fleck 1979).

The history of endocytobiology has been described in *Mosaic* by Fisher (1989), a magazine published by the National Science Foundation. The issue, which was dedicated to new research results in cell biology, launched endosymbiosis into the biological mainstream as an important mechanism of organelle origin and cell evolution. This article has aided communication among biologists and prompted the process of reintegra-

*Like most incipient symbioses this one between journals failed to permanently establish itself. However the scientific community of scholars did succeed in holding an International Congress on Symbiosis research at the Marine Biological Laboratory, Woods Hole, MA April 13–18, 1997. The proceedings are forthcoming in the Balaban journal, *Symbiosis,* 1998. A new International Symbiosis Society (ISS) was founded.

tion of their subfields. Considering the historical contributions of eastern European scientists, translation of Fisher's article into the Russian language is in order.

Conclusions

"Words become battle cries," wrote Ludwig Fleck (1919), describing the penchant of the scientist to fret about labels. Indeed, both endocytobiology and symbiogenesis are simultaneously neo-Lamarckian and Darwinian evolutionary ideas. Mitochondria, plastids, and other organelles began as bacteria; thus acquired characteristics, including their genomes, are inherited.

The Russian school recognized symbiogenesis as an evolutionary mechanism. Even British novelist and philosopher Samuel Butler described cells-inside-cells in eloquent literary terms at the turn of the century (Butler 1898). Furthermore, Darwin himself was a Lamarckian. He even anticipated symbiogenesis when he wrote, "We cannot fathom the marvelous complexity of an organic being; but on the hypothesis here advanced this complexity is much increased. Each living creature must be looked at as a microcosm—a little universe, formed of a host of self-propagating organisms, inconceivably minute and as numerous as the stars in heaven" (Darwin 1868, p. 453).

23

SCIENCE EDUCATION, USA:

Not Science, Not Yet Education
The Ecology Example

LYNN MARGULIS

Everybody Counts

Science education. We all say we want it. But the clever, hard-working students who flocked to physics in the post-Sputnik 60s and their colleagues who flooded medical school admission committees with applications have fled the halls of science. Research museums are computer-animating, merging, or storing their collections. Whereas in 1965, Peterson's Guide to Higher Education listed a hundred departments of botany or plant science, by 1995 far fewer remained. Oceanography is moribund; very few research vessels still go to sea. Thoughtful, critical young people now work for computer companies or film producers. Ambitious students who once became chemists or mining engineers now elect to study economics or management skills.

From the National Academy of Sciences' National Research Council (NRC) Report (1989) and the American Association for the Advancement of Science (AAAS 1989; Blackwell and Henkin 1989) come dire pronouncements (Brandwein and Passow 1988). That the United States lags behind Germany and Japan in mathematics and science education is not surprising, but that the quality of science education for young Greeks, Frenchmen, Spaniards, and Russians far exceeds ours is disturbing. Nearly half the graduate students in scientific fields in the United States are foreigners whose science background was obtained elsewhere. An attempt at the national level to counter these negative trends began in 1985 when the National Academy of Sciences (under the presidency of Dr. Frank Press) and The Smithsonian Institution (when Dr. Robert McAdams was Secretary) teamed up. Two years later, when they moved a small group of science educators into Smithsonian's Arts and Industries Building, Room 1201, in Washington, was launched the Sisyphysian work of the National Science Resource Center (NSRC) under the direction of physicist-educator Dr. Douglas Lapp. To quote Mark Twain (*Roughing It*, Penguin American Library, p. 384), few men in the world are gifted "with the pluck and perseverance necessary to follow up and hound such an undertaking to its completion." Although Mark Twain was referring to Mr. Sutro's eight-mile-long, two-thousand-foot–deep tunnel to extract silver ore from the Comstock mother lode, he might well have been referring to Doug Lapp and his close colleague and associate Sally Shuler's efforts to improve United States science education. (National Science Resource Center 1988)

What brings us, for the second time at least since Sputnik in 1959, to national malaise and desperation about science education improvement? Why does the United States show so poorly in all aspects of education involving quantitative science? Why, when society requires more, not less, science do our smart students flee from science and math? Is avoidance of math and science by American students irreversible? My informal comments here are limited to a single example: ecology education.

One aspect of limited science education stems from the ignorance of science in our elected officials, which of course reflects the ignorance of their constituency. Not only does elective office not require science education, but few with it ever elect to run for office. Each U.S. citizen, North or South, generates nearly ten pounds of garbage and trash per day, yet the 1988 presidential pollution debate either ignored all ecological issues or accusatively focused on ravaged Boston Harbor.

Lacking even passing familiarity with ecological principles, the public endures daily sound bites on popular topics such as "global warming," the "ozone hole," "fluorocarbons," "biodiversity," and "genetic engineering." This is media science. Basic scientific knowledge of U.S. journalists and citizens cannot be acquired by good intentions and fast money. "Briefings" are always too little and too late. Yet I doubt science ignorance should be blamed on the victim, the U.S. public. Education is needed; no elixir, no matter its cost, can produce a person well versed in science. Nothing substitutes for study.

What Makes Today's Ecologist?

An example of formal science education is how "ecology" or "environmental science" is learned by the U.S. citizen attending kindergarten through graduate school. Prior to college, "ecology study," if any, is memorization of textbook words about clouds and rocks. Even the choice of book is not controlled by the teacher. By college, "ecology" is taught as a specialized discipline, an option only for "biology majors." Usually these are third- or fourth-year students hoping to become pharmacists, dentists, or scientists.

Formal science education, for example, that in ecology, of course frightens any ordinary citizen. To become a professional ecologist today one essentially takes vows: one must declare to be a "biology major," a prerequisite to entry into the scientific priesthood where immense dedication of time is required to learn the language separating clergy from laity. For the study of life sciences, both high school and undergraduate requirements are rigorous. The aspiring student begins no later than his third year in high school to accumulate three to four years of foreign language, four to six years of biology, three to five years of chemistry, three to four years of math, five years of English, and at least one year of physics. These courses tend to be taught by overworked, underpaid teachers who were not trained and did not choose to teach them. Continuing in college, an aspiring professional is required to study biology each semester in a determined hierarchical order with few options. The ecology curricula, like nearly all courses of biological study, are essentially determined by medical school prerequisites.

Only rigid personalities, fearful flounderers, or extraordinarily dedicated college freshmen can contemplate the stretch of determined days, months, and years ahead. The potential citizen ecologist, such as the nature lover, the curious gardener, the philosopher of life, the biological anthropologist, the future physical therapist or nutritionist tends to face two extreme options: taking the plunge into the rigid waters of the chemistry or biology major, or enduring trivial and trivializing alternatives, such as "environmental studies" for nonscience majors. Most courses for "nonscience majors" are little more than lists of specialized vocabulary (for example, mitosis and meiosis, amino acids, osmosis, homeostasis, tracheids and trachea) scaled down from those in courses for "majors." Both the major and nonmajor, in my experience, enjoy few laboratory or field opportunities. Indeed, neither "serious scientist-aspirants" nor "environmental-studies amateurs" usually do *real* field or experimental science in formal classes. The standard teaching method is *still* the "chalk talk" of the medieval lecture theater.

The unstated scientific worldview of today is a tiny subset of the larger cultural philosophy. W.I. Thompson writes about the cultural fabric:

> . . . In our patriarchal imagination of what Laurie Anderson has called Big Science, we see the world as a collection of discrete individuals that own collectible things: egos contained in cars, wives and painting contained in houses, and kids contained in schools. . . . We are being asked to move out of our containers to enter into the evolutionary conversation to understand the biosphere and the emerging planetary culture as one in which Mankind (and I use the sexist term on purpose) as a defensive collection of competing and warring selves has come to its end.
>
> (Thompson, 1990)

Ecology students—whatever their former level of seriousness—are as embedded as the rest of us in Big Science assumptions. They are taught computer "model" interactions between "competing and warring" organisms directly from the sacred college text: the required reading. The "moving out of our containers to enter into the evolutionary conversation" is at odds with today's education. To "move out," the eager student needs experience, the recognition and direct sensing of his intimate relationship with the Earth's atmosphere, sediments, and vast array of living beings. Whereas the budding ecologist generally hungers for a direct relationship with nature, in cultural mockery she is treated to lessons in computing in climate controlled offices entirely devoid of non–*Homo sapiens* life. Even though she may "feel" that the borders of nations are most secure when they are in active ecological interchange with those of their neighbors, and that incessant trade and border crossing are social preludes to ecological health, she is told that these are political and economic questions—and hence "out of her field."

Philosophy and Life

The answers to nearly all the major philosophical questions are either found in or illuminated by the science of life, especially ecology, whose stated goal is the elucidation of the relationship of organisms to environment. Although academic ecologists stress mathematics and computer languages, detailed studies of thousands of plants, animals, and microbes clearly reveal that the language of life is carbon chemistry. Philosophical insights garnered from the life sciences are suppressed by the arbitrary pigeonholing of rigid academic traditions. What is our relation, as *Homo sapiens* mammals, with our environment? How much and what sort of land is required to ensure the health and growth of a person, a family? What is life? How did life begin? How did it evolve? What is sex? How did a two-sex system originate? What are the differences between humans and other primate species? How does human physiology work? These enlightening questions, of intrinsic interest, can not even be mused in the academic environment that requires "covering the material."

Both modern quantum physics and recent cognitive neurobiology tell us that the experimental apparatus cannot be ignored in accounting for scientific results. "Everything is seen by an observer" (Maturana 1987). Yet academic science courses still perpetuate the Cartesian concept of absolute objectivity. Science is taught as if there were an absolute separation between observer and observed. But if, as modern cognitive biology and physics imply, absolute objectivity is illusory, then the research programs of all the sciences, and perhaps especially the ecological sciences, lag behind. Because the older Newtonian, mechanistic paradigms are still firmly in place, our presumptive ecologist suffers from cognitive dissonance. Unstated, the mechanist view is perpetuated on the science student as indelibly as is his nationality, as we describe in "What Is Life?" (Margulis and Sagan 1996).

If we were to move ecology from the periphery of scientific discourse (where today it is accused of being "soft," of being "derived from physics and chemistry"), into the mainstream, perhaps we could improve the scientific outlook. Ecology is, of course, no more or less an absolute science than quantum physics. Rather, ecology is central because as live beings everything we learn is through our membership in a community of sensitive animals. The most important "object" of ecological study is our relationship as humans to some 30 million other species on this watery orbiting space island called Earth. Perhaps ecological science should move to the center of scientific education. Such a curricular revision would be indicative of the "science of life's" true position within the framework of human knowledge.

The irony is that the current curricula organization in U.S. education forces the curious student to face a ridiculous choice: whether to be a professional scientist or a scientific ignoramus. The academic institutions have molded not only most ecologists, but also chemists, physicists, and geologists, into just such crippled "specialists." Insidiously engineers, meteorologists, particle physicists, and many other scientists entirely lacking biology experience are considered "experts" and consulted to solve ecological problems. We cannot blame these scientists for this development; their ignorance is usually perpetuated against their own good instincts. These deficiencies result from the academic institutions whose histories of increasing specialization and departmentalization have walled off both access to knowledge about the living and practice with live organisms. No individual is at fault. We learn the hard way that while in our schools, universities, and research institutes we separate by force biology, chemistry, and geology, the planetary environment obeys no such strict "apartheid." The Earth's environment evolves and reacts as an ecosystem containing, but in no way limited to, human beings. As would-be planetary citizens we ignore this planetary bioplasm to our own detriment; our trash and garbage never go out—they only go around and around. Human beings, unlike cyanobacteria and grasses, are never productive. We are consumers of organic matter. Populations, intrinsically capable of unlimited growth, will always tend to expand and eventually be checked. We ignore these crucial messages from ecology to our peril.

Science is not primarily a profession, and ecology is no exception. Science is a way to find out about the world by inspection and direct sensation. Active observation and experimentation overrides memorization of vocabulary lists or textbook authorities. World-class ecologists interest us in ecosystems by studying coral reefs, ponds, deserts, forests, and salt marshes as part of their professional life. Why can't the questioning student be relieved of the imposition of educational excesses? Why can't serious scientific courses at the college level, while respecting quantitative inquiry, teach the science as liberal art? Why can't study of "natural history" regain its respectability?

A Course Called "Environmental Evolution"

For more than seventeen years at Boston University, and now since 1988 at the Biology and Geosciences Departments of the University of Massachusetts at Amherst, I, with colleagues, students, and friends, have practiced science as a liberal art through a learner-oriented course called Environmental Evolution. Active scientists, an international faculty, in one semester teach their own first scientific love. The program, which uses audiotaped lectures, slides, and other authentic materials of practicing scientists, focuses on the overall effect of more than three billion years of the history of life on planet Earth. Directed mainly to senior science majors, graduate students, and occasionally historians of science, science writers, and professional sociologists, the only prerequisite is a two-year (four semester) requirement of any science, such as physics, astronomy, chemistry, or geology. The talks, maps, and slides presented by engaging scientists are, to revive a word from the education debate of three decades ago, "relevant." The program involves no videos of people talking or computer linkages. Rather the excellent audiotapes, with their associated slides and, especially, "electric blackboard" presentations, are entirely accessible outside class. The "electrowriter" is a paper and pen device that must be seen to be believed. All class time is interactive. We instructors discuss the lectures and the lecturers with the students. The personal accessibility to lively idiosyncratic scientists (all scientists are idiosyncratic) augments interest far beyond that of any textbook.

The Interactive Lectures

Using the interactive lecture system (IAL) developed by the late Edwin Land and protégé, Stewart Wilson (now of the Polaroid Foundation) at the Polaroid Corporation, we cajoled first-rate scientists into talking to science students. In this course inherently interesting speakers, who connect their life work in meaningful and human ways to real-life experiences actually, via the IAL, converse with the student. Although the scientists are absent, their superb audiotaped electrowriter presentations—which can be stopped, slowed, and restarted—are lively, indeed fascinating. Many of the scientists we taped are at the summit of their respective fields. The teaching skills of three superb scholars—Prof. E.S. Barghoorn, Harvard University (1914–1984), T.S. Swain, Boston University (1916–1986), and Cyril Ponnamperuma (1923–1994)—have been preserved despite the fact that they are deceased.

The student listens to tapes of the scientist speaking, and views 35mm color slides, while via the electrowriter the handwriting of the lecturer appears in written form to complement the spoken address through an electronic ink-and-paper automatic recording device (the electrowriter itself). Thus, IAL scientists have not only their voices but the peculiarities of their handwriting is recorded for posterity. Our audiotaped lectures include Michael MacElroy, of Harvard University, on planetary atmospheres, and James E. Lovelock, Fellow of the Royal Society, on his Gaia hypothesis, the idea that the Earth's atmosphere and surface sediments are regulated by life. Also, Heinz Lowenstam, of the California Institute of Technology (1915–1994) has talked about how animals make minerals. The chemical origins of life have been addressed in different ways by David Deamer of the University of California/Santa Barbara and by Antonio Lazcano of Universidad Autónoma de México, Mexico. Stjepko Golubic and Paul Strother of Boston University have taken us to see microbial mats and stromatolites. I have explained symbiosis in cellular evolution, and Tony Swain has reviewed the coevolution of plants and the animals that eat them. Richard E. Schultes of Harvard has reviewed the chemistry and ethnography of hallucinogenic plants in his tapes, "Hallucinogenic Plants and Fungi of North America" and "Hallucinogenic Plants and Fungi of South America." Each student—only fifteen to eighteen are accepted for each class—listens to these lectures on his or her own schedule before class meetings. We also have collected other profoundly effective teaching aids: films of live microbes and related videotapes of lectures and television programs on subjects ranging from microbial behavior to global ecology, rocks, minerals, fossils, essential oil samples, and other natural "props."

Field Tripping

Field trips to nature preserves, science museums, aquaria, arboreta, and zoos are an integral part of our science education program. They are especially important for a budding ecologist from her precollege days through the graduate experience. These institutions are not simply toured; rather, their enormous and often unique resources are used as the basis for discussion, thought, and action. Zoo visits enhance the reality of readings on evolutionary mechanisms or mass extinctions. We have found that zoo and museum personnel warmly welcome science students behind the scenes. All are concerned that the most serious example of mass extinction since the Cretaceous demise of dinosaurs is in progress now. The current rates of poaching combined with the lack of effective habitat protection is causing even elephants, tigers, gorillas, and other favorite zoo animals to disappear from the face of the Earth. How devastating will the emotional effect of these extinctions be? The ecological and "geophysiological" effects of the extinctions of other, less charismatic species can be contemplated with trepidation. How will we live if planetary ecology is ignored and all nonhuman life is treated like a giant zoo?

Field trips generate questions that, in the context of supplementary readings and educational opportunities, are of immense importance to the future of humankind on this planet. For science to become a genuine liberal art, "formal training" must reject much of the professionized, computerized evolutionary speculation via population genetics and memorization of details of interest mostly to medical personnel and industrial executives. Our budding ecologists must experience live organisms. Everyone needs intimate familiarity with our food plants and animals. All students need some exposure to water chemistry measurements needed for determining quality and potability. Behind-the-scenes tours of botanical gardens, herbaria, water treatment plants, and natural history societies (for example, the National Wildlife, Audubon Society, Museum of Natural History) begin a process in which students no longer separate ecology as a discipline into a dull, specialized box. Awakening scientific interest means inaugurating genuine scientific experience.

In our IAL Environmental Evolution course, the students present oral reports on topics they find most interesting. They prepare their own slides, overhead-projector transparencies, posters, and maps. Unlike most science students, our "environmental evolutionists" take responsibility for the integration of knowledge into their personal lives. The high level of aesthetic criteria used in the presentations enhances the learning experience, making it far more memorable than less interactive media such as textbooks. The science of life is too important to be left to professional biologists, their publishers, and their money-making books. Ecology, wittingly or not, is an interactive discipline; we do not simply study life but live it. Literacy of science is intrinsic to making wise decisions. Scientific literacy, especially of life science, should be far more a prerequisite for public office than lawyer-like discursive abilities. Sadly, our culture behaves as if money were no longer a symbol but could actually be eaten and breathed—replacing the food, air, and other necessities that it represents. We teach the lie that money itself has intrinsic value. The number of people trying to get a little closer to the money by studying law and real estate is increasing far faster than the number that pursue careers in any of the sciences, especially ecology, in which employment opportunities are severely limited.

As a culture, we need to attract our critical, articulate, and enthusiastic young people to the sciences. Once attracted, they cannot help but become teachers of the young. By allowing part-time and casual involvement of everyone in scientific activities, the problem can at least be ameliorated. Science is an open way of knowing, not a tortuous path to middle-class security.

Epistemological Afterword

Dorion Sagan

Somewhere, and probably due to entropy and unplanned cell death I can't remember where, Nietzsche remarks that as you grow older you see the naïvete of your younger writings, until you grow older still, at which point you see the naïvete of *those* opinions. Another irreverent wit, Oscar Wilde (who also died in 1900), said one day that he had accomplished much: he had put in a comma in the morning and removed it again in the afternoon.

I mention such self-criticism in lieu of any conclusive statement about the present collection of essays, selected from the last decade and a half of collaboration with my workaholic mom. She is a true scientist and intellectual, and I am proud to have selected her, along with my dear recently departed father, as my parents. Although we do not always agree, I am truly lucky to have worked with her for the past sixteen years, especially considering her strong views on birth control, which even include a woman's right to commit infanticide. Thanks, Mom.

Now to the essays. I find them somewhat motley. Were I writing them today, I doubt any would appear in their present form. Artists and writers, like lovers, like to criticize what has passed, reserving their enthusiasm for current and future projects. Perfection ever fades, like money found in a dream that disappears upon waking. Still, much of what is said here strikes me as solid and sound.

Our Introduction, with its critique of religion, might have been extended to warn against the risk that science may crust over to become received wisdom. This is a particular danger in our era of relentlessly one-way communication via television, which craves pundits and experts and discourages nuance, ambiguity, or doubt. Whatever the populace's hunger for prophets and other authorities, the truth of science lies in the doing, in the open-mindedness of its approach, and not in erecting a fence between its (always provisional) answers and whatever data may come in later. This is where philosophy, much criticized in scientific circles, has a role to play: not as an alternative to science but rather as a rigorous voice, cognizant of history, external to science, and thus able to be objective about it in a way that a "scientific" view of science from within cannot be.

The essay on Oppenheimer shows that science alone, far from being the final answer, can lead to horrifying technological excesses. A sense of responsibility must guide our behavior, not technical feasibility. Lynn's essay "Red Shoes," shows that women can't get everything they want (although, as the song tells us, they may sometimes get what they need). Within some insular scientific disciplines, the difference between insiders and outsiders may actually be more consequential than that between men and women. Science is by nature egalitarian, even if it is not always practiced that way.

The stuff on symbiosis and individuality details our surprising multiple ancestries, the mixtures that are ourselves. If machines ever become conscious or humans ever achieve a true, decentralized democracy, we will no doubt find that we and our computers are retracing steps first taken billions of years ago by bacteria.

The Gaia essays attempt to negotiate the battleground of an evolving paradigm: the switch from geochemistry to geophysiology. These essays, like their subject, thus contain a sociological element that extends beyond the practice of science per se. Furthermore, considerable tension exists between competing visions of Gaia as a New Age idea quenching the thirst of those parched by the torches of Judeo-Christian belief and Gaia as an area of scientific research. The break between these two will never be as clean as scientific purists might like, and perhaps this is ultimately to society's advantage, if not to science's. The sullying of geophysiology by those questing for meaning may be looked on as poetic justice, if not karmic retribution, for science's inevitably unsuccessful attempts to isolate itself from the masses whose lives it so deeply affects.

The final essays, on evolution and evolutionists, continue in a politico-scientific vein. How can we best teach, and learn? What should graduate students be doing? Our civilization, so wonderfully successful at using science, must not be afraid to look at science itself critically. This is a far cry from simply trashing science because it doesn't tell us what we want to hear. However the ideas presented here are revised in the future, I believe the attitude of *Slanted Truths*—that truths in their superb surprise, are worth attaining, but we must take reports of them with a grain of salt—will remain valuable into the indefinite future.

REFERENCES

Abram, D. 1985. The perceptual implications of Gaia: The Gaia hypothesis suggests an alternative view of perception. *The Ecologist* 15:96–103.

Abram, D. 1996. *The Spell of the Sensuous.* New York: Pantheon Books.

Ahmadjian, V., and S. Paracer. 1986. *Symbiosis: An Introduction to Biological Associations.* Hanover, NH: University Press of New England.

American Association for the Advancement of Science. 1989. *Science for All Americans: A Project 2061 Report on Literacy Goals in Science, Mathematics and Technology.* Washington, D.C.: American Association for the Advancement of Science, 217 pp.

Amoore, J.E. 1961. Dependence of mitosis and respiration in roots upon oxygen tension. *Proc R Soc* B 154:109–129.

Avers, C.J. 1989. *Process and Pattern in Evolution.* New York: Oxford University Press.

Awramik, S.M. 1973. Stromatolites of the gunflint iron formation. Ph.D. diss., Harvard University Department of Geology, Cambridge, MA.

Awramik, S.M., J.W. Schopf, and M.J. Walter. 1983. The Warrawoona microfossils. *Precambrian Research* 20.

Ayala, F.J., and J.W. Valentine. 1979. *The Theory and Processes of Organic Evolution*. Menlo Park, CA: Benjamin/Cummings Publ.

Bada, J.L., and S.L. Miller. 1968. Ammonium ion concentration in the primitive ocean. *Science* 159:423–425.

Baltscheffsky, H., H. Jornvall, and R. Rigler (eds). 1986. *Molecular Evolution of Life*. Proceedings of a conference held at Sodergarn, Lidingo, Sweden 8–12 September 1985. Cambridge, England: Cambridge University Press.

Barghoorn, E.S. 1971. The oldest fossils. *Scientific Amer* 224:30–42.

Barghoorn, E.S., and J.W. Schopf. 1966. Microorganisms three billion years old from the Precambrian of South Africa. *Science* 152:758–763.

Barghoorn, E.S., and S.A. Tyler. 1965. Microorganisms from the Gunflint chert. *Science* 147:563–577.

Barlow, C., and T. Volk. 1990. Open systems living in a closed biosphere: A new paradox for the Gaia debate. *BioSystems* 23:371–384.

Bateson, W.B. 1928. "[S. Butler is the] most brilliant, and by far the most interesting of Darwin's opponents." From *Heredity and Variation in Modern Lights*. Cited in C.B. Bateson (ed). *William Bateson, F.R.S.* Cambridge: Cambridge University Press.

Bell, G. 1982. *The Masterpiece of Nature: The Evolution and Genetics of Sexuality*. Berkeley, CA: University of California Press.

Berman, M. 1989. *Coming to Our Senses: Body and Spirit in the Hidden History of the West*. New York: Simon & Schuster, p. 239.

Bermudes, D., and L. Margulis. 1987. Symbiont acquisition as neoseme: Origins of species and higher taxa. *Symbiosis* 4:185–198.

Blackwell, D., and L. Henkin. 1989. *Mathematics: Report of the Project 2061 Phase I Mathematics Panel*. Washington, D.C.: American Association for the Advancement of Science, 47 pp.

Bloom, H. 1995. *The Lucifer Principle: A Scientific Expedition into the Forces of History*. New York: Atlantic Monthly Press.

Botkin, D.B., and E.A. Keller. 1982. *Environmental Studies: The Earth as a Living Planet*. Columbus, OH: Charles E. Merrill Pubs.

Botkin, D.B., P.A. Jordan, S.A. Dominski, H.D. Lowendorf, and G.E. Hutchinson. 1973. Sodium dynamics in a northern ecosystem. *Proc Natl Acad Sci USA* 70:2745–2748.

Brandwein, P.F., and A.H. Passow (eds). 1988. *Gifted Young in Science*. Washington, D.C.: National Science Teachers Association, 422 pp.

Brock, T.D., P.J. Cook, H.P. Eugster et al. 1982. Sedimentary iron deposits, evaporites and phosphorites: State of the art report. In H.D. Holland and M. Schidlowski (eds). *Mineral Deposits and the Evolution of the Biosphere*. Berlin: Springer-Verlag, pp. 259–273.

Bunyard, P. 1995. *Gaia in Action*. Edinburgh: Floris Books.

Butler, S. 1898. *Life and Habit.* Reprinted 1923. New York: AMS Press.

Calvin, W.H. 1990. *The Cerebral Symphony: Seashore Reflections on the Structure of Consciousness.* New York: Basic Books.

Campbell, J. 1983. *The Way of the Animal Powers.* Vol. 1 of *Historical Atlas of World Mythology.* San Francisco: Harper & Row, p. 251.

Canuto, V.M., J.S. Levine, T.R. Augustsson, and C.L. Imhoff. 1982. UV radiation from the young sun and oxygen and ozone levels in the prebiological palaeoatmosphere. *Nature* 296:816–820.

Charlson, R.J., J.E. Lovelock, M.O. Andreae, and S.G. Warren. 1987. Oceanic phytoplankton, atmospheric sulphur, cloud albedo and climate. *Nature* 326: 655–661.

Clark, R.W. 1984. *The Survival of Charles Darwin: A Biography of a Man and an Idea.* New York: Random House.

Cloud, P.E., Jr. 1968. Atmospheric and hydrospheric evolution on the primitive earth. *Science* 160:729–736.

Cloud, P.E., Jr. 1983. The biosphere. *Scientific American* 249:176–189.

Cohen, R.S., and T. Schnelle. 1986. *Materials on Ludwik Fleck.* Boston: D. Reidel Pubs.

Collocott, T.C. (ed). 1972. *Dictionary of Science and Technology.* New York: Barnes & Noble.

Darwin, C. 1859. *On the Origin of Species by Means of Natural Selection.* London: J. Murray.

Darwin, C. 1868. *The Variation of Animals and Plants under Domestication.* Vol. 2. New York: Organe Judd.

Dawkins, R. 1976. *The Selfish Gene.* Oxford: Oxford University Press.

Dawkins, R. 1982. *The Extended Phenotype: The Gene as the Unit of Expression.* Oxford: W.H. Freeman & Co.

Dawkins, R. 1986. *The Blind Watchmaker.* Harlow, England: Longman.

De Bary, H.A. 1879. *Die Erscheinung der Symbiose.* Strasburg, Germany: R.J. Trubner.

Dévai, I., L. Felföldy, I. Wittner, and S. Plósz, 1988. New aspects of the phosphorus cycle in the hydrosphere. *Nature* 333:343–345.

Dobell, C. 1913. Observations on the life-history of Cienkowski's "*Arachnula.*" *Arch Protistenkund* 31:317–353.

Doolittle, W.F. 1981. Is nature really motherly? *CoEvolution Q* 29:58–63.

Dover, G.A. 1988a. Evolving the improbable. *TREE* 3:84.

Dover, G.A. 1988b. The new genesis. In D.L. Hawksworth (ed). *Prospects in Systematics.* Oxford: Oxford University Press, pp. 151–168.

Dyer, B. 1989. Symbiosis and organismal boundaries. *Am Zool* 29:1085–1093.

Dyer, B.D., and R. Obar (eds). 1985. *The Origin of Eukaryotic Cells: Benchmark Papers in Systematic and Evolutionary Biology.* New York: Van Nostrand Reinhold.

Edelman, G.M. 1985. Neural Darwinism: Population thinking and higher brain function. In M. Shafto (ed). *How We Know, Nobel Conference XX.* St. Peter, MN: Gustavus Adolphus College, pp. 1–30.

Edelman, G.M. 1987. *Neural Darwinism: The Theory of Group Selection.* New York: Basic Books.

Egami, F. 1974. Minor elements and evolution. *J Mol Evol* 4:113–120.

Ehrlich, P.R., and R.W. Holm. 1963. *The Process of Evolution.* New York: McGraw-Hill.

Elias, N. 1978. *The Civilizing Process: The History of Manners.* Translated by E. Jephcott. New York: Urizen Books, pp. 252–253.

Ferris, J., and L. Nicodem. 1974. In K. Dose, S.W. Fox, C.A. Deborin, and T.E. Pavlovskaya (eds). *Origins of Life and Evolutionary Biochemistry.* New York: Plenum Press, pp. 107–117.

Filipelli, G.M., and M.L. Delaney. 1992. Similar phosphorus fluxes in ancient phosphorite deposits and a modern phosphogenic environment. *Geology* 20:707–712.

Fisher, A. 1989. Endocytobiology: The wheels within wheels in the superkingdom Eucaryotae. *Mosaic* 20:2–13.

Fleck, L. 1979. *Genesis and Development of a Scientific Fact.* Chicago: University of Chicago Press.

Fleischaker, G.R. 1988. *Autopoiesis: The System, Logic and Origin of Life.* Boston, MA: Boston University Professors Program, The Graduate School.

Foucault, M. 1977. What is an author? In D.F. Bouchard (ed). *Language, Counter-Memory and Practice: Selected Essays and Interviews.* Ithaca, NY: Cornell University Press, p. 124.

Futuyma, D.J. 1985. Is Darwinism dead? *The Science Teacher* 52 (Jan.):16–21.

Futuyma, D.J. 1986. *Evolutionary Biology,* 2nd ed. Sunderland, MA: Sinauer Associates.

Garrels, R.M., A. Lerman, and F.T. MacKenzie. 1981. Controls of atmospheric oxygen: Past, present, and future. *Am Scientist* 61:306–315.

Gillespie, C.C. 1969. *Genesis and Geology: A Study in the Relations of Scientific Thought, Natural Theology and Social Opinions in Great Britain, 1790–1850.* Cambridge, MA: Harvard University Press.

Glashow, S. 1988. *Interactions, A Journey through the Mind of a Particle Physicist and the Matter of this World.* New York: Warner.

Goff, L.J., and A.W. Coleman. 1987. Nuclear transfer from parasite to host: A new regulatory mechanism of parasitism. In J.J. Lee and J.F. Fredrick (eds). *Endocytobiology III.* New York: New York Academy of Sciences, pp. 402–423.

Gold, P.E. 1987. Sweet memories. *Am Scientist* 75:151–155.

Goody, R., and J.C.G. Walker. 1972. *Atmospheres.* Englewood Cliffs, NJ: Prentice Hall.

Gould, S.J. 1989. *Wonderful Life: The Burgess Shale and the Nature of History.* New York: W.W. Norton.

Gregory, P.H. 1973. *Microbiology of the Atmosphere,* 2nd ed. New York: John Wiley & Sons.

Grinevald, J. 1988. A history of the idea of the biosphere. In P. Bunyard and E. Goldsmith (eds). *Gaia: The Thesis, the Mechanisms and the Implications. Proceedings of the First Annual Camelford Conference on the Implications of the Gaia Hypothesis.* Cornwall, UK: Quintrell & Co. Reprinted 1995 in Bunyard, P. (ed) *Gaia in Action: Science of the Living Earth.* Edinburgh: Floris Books.

Grinevald, J. 1996. Sketch for a history of the biosphere. In P. Bunyard (ed). *Gaia in Action: Science of the Living Earth.* Edinburgh: Floris Books, pp. 34–53.

Guerrero, R., C. Pedrós-Alió, I. Esteve, J. Mas, D. Chase, and L. Margulis. 1986. Predatory prokaryotes: Predation and primary consumption evolved in bacteria. *Proc Natl Acad Sci USA* 83:2138–2142.

Habermas, J. 1987. *The Philosophical Discourse of Modernity.* Translated by F. Lawrence. Cambridge, MA: MIT Press, p. 137.

Hall, J.L., Z. Ramanis, and D.J.L. Luck. 1989. Basal body/centriolar DNA: Molecular genetic studies in *Chlamydomonas. Cell* 59:121–132.

Haynes, R.H. 1990. Ecce ecopoiesis: Playing God on Mars. In D. MacNiven (ed). *Moral Expertise: Studies in Practical and Professional Ethics.* London: Routledge, pp. 161–183.

Haynes, R.H. 1992. How might Mars become a home to humans? *Gaia Sci* 2:7–9.

Hinkle, G.J. 1988. Marine salinity: A Gaian phenomenon? In P. Bunyard and E. Goldsmith (eds). *Gaia: The Thesis, the Mechanism and Its Implications. Proceedings of the First Annual Camelford Conference on the Implications of the Gaia Hypothesis.* Cornwall, England: Wadebridge Ecological Centre, pp. 91–98. Reprinted 1995 in Bunyard, P. (ed) *Gaia in Action: Science of the Living Earth.* Edinburgh: Floris Books.

Hitchcock, D.R., and J.E. Lovelock. 1967. Life detection by atmospheric analysis. *Icarus* 7:149–159.

Honegger, R. 1991. Fungal evolution: Symbiosis and morphogenesis. In L. Margulis and R. Fester (eds). *Evolution and Speciation: Symbiosis As a Source of Evolutionary Innovation.* Cambridge, MA: MIT Press, pp. 319–343.

Hori, H. 1982. The phylogenetic structure of the metabacteria. *Zbl Bakt Hyg I Abt Orig.* C3:18–30.

Hughes, A.J., and D.M. Lambert. 1984. Functionalism, structuralism and "Ways of Seeing." *J Theoret Biol* 3:787–800.

Hughes, J.D. 1983. Gaia: An ancient view of our planet. *The Ecologist: J Post-Industr Age* 13:54–60.

Hughes, J.D. 1994. "[Gaia], oldest of gods, mother of gods, humans and every living thing." *Homeric Hymns* 30.1. "Mistress, from you come our fine children and bountiful harvests; yours is the power to give mortals life and to take it away." *Hymn. Hom.* 14.4–5. In *Pan's Travail: Environmental Problems of the Ancient Greeks and Romans*. Baltimore, MD: Johns Hopkins University Press, p. 47.

Hutchinson, G.E. 1954. *Biogeochemistry of Vertebrate Excretion*. New York: American Museum of Natural History.

Huxley, J. 1912. *The Individual in the Animal Kingdom*. New York: G.P. Putnam's Sons, p. 125.

Jacob, F. 1973. *The Logic of Life: A History of Heredity*, New York: Pantheon, p. 25.

Jeon, K. 1991. Amoeba and X-bacteria: Symbiont acquisition and possible species change. In L. Margulis and R. Fester (eds). *Evolution and Speciation: Symbiosis as a Source of Evolutionary Innovation*. Cambridge, MA: MIT Press.

Kaveski, S., D.C. Mehos, and L. Margulis. 1983. There's no such thing as a one celled plant or animal. *The Science Teacher* 50:34–36, 41–43.

Keller, E.F., and E.A. Lloyd (eds). 1991. *Keywords in Evolutionary Biology*. Cambridge, MA: Harvard University Press.

Kellogg, W.W., and S.H. Schneider. 1974. Climate stabilization: For better or for worse? *Science* 186:1163–1172.

Kendrick, B. 1985. *The Fifth Kingdom*, 2nd ed. Waterloo, Canada: Mycologue Publ.

Khakhina, L.N. 1979. *Concepts of Symbiogenesis* (in Russian). Leningrad: Akademie Nauk, USSR (Soviet Academy of Sciences). English translation: Yale University Press, New Haven, CT, 1992.

Kimura, M. 1983. *The Neutral Theory of Molecular Evolution*. New York: Cambridge University Press.

Klossowski, P. 1991. *Sade My Neighbor*. Translated by A. Lingis. Evanston, IL: Northwestern University Press.

Kluger, J. 1992. Mars, in Earth's image. *Discover* 13:70–75.

Koshland, D.E., Jr. 1992. A response regulator model in a simple sensory system. *Science* 196:1055–1056.

Kropotkin, P. 1902. *Mutual Aid: A Factor of Evolution*. Reprinted 1955. Boston: Extending Horizons Books.

Lacan, J. 1977. The mirror stage as formative in the function of the I. In *Écrites: A Selection*. Translated by A. Sheridan. New York: W.W. Norton, pp. 1–7.

Lake, J.A. 1988. Origin of the eukaryotic nucleus determined by the rate invariant analysis of rRNA sequences. *Nature* 331:184–186.

Lambert, D.M., C.D. Miller, and T.J. Hughes. 1986. On the classic case of natural selection. *Rivista Biol (Biol Forum)* 79:11–49.

Lapo, A.V. 1988. *Traces of Bygone Biospheres*. Moscow: MIR. Translated by V. Purto. Oracle, AZ: Synergistic Press.

Lee, J.J., and J.F. Fredrik. 1987. Endocytobiology III. *Ann NY Acad Sci* 503:1–590.

Leenhardt, M. 1979. *Do Kamo.* Translated by B.M. Gluati. Chicago: University of Chicago, p. 22.

Levine, J.S. 1989. Photochemistry of biogenic gases. In M.B. Rambler, L. Margulis, and R. Fester (eds). *Global Ecology: Towards a Science of the Biosphere.* San Diego, CA: Academic Press, pp. 51–74.

Lewis, D.H. 1973a. The relevance of symbiosis to taxonomy and ecology with particular reference to mutualistic symbiosis and the exploitation of marginal habitats. In V.H. Heywood (ed). *Taxonomy and Ecology.* New York: Academic Press, pp. 151–172.

Lewis, D.H. 1973b. Concepts in fungal nutrition and the origin of biotrophy. *Biol Rev Camb Phil Soc* 48:261–278.

Lovelock, J.E. 1972. Gaia as seen through the atmosphere. *Atmospher Environ* 6:579–580.

Lovelock, J.E. 1979. *Gaia: A New Look at Life on Earth.* Oxford: Oxford University Press.

Lovelock, J.E. 1983a. Gaia as seen through the atmosphere. In P. Westbroek and E.W. de Joeng (ed). *The Fourth International Symposium on Biomineralization.* Dordrecht, Holland: Reidel Pub.

Lovelock, J.E. 1983b. Daisy World: A cybernetic proof of the Gaia hypothesis. *CoEvolution Q* 31:66–72.

Lovelock, J.E. 1988. *The Ages of Gaia.* New York: W.W. Norton.

Lovelock, J.E., and J.P. Lodge. 1972. Oxygen in the contemporary atmosphere. *Atmos Environ* 6:575–578.

Lovelock, J.E., and L. Margulis. 1974a. Atmospheric homeostasis by and for the biosphere: The Gaia hypothesis. *Tellus* 26:2–10.

Lovelock, J.E., and L. Margulis. 1974b. Homeostatic tendencies of the Earth's atmosphere. *Origins of Life* 1:12–22.

Lovelock, J.E., and L. Margulis. 1976. Is Mars a spaceship too? *Natur Hist Mag* 85:86–90.

Lovelock, J.E., and M. Whitfield. 1982. Life span of the biosphere. *Nature* 296:561–563.

Lovelock, J.E., R.J. Maggs, and R.A. Rasmussen. 1972. Atmospheric dimethyl sulfide and the natural sulfur cycle. *Nature* 237:452–453.

Lumiére, A. 1919. *Le Myth des Symbiotes.* Paris: Masson.

Madigan, M., J. Martinko, and J. Parker. 1996. *Brock's Biology of Microorganisms.* Englewood Cliffs, NJ: Prentice Hall.

Madigan, M., J. Martinko, and J. Parker (eds). 1997. *Brock's Biology of Microorganisms.* Upper Saddle River, NJ: Prentice-Hall.

Margulis, L. 1976. A review: Genetic and evolutionary consequences of symbiosis. *Exp Parasitol* 39:277–349.

Margulis, L. 1982. *Early Life*. Boston: Jones and Bartlett.

Margulis, L. 1988. Serial endosymbiotic theory (SET): Undulipodia, mitosis and their microtubule systems preceded mitochondria. *Endocytobiosis and Cell Res* 5:133–162.

Margulis, L. 1991a. Big trouble in biology: Physiological autopoiesis vs. mechanistic neodarwinism. In J. Brockman (ed). *Doing Science: Reality Club #2*. New York: Prentice Hall, pp. 211–235.

Margulis, L. 1991b. Symbiogenesis and symbionticism. In L. Margulis and R. Fester (eds). *Symbiosis as a Source of Evolutionary Innovation: Speciation and Morphogenesis*. Cambridge, MA: MIT Press, pp. 1–14.

Margulis, L. 1991c. Symbiosis in evolution: Origins of cell motility. In S. Osawa and T. Honjo (eds). *Evolution of Life: Fossils, Molecules and Culture*. Tokyo: Springer-Verlag, pp. 305–324.

Margulis, L. 1993. *Symbiosis in Cell Evolution*, 2nd ed. New York: W.H. Freeman.

Margulis, L., and D. Bermudes. 1985. Symbiosis as a mechanism of evolution: Status of cell symbiosis theory. *Symbiosis* 1:104–124.

Margulis, L., and D. Bermudes. 1988. Symbiosis and evolution: A brief guide to recent literature. In S. Scannerini, D.C. Smith, P. Bonfante-Fasolo, and V. Gianinazzi-Pearson (eds), *Cell-to-Cell Signals in Plant, Animal, and Microbial Symbiosis*. Berlin: Springer-Verlag, pp. 159–165.

Margulis, L., and R. Fester (eds). 1991. *Symbiosis as a Source of Evolutionary Innovation: Speciation and Morphogenesis*. Cambridge, MA: MIT Press.

Margulis, L., and G. Hinkle. 1991. The biota and Gaia: One hundred and fifty years of support for environmental sciences. In S.H. Schneider and P.J. Boston (eds). *Scientists on Gaia*. Cambridge, MA: MIT Press, pp. 11–18.

Margulis, L., and J.E. Lovelock. 1974. Biological modulation of the Earth's atmosphere. *Icarus* 21:471–489.

Margulis, L., and D. Sagan. 1990. *Origins of Sex*. New Haven: Yale University Press.

Margulis, L., and D. Sagan. 1996. *What Is Life?* New York: Simon & Schuster, pp. 207.

Margulis, L., and D. Sagan. 1997. *Microcosmos: Four Billion Years of Evolution from Our Bacterial Ancestors*. Berkeley, CA: University of California Press.

Margulis, L., and K.V. Schwartz. 1997. *Five Kingdoms*, 3rd ed. New York: W.H. Freeman Co.

Margulis, L., and J. Stolz. 1983. Microbial systematics and a Gaian view of the sediments. In P. Westbroek and E. de Jong (eds). *Biomineralization and Biological Metal Accumulation*. Dordrecht: Netherlands, Reidel, pp. 27–53.

Margulis, L., and D. West. 1993. Gaia and the Colonization of Mars. *GSA Today*, November, pp. 278–291.

Margulis, L., J.O. Corliss, M. Melkonian, and D.J. Chapman (eds). 1990. *Handbook of Protoctista: The Structure, Cultivation, Habitats and Life Cycles of the Eukaryotic Microorganisms and Their Descendants Exclusive of Animals, Plants and Fungi.* Boston: Jones and Bartlett.

Margulis, L., M. Enzien, and H.I. McKhann. 1990. Revival of Dobell's "chromidia" hypothesis: Chromatin bodies in the amoebomastigote *Paratetramitus jugosus. Biol Bull* 178:300–304.

Margulis, L., L. Lopez Baluja, S.M. Awramik, and D. Sagan. 1986. Community living long before man. *Science of Total Earth* 55:379–400.

Margulis, L., H.I. McKhahn, and L. Olendzenski. 1993. *Illustrated Glossary of the Protoctista.* Boston: Jones and Bartlett.

Margulis, L., J.C.G. Walker, and M.B. Rambler. 1976. A reassessment of the roles of oxygen and ultraviolet light in Precambrian evolution. *Nature* 264:620–624.

Maturana, H. 1987. Everything is said by an observer. In W.I. Thompson (ed). *Gaia: A Way of Knowing.* New York: Lindisfarne Press.

Maturana, H.R., and F.J. Varela. 1973. Autopoiesis: The organization of the living. In H.R. Maturana and F.J. Varela (eds). *Autopoiesis and Cognition.* Boston: D. Reidel.

Maturana, H.R., and F.J. Varela (eds). 1980. *Autopoiesis and Cognition—The Realization of the Living.* Vol. 42 of *Boston Studies in the Philosophy of Science.* Dordrecht, Holland: D. Reidel Publishing Co.

Maynard-Smith, J. 1978a. *Evolution of Sex.* Cambridge: Cambridge University Press.

Maynard-Smith, J. 1978b. Optimization theory in evolution. *Ann Rev Ecol Systemat* 9:31–56.

Maynard-Smith, J. (ed). 1983. *Evolution Now: A Century after Darwin.* San Francisco: W.H. Freeman.

McFall-Ngai, M.J. 1991. Luminous bacterial symbioses in fish evolution: Adaptive radiation among the leiognathid fishes. In L. Margulis and R. Fester (eds). *Symbiosis as a Source of Evolutionary Innovation: Speciation and Morphogenesis.* Cambridge, MA: MIT Press.

McKay, C.P. 1987. Terraforming: Making an Earth of Mars. *Planetary Report* 7:26–27.

McKay, C.P., O.B. Toon, and J.F. Kasting. 1991. Making Mars habitable. *Nature* 352:489–496.

McMenamin, M.A.S., and D.L.S. McMenamin. 1990. *The Emergence of Animals: The Cambrian Breakthrough.* New York: Columbia University Press.

Mehos, D.C. 1992. The defeat of Ivan E. Wallin and his symbiotic theory of evolution in Khakhina q. v. 1992.

Minkoff, E.C. 1983. *Evolutionary Biology.* Reading, MA: Addison-Wesley Publ.

Miyashita, Y., and H.S. Chang. 1988. Neuronal correlate of pictorial short-term memory in the primate temporal cortex. *Nature* 331:68–70.

Mortlock, R.P. (ed). 1984. *Microorganisms as Model Systems for Studying Evolution.* Monographs in Evolutionary Biology. New York: Plenum Press.

Nardon, P., and A.-M. Grenier. 1991a. Serial endosymbiosis and weevil evolution: The role of symbiosis. In L. Margulis and R. Fester (eds). *Symbiosis as a Source of Evolutionary Innovation.* Cambridge: MIT Press, pp. 153–169.

Nardon, P., and A.-M. Grenier. 1991b. Symbiosis in weevil evolution. In L. Margulis and R. Fester (eds). *Evolution of Speciation: Symbiosis as a Source of Evolutionary Innovation.* Cambridge, MA: MIT Press.

National Research Council. 1989. *Everybody Counts: A Report to the Nation on the Future of Mathematics Education.* Washington, D.C.: National Academy Press.

National Research Council. 1993. *Solid-Earth Sciences and Society—Summary of Global Overview.* Washington, D.C.: National Academy Press.

National Science Resource Center. 1988. *Science for Children: Resources for Teachers.* Washington, D.C.: National Academy Press.

Newman, M. 1978. Evolution of the solar constant. In C. Ponnam Peruma and L. Margulis (eds). *Limits to Life: Proceedings of the Fourth College Park Colloquium on Chemical Evolution*, University of Maryland, October 18–20, 1978. Dordrecht: D. Reidel.

Newman, M.J. 1980. Evolution of the solar "Constant." In C. Ponnamperuma and Lynn Margulis (eds). *Limits to Life.* Dordrecht, Holland: Reidel Pub. Co.

Oster, L. 1973. *Modern Astronomy.* San Francisco: Holden Day.

Pagel, W. 1951. William Harvey and the purpose of circulation. *Isis* 42:22–38.

Patterson, C. 1983. Cladistics. In J. Maynard-Smith (ed). *Evolution Now: A Century after Darwin.* San Francisco: W.H. Freeman.

Pert, C., and N. Griffiths-Marriott. 1988. Bodymind. *Woman of Power* 11:22–25.

Portier, P. 1918. *Les Symbiotes.* Paris: Masson et Cie.

Raikov, I.B. 1982. *The Protozoan Nucleus: Morphology and Evolution.* Revised from the 1978 Russian edition. New York: Springer-Verlag.

Rasool, I. (ed). 1974. *The Lower Atmosphere.* New York: Plenum Press.

Reig, O.A. 1987. Notes on biological progress, the changing concepts of anagenesis and macroevolution. *Genetica Iberica* 39:473–520.

Rothschild, L. 1990. Earth analogs for Martian life: Microbes in evaporites, a new model system for life on Mars. *Icarus* 88:246–260.

Sagan, C. 1970. *Planets and Explorations.* Condon Lectures. Eugene, OR: Oregon State System.

Sagan, C., and G. Mullen. 1972. Earth and Mars: Evolution of atmosphere and surface temperatures. *Science* 177:52–56.

Sagan, D. 1987. Biosphere II: Meeting ground for ecology and technology. *Environmentalist* 7:271–281.

Sagan, D. 1988. If the Earth is alive: The Gaia hypothesis may be biology's grand unified theory. *Earthwatch* 7:14–15.

Sagan, D. 1990a. *Biospheres. Metamorphosis of Planet Earth.* New York: McGraw-Hill Publ.

Sagan, D. 1990b. What Narcissus saw: The Oceanic "I"/"eye." In J. Brockman (ed). *Speculations: The Reality Club* 1. Englewood Cliffs, NJ: Prentice Hall, pp. 245–266.

Sagan, D. 1992. Metametazoa: Biology and Multiplicity. In J. Crary and S. Kwinter (eds). *Incorporations, Zone 6.* Cambridge, MA: MIT Press.

Sagan, D., and L. Margulis. 1983. The Gaia perspective of ecology. *Ecologist* 13:160–167.

Sagan, D., and L. Margulis. 1987. Bacterial bedfellows. *Nat Hist* 96:26–33.

Sagan, D., and L. Margulis. 1993. *Garden of Microbial Delights: A Practical Guide to the Subvisible World.* Dubuque, IA: Kendall-Hunt Publ.

Sapp, J. 1995. *Evolution by Association: History of Symbiosis Research.* NY: Oxford University Press.

Schenk, H., and W. Schwemmler. 1983. *Endocytobiology II: Intracellular Space as Oligogenetic System.* New York: Walter de Gruyter.

Schiff, J.A., and H. Lyman. 1982. *On the Origins of Chloroplasts.* New York: Elsevier.

Schopf, J.W. 1970. Precambrian microorganisms and evolutionary events prior to the origin of vascular plants. *Biol Rev* 45:319–352.

Schopf, J.W. (ed). 1983. *Precambrian Paleobiology Research Group Report.* Princeton, NJ: Princeton University Press.

Schopf, J.W. 1994. The oldest known records of life: Early Archaean stromatolites, microfossils and organic matter. In S. Bengston (ed). *Early Life on Earth.* New York: Columbia University Press, pp. 193–207.

Schwemmler, W. 1989. *Symbiogenesis: A Macro-mechanism of Evolution.* Berlin: Walter de Gruyter.

Schwemmler, W., and H.E.A. Schenk (eds). 1980. *Endocytobiology, Endosymbiosis and Cell Biology: A Synthesis of Recent Research.* New York: Walter de Gruyter.

Shimony, A. 1989a. The nonexistence of a principle of natural selection. *Biol Philos* 4:255–273.

Shimony, A. 1989b. Reply to Sober. *Biol Philos* 4:280–286.

Shimony, A. 1989c. The theory of natural selection is a null theory. In D. Constantini and R. Cook (eds). *Statistics and Sciences.* Norwell, MA: Kluwer/Reidel.

Shukla, J., and Y. Mintz. 1982. Influence of the land-surface evapotranspiration on the Earth's climate. *Science* 215:1498–1501.

Simmons, J. 1996. *The Scientific 100: A Ranking of Scientists Past and Present.* Seacaucus, NJ: Carol Publishing.

Simpson, G.G. 1960. In S. Tax (ed). *Evolution after Darwin.* Vol. 1. Chicago, IL: University of Chicago Press, pp. 177–180.

Smith, D.C., and A.E. Douglas. 1987. *The Biology of Symbiosis.* London: Edward Arnold.

Snell, B. 1960. *The Discovery of the Mind.* Translated by T.C. Rosenmeyer. New York: Harper Torchbooks, p. 8.

Sonea, S. 1987. Bacterial viruses, prophages, and plasmids reconsidered. *Ann NY Acad Sci* 503:251–260.

Sonea, S. 1988. A bacterial way of life. *Nature* 331:216.

Sonea, S. 1991a. Bacterial evolution without speciation. In L. Margulis and R. Fester (eds). *Symbiosis as a Source of Evolutionary Innovation: Speciation and Morphogenesis.* Cambridge, MA: MIT Press, pp. 95–105.

Sonea, S. 1991b. Bacterial symbioses and the two contrasting types of evolution. In L. Margulis and R. Fester (eds). *Symbiosis as a Source of Evolutionary Innovation: Speciation and Morphogenesis.* Cambridge, MA: MIT Press.

Sonea, S., and M. Panisset. 1983. *A New Bacteriology.* Boston: Jones and Bartlett.

Stadtman, T. 1974. Selenium biochemistry. *Science* 183:915–922.

Starr, M.P. 1975. A generalized scheme for classifying organismic associations. In D.H. Jennings and D.L. Lee (eds). *Symbiosis: Symposia of the Society for Experimental Biology.* New York: Cambridge University Press, pp. 1–20.

Starr, M.P., H. Stolp, H.G. Trüper, A. Balows, and H.G. Schlegel. (eds). 1981. *The Prokaryotes: A Handbook on Habitats, Isolation and Identification of Bacteria.* Berlin: Springer-Verlag.

Stolz, J.E. 1980. Magnetotactic bacteria. In S. Awramik (ed).

Stolz, J.F. 1993. Magnetosomes. *J Gen Microbiol* 139:1663–1670.

Suzuki D., and P. Knudtson. 1992. *Wisdom of the Elders: Honouring Sacred Visions of Nature.* New York: Bantam Press.

Tennyson, A.B. 1898. In Memoriam. Stanza IV. In H.E. Scudder (ed). *Cambridge Edition of the Poets.* Cambridge: Houghton-Mifflin, p. 164.

Thompson, W.I. 1981a. *The Time Falling Bodies Take to Light.* New York: St. Martin's Press.

Thompson, W.I. 1981b. Walking out on the university. In *Passages about Earth: An Exploration of the New Planetary Culture.* New York: Harper & Row.

Thompson, W.I. 1990. *Imaginary Landscape: Making Worlds of Myth and Science.* New York: St. Martin's Press, p. 91.

Todd, N. 1970. Karyotypic fissioning and carnivore evolution. *J Theoret Biol* 26:445–480.

Twain, M. 1984. *Roughing It.* New York: Literary Classics, p. 384.

Van Beneden, P.J. 1873. Un mot sur la vie sociale des animaux inferieurs. *Bull Acad Roy Belgique,* serie 2, 28:621–648.

Varela, F.G., H.R. Maturana, and R. Uribe. 1974. Autopoiesis: The organization of living systems, its characterization and a model. *BioSystems* 5:187–196.

Vernadsky, V.I. 1929. *La Biosphère.* Expurgated version translated under the title *The Biosphere.* 1986. Oracle, AZ: Synergetic Press. Translation from the Russian 1926 edition to be published *in toto,* 1998, The Biosphere Copernicus Books, Springer-Verlag.

Vernadsky, V.I. 1944. Problems of biogeochemistry. Translated by George Vernadsky. Edited and condensed by G. Evelyn Hutchinson. *Trans Connect Acad Arts Sci* 35:483–517.

Vernadsky, V.I. 1945. The biosphere and the noösphere. *The Am Scientist* 33:1–12.

Vetter, R. 1991. Symbiosis and the evolution of novel trophic strategies: Thiotrophic organisms at hydrothermal vents. In L. Margulis and R. Fester (eds). *Symbiosis as a Source of Evolutionary Innovation: Speciation and Morphogenesis.* Cambridge, MA: MIT Press, pp. 219–245.

Vorontsov, N.N. 1980. The synthetic theory of evolution: Its sources, basic postulates, and unsolved problems. *Zhurnal Vses Khim Obva im Mendeleeva* 25:295–314.

Waddington, C.H. 1976. Concluding remarks. In E. Jantsch and C.H. Waddington (eds). *Evolution and Consciousness.* Reading, MA: Addison Wesley, pp. 243–250.

Wakeford, T., and M. Walter (eds). 1995. *Science for the Earth.* Chichester, UK: John Wiley & Sons, pp. 19–37.

Wallin, I.E. 1927. *Symbionticism and the Origin of Species.* Baltimore, MD: Williams & Wilkins.

Walter, M.J. (ed). 1976. *Stromatolites: Developments in Sedimentology.* Vol. 20. New York: Elsevier.

Walter, M.R. 1976. Stromatolites: The main source of information on the evolution of the early benthos. In *Early Life on Earth,* pp. 270–286.

Watson, A.J., and J.E. Lovelock. 1983. Biological homeostasis of the global environment: The parable of "Daisy World." *Tellus* 35b:284–289.

Watson, A.J., J.E. Lovelock, and L. Margulis. 1978. Methanogenesis, fires, and the regulation of atmospheric oxygen. *BioSystems* 10:293–298.

Westbroek, P. 1991. *Life as a Geological Force.* New York: W.W. Norton.

Wheery, W.B. 1913. Studies on the biology of an amoeba of the limax group. *Vahlkampfia* sp. No. I. *Arch Protistenkund* 31:77–94.

White, F. 1987. *The Overview Effect: Space Exploration and Human Evolution.* Boston, MA: Houghton Mifflin.

Williams, G.C. 1992. Gaia, nature worship and biocentric fallacies. *Q Rev Biol* 67:479–486.

Wilson, E.O. 1970. *The Insect Societies.* Cambridge, MA: Harvard University Press.

Woese, C. 1987. Bacterial evolution. *Microbiol Rev* 51:221–271.

Younger, K.B., S. Banerjee, J.K. Kelleher, M. Winston, and L. Margulis. 1972. Evidence that the synchronized production of new basal bodies is not associated with DNA synthesis in *Stentor coeruleus. J Cell Sci* 11:621–637.

Professional Literature on Gaia

1965 Lovelock, J.E. A physical basis for life detection experiments. *Nature* 207:568–569.

1967 Hitchcock, D.R., and J.E. Lovelock. Life detection by atmospheric analysis. *Icarus* 7:149–159.

1967 Lovelock, J.E., and D.R. Hitchcock. Detecting planetary life from Earth. *Science J* (April):2–4.

1968 Lovelock, J.E., and C.E. Griffen. Planetary atmospheres: Compositional and other changes associated with the presence of life. In D.L. Tiffany and E. Galtzeff (eds). *Advanced Space Experiments.* Vol. 25. Washington, D.C.: American Astronomical Society.

1972 Holland, H.D. The geologic history of seawater—An attempt to solve the problem. *Geochimi Cosmochim Acta* 36:637–651.

1972 Lovelock, J.E., and J.P. Lodge. Oxygen—The contemporary atmosphere. *Atmospher Environ* 6:575–578.

1972 Lovelock, J.E. Gaia as seen through the atmosphere. *Atmospher Environ* 6:579–580.

1974 Lovelock, J.E., and L. Margulis. Atmospheric homeostasis by and for the biosphere: The Gaia hypothesis. *Tellus* 26:2–10.

1974 Margulis, L., and J.E. Lovelock. Biological modulation of the Earth's atmosphere. *Icarus* 21:471–489.

1974 Lovelock, J.E., and L. Margulis. Homeostatic tendencies of the Earth's atmosphere. *Origins of Life* 5:93–103.

1975 Margulis, L., and J.E. Lovelock. The atmosphere as circulatory system of the biosphere—The Gaia hypothesis. *CoEvol Q* 6:30–41.

1975 Lovelock, J.E. Thermodynamics and the recognition of alien biospheres. *R Soc Lon Proc* Ser. B 189:167–181.

1976 Margulis, L., J.C.G. Walker, and M. Rambler. Reassessment of roles of oxygen and ultraviolet light in PreCambrian evolution. *Nature* 264:620–624.

1977 Margulis, L., and J.E. Lovelock. The view from Mars and Venus. *Sciences* (March-April):10–13.

1978 Margulis, L., and J.E. Lovelock. The biota as ancient and modern modulator of the Earth's atmosphere. *Pageoph* 116:239–243.

1978 Dutsch, H.U. (ed). *Influence of the Biosphere on the Atmosphere.* Basel: Birkauser. (Reprinted from *Pure Appl Geophys* 116:213–582.)

1978 Watson, A., J.E. Lovelock, and L. Margulis. Methanogenesis, fires and the regulation of atmospheric oxygen. *BioSystems* 10:293–298.

1980 Watson, A.J., J.E. Lovelock, and L. Margulis. What controls atmospheric oxygen? *BioSystems* 12:123–125.

1981 Dastoor, M., K.H. Nealson, and L. Margulis (eds). *Interaction of the Biota with the Atmosphere and Sediments.* Washington, D.C.: NASA Workshop Report, Oct. 18–19, 1979.

1981 Doolittle, W.F. Is nature really motherly? *CoEvol Q* 29:58–63.

1981 Margulis, L., K.H. Nealson, and I. Taylor (eds). *Planetary Biology and Microbial Ecology: Biochemistry of Carbon and Early Life.* NASA Technical Memorandum 86043. Summer Program Research Report, 1980.

1981 Margulis, L., and J.E. Lovelock. Atmospheres and evolution. In J. Billingham (ed). *Life in the Universe.* Cambridge, MA: MIT Press, pp. 79–100.

1982 Brock, T.D., P.J. Cook, H.P. Eugster, A.M. Goodwin, H.L. James, L. Margulis, K.H. Nealson, J.O. Nriagu, A.F. Trendall, and M.R. Walter. Sedimentary iron deposits, evaporites and phosphorites: State of the art report. In H.D. Holland and M. Schidlowski (eds). *Mineral Deposits and the Evolution of the Biosphere.* Berlin: Springer-Verlag, pp. 259–273.

1982 Lovelock, J.E., and A.J. Watson. The regulation of carbon dioxide and climate: Gaia or geochemistry. *J Planet Sci* 30:795–802.

1982 Lovelock, J.E., and M. Whitfield. The life span of the biosphere. *Nature* 296:561–563.

1982 Margulis, L. The biological point of view: The effect of life on the planet. In A. Brahic (ed). *Formation of Planetary Systems.* Toulouse, France: Centre d'Études Spatiales, Capaude-Editions, pp. 891–893.

1983 Lovelock, J.E. Gaia as seen through the atmosphere. In P. Westbroek and E. de Jong (eds). *Biomineralization and Biological Metal Accumulation.* Dordrecht, Netherlands: Reidel, pp. 15–25.

1983 Margulis, L., and J. Stolz. Microbial systematics and a Gaian view of the sediments. In P. Westbroek and E. de Jong (eds). *Biomineralization and Biological Metal Accumulation.* Dordrecht, Netherlands: Reidel, pp. 27–53.

1983 Watson, A., and J.E. Lovelock. Biological homeostasis of the global environment: The parable of Daisyworld. *Tellus* 35:284–289.

1985 Sagan, D. (ed). *Planetary Biology and Microbial Ecology: The Global Sulfur Cycle.* NASA Technical Memorandum. Summer Program Research Report, June-August, 1984.

1986 Lovelock, J.E. Geophysiology: A new look at Earth science. *Am Meteorol Soc Bull* 67:392–397.

1987 Charlson, R.J., J.E. Lovelock, M.O. Andreae, and S.G. Warren. Oceanic phytoplankton, atmospheric sulphur, cloud albedo and climate. *Nature* 326:655–661.

1987 Lovelock, J.E. Ecopoiesis of Daisy World. In J.M. Robson (ed). *Origin and Evolution of the Universe—Evidence for Design?* Ottawa, Canada: Royal Society of Canada, pp. 153–166.

1987 Lovelock, J.E. *Gaia: A New Look at Life on Earth,* 2nd ed. New York: Oxford University Press.

1987 Margulis, L. Early life: The microbes have priority. In W.I. Thompson (ed). *Gaia: A Way of Knowing: Political Implications of the New Biology.* Great Barrington, MA: Lindisfarne Press, pp. 98–109.

1988 Lovelock, J.E. *The Ages of Gaia.* New York: W.W. Norton.

1989 Lovelock, J.E. Geophysiology: Royal Society of Edinburgh. *Earth Sci* 80:169–175.

1989 Lovelock, J.E. Geophysiology, the science of Gaia. *Rev Geophys* 27:215–222.

1989 Lovelock, J.E. The First Leslie Cooper Memorial Lecture: Gaia. *J Marine Biol* 69:746–758.

1989 Margulis, L., and J.E. Lovelock. Gaia and geognosy. In M.B. Rambler, L. Margulis, and R. Fester, (eds). *Global Ecology: Towards a Science of the Biosphere.* Boston: Academic Press, pp. 1–30.

1990 Hinkle, G., and L. Margulis. Global ecology and the Gaia hypothesis [Special issue: Ecology for tomorrow]. *Physiol Ecol Jpn* 27:53–62.

1991 Lovelock, J.E. Geophysiology of the oceans. In R.F.C. Mantoura, J.-M. Martin, and R. Wollast (eds). *Ocean Margin Processes in Global Change* (Dahlem Conference). London: John Wiley & Sons, pp. 419–431.

1991 Margulis, L., and G. Hinkle. The biota and Gaia: 150 years of support for environmental sciences. In S.H. Schneider and P.J. Boston (eds). *Scientists on Gaia.* Cambridge, MA: MIT Press, pp. 11–18.

1992 Margulis, L., and L. Olendzenski (eds). *Environmental Evolution: The Effect of the Origin and Evolution of Life on Planet Earth.* Cambridge, MA: MIT Press.

1992 Lovelock, J.E. A numerical model for biodiversity. *R Soc Lon Phil Trans* ser. B 338:383–391.

1992 Lovelock, J.E. Geophysical aspects of biodiversity. In O.T. Solbrig, H.M. Van Emden, and P.G.W.J. Van Oordt (eds). IUBS Monograph 8. Paris: International Union of Biological Sciences, pp. 57–70.

1992 Schneider, S.H., and P. Boston (eds). *Scientists on Gaia.* Cambridge, MA: MIT Press, pp. xxii, 431.

1993 Lovelock, J.E. *Las Edades de Gaia: Una Biografía de Nuestro Planeta Vivo.* Presentacion de Ricardo Guerrero. Barcelona, Spain: Fundacío la Caixa, Museu de la Ciéncia.

Morowitz, H.J. *Beginning of Cellular Life: Metabolism recapitulates biogenesis.* New Haven, CT: Yale University Press.

1994 Bengtson, S. (ed). *Early Life on Earth.* Nobel Symposium #84. NY: Columbia University Press.

Sagan, D., and L. Margulis. Gaian views. pp. 3–9 in C.K. Chapple (ed). *Ecological Prospects: Scientific, Religious and Aesthetic Perspective.* Albany, NY: SUNY Press.

1994 Barlow, C. (ed). *Evolution Extended: Biological Debates on the Meaning of Life.* Cambridge, MA: MIT Press.

1995 Dixon, B. *Unseen Power: How Microbes Rule the World.* NY: Oxford University Press.

1995 Matthews, C., and R. Varghese (eds). *Cosmic Beginnings and Human Ends.* Chicago, IL: Open Court Publishers.

1995 Rolston, H. (ed). *Biology, Ethics and the Origins of Life.* Boston, MA: Jones and Bartlett.

1995 Makofske, W.J. and E. Karlin (eds). *Technology and Global Environmental Issues.* NY: HarperCollins.

Margulis, L. Gaia Is a Tough Bitch. In J. Brockman (ed). *The Third Culture.* NY: Simon & Schuster.

1996 Broecker, W.S. The biosphere and me. *GSA Today* 6:1–7.

Bungard, P. (ed). *Gaia in Action. Science of the Living Earth.* Edinburgh: Floris Books.

Popular Literature on Gaia

1979 Lovelock, J.E. *Gaia: A New Look at Life on Earth.* Oxford, England: Oxford University Press.

1980 Margulis, L., and J. E. Lovelock. After Viking: Life on Earth. *Sciences,* (November): 20:24–26.

1980 Margulis, L., and J.E. Lovelock. L'atmosphère est-elle le système circulatoire de la biosphère? L'hypothèse Gaia. *CoEvolution* 1:20–31.

1982 Dawkins, R. *The Extended Phenotype.* New York: W.H. Freeman.

1982 Lovelock, J.E. From gas chromatography to Gaia. *Chromatographia* 16:26–31.

1983 Margulis, L., and J.E. Lovelock. Le petit monde des pâquerettes: Un modèle quantitatif de Gaia. *CoEvolution* 11:48–52.

1983 Sagan, D., and L. Margulis. The Gaian perspective of ecology. *Ecologist* 13:160–167.

1984 Sagan, D., and L. Margulis. Gaia and philosophy. In L.S. Rouner (ed). *On Nature.* Notre Dame, IN: University of Notre Dame Press, pp. 60–75.

1985 Lovelock, J.E., and J. Allaby. *The Greening of Mars.* London: Allan & Unwin.

1986 Lovelock, J.E. Gaia: The world as a living organism. *New Scientist,* December 18, 25–28.

1986 Margulis, L., and D. Sagan. *Microcosmos: Four Billion Years of Evolution from Our Microbial Ancestors.* New York: Summit Books. French translation, A. Michel, Paris, 1989; Italian translation, A. Mondadori, Milan, 1989; Japanese translation, Tokyo Kagaku Dozin, Tokyo, 1989; Danish translation, Nysyn, Munksgaard, 1990; Portugese translation: Edicoes 70, Rio de Janeiro, 1990; paperback, Berkeley: University of California Press, 1997.

1986 Margulis, L., L. Lopez Baluja, S.M. Awramik, and D. Sagan. Community living long before man. In D. Botkin and A.A. Orio (eds). *Man's Effect on the Global Environment: Science of the Total Environment,* Vol. 56, pp. 379–397.

1987 Sagan, D., and Margulis, L. Gaia and the evolution of machines. *Whole Earth Review* 55:15–21.

1988 Lovelock, J.E. *The Ages of Gaia: A Biography of Our Living Earth.* New York: W.W. Norton.

1988 Margulis, L. Jim Lovelock's Gaia. In P. Bunyard and E. Goldsmith (eds). *Gaia, the Thesis, the Mechanisms and the Implications.* Cornwall, England: Wadebridge Ecological Centre, pp. 50–65.

1988 Sagan, D. What Narcissus saw: The oceanic "I"/eye. In *The Reality Club 1.* New York: Prentice Hall, pp. 248–266.

1988 Sagan, D., and L. Margulis. Gaia and biospheres. In P. Bunyard and E. Goldsmith (eds). *Gaia, the Thesis, the Mechanisms and the Implications.* Cornwall, England: Wadebridge Ecological Centre, pp. 237–242.

1990 Joseph, L.E. *Gaia, the Growth of an Idea.* New York: St. Martin's Press (paperback: 1991, Arkana).

1990 Lovelock, J.E. Hands up for the Gaia hypothesis. *Nature* 344:100–102.

1991 Barlow, C. (ed). *From Gaia to selfish genes: Selected writings in the life sciences.* Cambridge, MA: MIT Press.

1991 Lovelock, J.E. *Healing Gaia: Practical Medicine for the Planet.* United Kingdom: Gaia Books; New York: Harmony Books.

1991 Margulis, L. Gaia, a new look at the Earth's systems. In W.J. Makofske, H. Horowitz, E.F. Karlin, and P. McConnell (eds). *Technology, Development and the Global Environment.* Mahwah, NJ: Ramapo College, Institute for Environmental Studies, School of Theoretical and Applied Science, pp. 299–305.

1991 Margulis, L., and R. Guerrero. Two plus three equal one: Individuals emerge from bacterial communities. In W.I. Thompson (ed). *Gaia 2. Emergence: The New Science of Becoming.* Hudson, NY: Lindisfarne Press, pp. 50–67.

1991 Sagan, D. *Biospheres: Metamorphosis of Planet Earth:* New York, McGraw-Hill (paperback edition: New York: Bantam Books).

1991 Westbroek, P. *Life as a Geological Force.* New York: W.W. Norton.

1992 Lovelock, J.E. The Gaia hypothesis. In *Environmental Evolution: Effects of the Origin and Evolution of Life on Planet Earth.* Cambridge, MA: MIT Press.

1993 Barlow, C. *From Gaia to Selfish Genes.* Cambridge, MA: MIT Press.

1993 Levine, L. Gaia: Goddess and idea. *BioSystems* 31:100–200.

1995 Margulis, L. *Gaia Is a Tough Bitch in the Third Culture.* Edited by John Brochman. NY: Simon & Schuster, pp. 129–146.

1996 Bunyard, P, (ed). *Gaia in Action: Science for the Living Earth.* Edinburgh: Floris Books.

Contributors

Michael F. Dolan
Organismal and Evolutionary Biology Graduate Program
University of Massachusetts
Amherst, MA 01003

Ricardo Guerrero
Department of Microbiology
University of Barcelona
Apartado Postal 16-009
Barcelona, Spain 08080

Gregory Hinkle
Department of Biology
University of Massachusetts
North Dartmouth, MA 02747-2300
and
Laboratory of Molecular Evolution
Marine Biological Laboratory
Woods Hole, MA 02543

James E. Lovelock
Coombe Mill Experiment Station
St. Giles on the Heath
Launceston, Cornwall
United Kingdom PL1S 9RY

Mark A. McMenamin
Mount Holyoke College
South Hadley, MA 01075-1617

Lynn Margulis
Sciencewriters
PO Box 671
Amherst, MA 01004-0671

Dorion Sagan
Sciencewriters
PO Box 671
Amherst, MA 01004-0671

Oona West
Organismal and Evolutionary Biology Graduate Program
University of Massachusetts
Amherst, MA 01003

Glossary

Alchemy: proto-chemistry, practiced in ancient and medieval times, in which natural processes are expressed in mythic and allegorical symbolism, sometimes in an effort to make gold from other substances.

Animal: diploid organism that develops from a multicellular blastula, the animal embryo, which is the product of fertilization of eggs and sperm, generally heterotrophic by phagotrophy (ingestive nutrition).

Archaebacteria (=Archaea): a subkingdom of bacteria with a distinctive ribosome RNA structure and other chemical features. Includes methane-making, high salt and high heat and acid-loving bacteria.

Atmospheric anomalies: concentrations of atmospheric gases in combinations which differ significantly from stable mixtures determined by experimental and theoretical chemical expectations.

ATP (adenosine triphosphate): a phosphorus-containing organic compound which functions to deliver chemical energy in all living cells.

Autopoiesis: self-making and self-maintaining properties of living systems; unlike mechanistic systems they produce and maintain their own boundaries.

Bacteria: unicellular microorganisms and their multicellular descendants bounded by cell membranes, but lacking chromosomes bounded by nuclear membranes. Comprising a kingdom referred to by various names (Bacteria, Monera, or Prokaryotae), they are chemically and metabolically more diverse than any other form of life on Earth.

Biochips: a computer chip based not on silicon, but on complex organic compounds.

Biodiversity: the variety of species, or genetically distinctive plants, animals, fungi, protoctists, and bacteria in any natural environment

Bridgewater Treatises: the Earl of Bridgewater (England) in the 1820s commissioned these studies by leading thinkers of the day expressly to show evidence of divine plan and providence in the natural sciences.

Calonymphids: a family of the protoctists, restricted to termite digestive tracts studied by H. Kirby. They lack mitochondria and in them the multiple nuclei and undulipodia reproduce independently of the rest of the cell.

Cartesian dualism: the philosophical perspective, deriving from the 17th century philosopher René Descartes, in which reality is considered to be composed of two substances; one of mind, and the other of matter.

Centrioles: reproducing cylindrical structures in many nucleated cells, conspicuously in animals, during cell division. Centrioles, precursors to kinetosomes, are not found in plant cells, except prior to sperm formation. Whereas they are abscent in fungi, their position and behavior vary widely in protoctists.

Centromere: kinetochore, the specialized reproductive DNA-protein structure by which a chromosome attaches to the spindle in cell reproduction (mitosis) and insures its accurate distribution to offspring cells.

Chemotactic bacteria: prokaryotic microorganisms which swim or glide in response to chemical stimuli; e.g., the detection of chemical gradients in their ambient environment.

Chloroplasts: chlorophyll-containing membrane-bounded structures inside plant, algal, and occasional animal cells which function in photosynthesis.

Coevolution: a process of complementary evolution, in which two species or types of life exert evolutionary influence on each other, often in symbiotic relations.

Community: the set of all organisms of different species living in the same place at the same time.

Complexity: broadly, the measure of the number and form of interacting parts in an embedded structure; sometimes expressed as the bit-length of an algorithmic procedure sufficient to the replication of the structure.

Convergence: the tendency of organisms to evolve similar metabolism, behaviors, and structures under similiar environmental pressures, despite the fact that they have different recent ancestors.

Cybernetic systems: self-governing systems, in which system output, via feedback loops, is used to control system function.

Cybersymbiosis: the evolution of machines and human beings in future life forms.

Daisy World model: mathematical model of temperature modulation by biota (dark and light daisies), which indicates a possible mechanism for global homeorrhesis.

Deconstruction: a style of textual criticism, developed by Jacques Derrida, in which texts supposedly closed, intended, or decided, are through juxtaposition shown to be undecided, open, and perhaps undecidable.

Deoxyribonucleic acid (DNA): material substances of the genes, long-chain molecules comprising nucleoids of prokaryotes and chromosomes of eukaryotes through which hereditary characteristics are potentially determined.

Differentiation: cellular specialization in many-celled organisms.

Ecopoiesis: the settlement of a formerly uninhabited plantetary (carbon, nitrogen, etc.) surface with viable living systems.

Ecosystem: the system of cyclic flows of matter and energy (light and chemical) characterizing the relations of a community of biological organisms with its physical environment.

Emergence: the appearance of characteristics or a system that could not have been predicted on the basis of known properties of its components alone.

Eukaryote: an organism consisting of cells in which genetic material is contained within a distinct membrane-bounded nucleus. Except for bacteria, all organisms (animals, plants, fungi, and protoctists) are eukaryotes.

Endocytobiology: the study of intracellular space as a community of microorganisms; its objects of study are all intracellular symbionts.

Epistemology: the philosophical study of the grounds of our knowledge.

Evolution: the process, so far (over a period of 3,000 million years), of biological change on Earth.

Feedback: a self-controlling system function, in which the output of some process is used as input to that same process.

Flagella: rotary-motor based flagellin-protein fibers of swimming bacteria. External to cell membrane; used in locomotion.

Fractal theory: the study of systems exhibiting self-similarity, in which characteristic features are found repeated at different levels of scale.

Fungi: eukaryotic organisms that develop from certain kinds of spores (e.g., ascospores, basidiospores); they lack both embryos and undulipodia at all stages of their life cycle.

Gaia: after the Greek mother of the Titans, Gaia; the organismal self-maintaining environmental regulatory system extending within 20 kilometers of the Earth's surface, comprised of more than 30,000,000 extant species.

Gaia hypothesis: theory of British scientist James E. Lovelock, that the Earth, including both its biotic and abiotic components, functions as a single self-regulating system, in which the growth and activities of living organisms in response to the environment regulate the reactive gas composition, acidity-alkalinity and temperature bringing about changes that make the Earth continuously habitable.

Gene: the biological unit of heredity, composed of long-chain DNA molecules, found in the nucleoid (bacteria) and chromosomes (of eukaryotes).

Green politics: an international political movement concerned primarily with environmental and ecological issues.

Group selection: the idea that evolution favors populations of individuals that act together rather than independent individuals.

Homeorrhesis, Homeorrhetic regulation: the regulation of aspects of a system (such as temperature, composition) around moving, rather than fixed-from-the-outside, set-points.

Homeostasis: regulation of a system around fixed set-points. Engineering cybernetic systems are regulated from the outside whereas homeostatic ones are physiological, regulated from the inside.

Homuncular inner self: a persistent regressive structure employed, in explanations of consciousness in which thought is not explained but assumed to inhere at lower levels .

Hormones: biologically-produced chemicals secreted into the body fluids which regulate functions of the body of the organism at a distance from the source of the secretion.

Human Genome Project: an international project, begun in 1988, to determine the sequence of nucleotides in the entire set of DNA in humans. The genome of *Homo sapiens* includes over 100,000 genes.

Incompleteness theorems: the results of K. Godel in which he demonstrated that axiomatic systems must contain true statements not demonstrable as theorems.

Interactive lecture system: learning program of higher education, in which audiotapes, transparencies, and an "electrowriter" which reproduces handwriting, are employed when leading scientists present their own ideas directly to students.

Karyotypic fission: a process inferred in mammalian evolution in which chromosomal kinetochores (centromeres) reproduce once independently of the chromosomes leading to half-size, double-number chromosomes in egg or sperm cells.

Kinetocore: See Centromeres

Kinetosomes: basal bodies; centrioles after their shafts emerge. Intracellular organelles at the base of the undulipodia necessary for the development of undulipodia.

Kingdom: the highest, most inclusive, most general group (taxon) in the classification of organisms. The two-kingdom, plant and animal system has largely been replaced by the five-kingdom scheme.

Locomotion: the ability of any organism to move itself about (swim, glide, walk, fly) in a particular direction by use of locomotory organelles such as flagella, undulipodia or organs such as legs or wings.

Mechanistic philosophy: the Cartesian-Newtonian perspective which became the basis of classical scientific theory in which all natural phenomena are considered to be determined by the effects of matter in motion.

Med mesign: concept of the Chewong peoples of Malaysia, meaning "different eye"; each type of creature sees the world in its own way.

Metabolism: the sum total of chemical reactions occuring within and maintaining a living organism.

Methane anomaly: the variance, of some 30 orders of magnitude, between measured and chemically predicted levels of methane in the Earth's atmosphere.

Mitochondria: structures (membrane-bounded organelles) within the most eukaryotic cells responsible for the cell's energy production via oxygen respiration.

Neo-darwinism: biological theory which asserts that new organisms and organs evolve through the accumulation and genetic transmission of random mutations in DNA.

Noosphere: term employed by Vernadsky and Teilhard to describe the human and technological planetary layer forming in conjunction with the biosphere.

Organelle: generic term for distinctive structure inside any cell as seen with a microscope.

Plasmids: gene-bearing small circles or strands of DNA that can replicate independently of the rest of the cell's DNA, which may contain one or two of the genes vital for conditional survival.

Polyenergy: an increase in the number of homologous genomes in a nucleus.

Population: the set of all organisms of the same species living in a given place at the same time.

Plastids: generic term for photosynthetic organelles and their derivatives in plants and protoctists (all algae), e.g., chloroplasts, proplastids.

Prokaryote: bacterial organisms composed of cells lacking a membrane-bounded nucleus.

Protoctists: eukaryotic unicellular or small multicellular organisms (protists) and their large descendants which possess nuclei and evolved by bacterial symbiosis. They cannot be classified as animals or plants (lacking embryos), or as fungi (lacking fungal spores); protoctista.

Propagule: any cellular structure produced by organisms, and capable of survival, dissemination, and future growth. Usually resistant to extreme temperatures, desiccation or starvation relation to the growing form of the same organism (e.g., spores, cysts)

Serial endosymbiosis theory (SET): the theory that all eukaryotic cells with nuclei are composites formed from a sequence of mergers of different kinds of bacteria.

Sociobiology: a branch of neo-Darwinist biological theory in which social behaviors of animals, including people, can be inferred from genetic information.

Spirochetes: flexible, fast-swimming corkscrew-shaped heterotrophic bacteria which have their flagella outside their cell membranes, but inside their outer membranes, i.e. in their periplasm.

Symbiogenesis: concept of evolutionary change in which long-term symbioses lead to new behaviors, metabolism, organelles, organs and eventually even species.

Symbiosis: prolonged physical association between two or more different kinds of organisms. Many symbioses are obligatory (i.e., the participants cannot survive without the interaction), others are casual.

Symbiotrophy: mode of nutrition in which a heterotrophic symbiont derives both its carbon and its energy from a living partner.

Synchronicity: coincidences with such deep significance that one concludes they are more than mere coincidences, after Jung.

Taxonomy: the indentification, naming and placement of organisms into categories (kindoms, families, species) today on the basis of their evolutionary history.

Terraformation: the recreation of an Earth-like environment on another planetary surface.

Trichomonads: a class of protists that lack mitochondria, they evolved from small swimmers with one nucleus connected to four undulipodia (such as *Trichomonas*.) The nucleus-undulipodia system is capable of independent reproduction that occurs to form multinucleate descendants, like members of the calonymphid family.

Tropy: locomotion, morphogenetic movement or growth toward or away from an external stimulus; as in phototropy (light-seeking). Usually a suffix.

Trophy: mode of nutrition as in photoautotrophy (light as energy source and carbon dioxide as carbon source) or heterotrophy (organic chemicals as sources of energy and carbon).

Uncertainty principle: in modern physical theory, the principle enunciated by Heisenberg, in which limits on the simultaneous determination of the values of complementary variables (e.g., the position and momentum of a particle) are established.

Undulipodium: generic term for motility structure of eukaryotes. Microtubule-based tubulin (and hundreds of other proteins). Cell appendage showing a 9-fold symmetry. Each develops from its centriole-kinetosome. Includes sperm tails, cilia or eukaryotic "flagella" and many other intracellular organelles used in sensory perception, cell locomotion and feeding.

Vahlkampfids: small shelless "monopodial" or "one-foot" amoebae. They tend to slowly move forward rather than simultaneously in many directions or form an exuberance of spines as other amoebae do.

Index